Silicon Carbide
Power Devices

Silicon Carbide
Power Devices

B JAYANT BALIGA

North Carolina State University, USA

World Scientific

NEW JERSEY • LONDON • SINGAPORE • BEIJING • SHANGHAI • HONG KONG • TAIPEI • CHENNAI

Published by

World Scientific Publishing Co. Pte. Ltd.

5 Toh Tuck Link, Singapore 596224

USA office: 27 Warren Street, Suite 401-402, Hackensack, NJ 07601

UK office: 57 Shelton Street, Covent Garden, London WC2H 9HE

Library of Congress Cataloging-in-Publication Data
Baliga, B. Jayant, 1948–
   Silicon carbide power devices / B. Jayant Baliga.
   p. cm.
   ISBN-13 978-981-256-605-8
   ISBN-10 981-256-605-8
   Includes bibliographical references and index.
   1. Silicon carbide--Electric properties. 2. Semiconductors. I. Title.
   TK7871.15.S56
   621.3815'2--dc22

                              2006283987

British Library Cataloguing-in-Publication Data
A catalogue record for this book is available from the British Library.

First published 2005 (Hardcover)
Reprinted 2016 (in paperback edition)
ISBN 978-981-3203-23-5

Dedication

The author would like to dedicate this book to his family – his wife, Pratima and his sons, Avinash and Vinay - for their support of his commitment to create a world-wide effort on the development and commercialization of power devices based upon wide band-gap semiconductors.

Preface

Power semiconductor devices are a key component of all power electronic systems. It is estimated that at least 50 percent of the electricity used in the world is controlled by power devices. With the wide spread use of electronics in the consumer, industrial, medical, and transportation sectors, power devices have a major impact on the economy because they determine the cost and efficiency of systems. After the initial replacement of vacuum tubes by solid state devices in the 1950s, semiconductor power devices have taken a dominant role with silicon serving as the base material.

Bipolar power devices, such as bipolar transistors and thyristors, were first developed in the 1950s. Their power ratings and switching frequency increased with advancements in the understanding of the operating physics and availability of more advanced lithography capability. The physics underlying the current conduction and switching speed of these devices has been described in several textbooks[1,2]. Since the thyristors were developed for high voltage DC transmission and electric locomotive drives, the emphasis was on increasing the voltage rating and current handling capability. The ability to use neutron transmutation doping to produce high resistivity n-type silicon with improved uniformity across large diameter wafers enabled increasing the blocking voltage of thyristors to over 5000 volts while being able to handle over 2000 amperes of current in a single device. Meanwhile, bipolar power transistors were developed with the goal of increasing the switching frequency in medium power systems. Unfortunately, the current gain of bipolar transistors becomes low when it is designed for high voltage operation at high current density. The popular solution to this problem, using the Darlington configuration, had the disadvantage of increasing the on-state voltage drop resulting in an increase in the power dissipation. In addition to the large control currents required for bipolar transistors, they suffered from second breakdown failure modes. These issues produced a cumbersome design with snubber networks, which raised the cost and degraded the efficiency of the power control system.

In the 1970s, the power MOSFET product was first introduced by International Rectifier Corporation. Although initially hailed as a replacement for all bipolar power devices due to its high input impedance and fast switching speed, the power MOSFET has successfully cornered the market for low voltage (< 100 V) and high switching speed (> 100 kHz) applications but failed to make serious inroads in the high voltage arena. This is because the on-state resistance of power MOSFETs increases very rapidly with increase in the breakdown voltage. The resulting high conduction losses, even when using larger more expensive die, degrade the overall system efficiency.

In recognition of these issues, I proposed two new thrusts in 1979 for the power device field. The first was based upon the merging of MOS and bipolar device physics to create a new category of power devices[3]. My most successful innovation among MOS-Bipolar devices has been the *Insulated Gate Bipolar Transistor (IGBT)*. Soon after commercial introduction in the early 1980s, the IGBT was adopted for all medium power electronics applications. Today, it is manufactured by more than a dozen companies around the world for consumer, industrial, medical, and other applications that benefit society. The triumph of the IGBT is associated with its huge power gain, high input impedance, wide safe operating area, and a switching speed that can be tailored for applications depending upon their operating frequency.

The second approach that I suggested in the early 1980s for enhancing the performance of power devices was to replace silicon with wide band gap semiconductors. The basis for this approach was an equation that I derived relating the on-resistance of the drift region in unipolar power devices to the basic properties of the semiconductor material. This equation has since been referred to as *Baliga's Figure of Merit (BFOM)*. In addition to the expected reduction in the on-state resistance with higher carrier mobility, the equation predicts a reduction in on-resistance as the inverse of the cube of the breakdown electric field strength of the semiconductor material.

In the 1970s, there was a dearth of knowledge of the impact ionization coefficients of semiconductors. Consequently, an association of the breakdown electric field strength was made with the energy band gap of the semiconductor[4]. This led to the conclusion that wide band gap semiconductors offer the opportunity to greatly reduce the on-state resistance of the drift region in power devices. With a sufficiently low on-state resistance, it became possible to postulate that unipolar power devices could be constructed from wide band gap semiconductors with lower on-state voltage drop than bipolar devices made out of silicon. Since unipolar devices exhibit much faster switching speed than bipolar devices because of the absence of minority carrier stored charge, wide band gap based power devices offered a much superior alternative to silicon bipolar devices for medium and high power applications. Device structures that were particularly suitable for development were identified as Schottky rectifiers to replace silicon P-i-N rectifiers, and power Field Effect Transistors to replace the bipolar transistors and thyristors prevalent in the 1970s.

The first attempt to develop wide band gap based power devices was undertaken at the General Electric Corporate Research and Development Center, Schenectady, NY, under my direction. The goal was to leverage a 13-fold reduction in specific on-resistance for the drift region predicted by the BFOM for Gallium Arsenide. A team of 10 scientists was assembled to tackle the difficult problems of the growth of high resistivity epitaxial layers, the fabrication of low resistivity ohmic contacts, low leakage Schottky contacts, and the passivation of the GaAs surface. This led to an enhanced understanding of the breakdown strength[5] for GaAs and the successful fabrication of high performance Schottky rectifiers[6] and MESFETs[7]. Experimental verification of the basic thesis of the analysis represented by BFOM was therefore demonstrated during this period. Commercial GaAs based Schottky rectifier products were subsequently introduced in the market by several companies.

In the later half of the 1980s, the technology for the growth of silicon carbide was developed with the culmination of commercial availability of wafers from CREE Research Corporation. Although data on the impact ionization coefficients of SiC was not available, early reports on the breakdown voltage of diodes enabled estimation of the breakdown electric field strength. Using these numbers in the BFOM predicted an impressive 100-200 fold reduction in the specific on-resistance of the drift region for SiC based unipolar devices. In 1988, I joined North Carolina State University and subsequently founded the *Power Semiconductor Research Center (PSRC)* - an industrial consortium – with the objective of exploring ideas to enhance power device performance. Within the first year of the inception of the program, SiC Schottky barrier rectifiers with breakdown voltage of 400 volts were successfully fabricated with on-state voltage drop of about 1 volt and no reverse recovery transients[8]. By improving the edge termination of these diodes, the breakdown voltage was found to increase to 1000 volts. With the availability of epitaxial SiC material with lower doping concentrations, SiC Schottky rectifiers with breakdown voltages over 2.5 kV have been fabricated at PSRC[9]. These results have motivated many other groups around the world to develop SiC based power rectifiers. In this regard, it has been my privilege to assist in the establishment of national programs to fund research on silicon carbide technology in the United States, Japan, and Switzerland-Sweden. Meanwhile, accurate measurements of the impact ionization coefficients for 6H-SiC and 4H-SiC in defect free regions were performed at PSRC using an electron beam excitation method[10]. Using these coefficients, a BFOM of over 1000 is predicted for SiC providing even greater motivation to develop power devices from this material.

Although the fabrication of high performance, high voltage Schottky rectifiers has been relatively straight-forward, the development of a suitable silicon carbide MOSFET structure has been more problematic. The existing silicon power D-MOSFET and U-MOSFET structures do not directly translate to suitable structures in silicon carbide. The interface between SiC and silicon dioxide, as a gate dielectric, needed extensive investigation due to the large density of traps that prevent the formation of high conductivity inversion layers. Even after overcoming this hurdle, the much higher electric field in the silicon dioxide when compared with silicon devices, resulting from the much larger electric field in the underlying SiC, leads to reliability problems. Fortunately, a structural innovation, called the ACCUFET, to overcome both problems was proposed and demonstrated at PSRC[11]. In this structure, a buried P^+ region is used to shield the gate region from the high electric field within the SiC drift region. This concept is applicable to devices that utilize either accumulation channels or inversion channels. Devices with low specific on-resistance have been demonstrated at PSRC using both 6H-SiC and 4H-SiC with epitaxial material capable of supporting over 5000 volts[12]. This device structure has been subsequently emulated by several groups around the world.

Although many papers have been published on silicon carbide device structures and process technology, no comprehensive book written by a single author is available that provides a unified treatment of silicon carbide power device structures. This book has been prepared to fill this gap. The emphasis in the book is on the physics of operation of the devices elucidated by extensive two-dimensional numerical analysis.

The simulations were done for 4H-SiC, rather than 6H-SiC, because its larger breakdown strength results in superior device performance. This analysis provides general guidelines for understanding the design and operation of the various device structures. For designs that may be pertinent to specific applications, the reader should refer to the papers published in the literature, the theses of my M.S. and Ph.D. students, as well as the work reported by other research groups. Comparison with silicon devices is provided to enable the reader to understand the benefits of silicon carbide devices.

In the introduction chapter, the desired characteristics of power devices are described with a broad introduction to potential applications. The second chapter provides the properties of silicon carbide that have relevance to the analysis and performance of power device structures. Issues pertinent to the fabrication of silicon carbide devices are reviewed here because the structures analyzed in the book have been constructed with these process limitations in mind. The third chapter discusses breakdown voltage, which is the most unique distinguishing characteristic for power devices, together with edge termination structures. This analysis is pertinent to all the device structures discussed in subsequent chapters of the book.

The fourth chapter provides a brief analysis of P-i-N rectifiers followed by a detailed analysis of the Schottky rectifier structure in chapter five. The sixth chapter introduces the concept and explains the benefits of shielding the Schottky contact. This approach is essential for mitigating the generation of high leakage current in silicon carbide Schottky rectifiers arising from Schottky barrier lowering and tunneling currents. The next chapter describes power JFET and MESFET structures with planar and trench gate regions. These normally-on structures can be used to construct the *Baliga-Pair* circuit[13], described in chapter eight, which has been shown to be ideally suitable for motor control applications. This represents a near term solution to taking advantage of the low specific on-resistance of the silicon carbide drift region while awaiting resolution of reliability issues with the gate oxide in silicon carbide power MOSFET structures.

Chapter nine provides a description of silicon carbide power MOSFET structures with emphasis on problems associated with simply replicating structures originally developed in silicon technology. Innovative approaches to prevent high electric field in the gate oxide of silicon carbide power MOSFETs are then discussed in chapter ten. In this chapter, accumulation-mode structures are shown to provide the advantages of larger channel mobility and lower threshold voltage leading to a reduction in the specific on-resistance. In chapter eleven, issues with adopting the silicon UMOSFET structure to silicon carbide are enunciated followed by analysis of solutions to these problems in chapter twelve. Once again, methods for shielding the gate oxide are shown to enable reduction of the electric field in the gate oxide to acceptable levels. The shielding is also demonstrates to ameliorate the base reach-through problem allowing a reduction of the base width and hence the channel resistance contribution.

The application of the charge-coupling concept to silicon carbide is explored in chapter thirteen to reduce the resistance of the drift region. This approach is applicable to very high voltage structures with blocking voltage capability exceeding 3000 volts. In chapter fourteen, the operation of the integral diode within the silicon carbide power MOSFET structures is analyzed for utilization as a fly-back rectifier in motor control applications. A unique mode of current conduction in the silicon carbide power MOSFET in the third quadrant of operation is identified here which allows for a reduction of the on-state voltage drop of the integral diode.

Although the emphasis in this book has been on discrete vertical silicon carbide structures, for the sake of completeness, the fifteenth chapter describes high voltage lateral silicon carbide structures that are suitable for integration with CMOS circuits. In the concluding sixteenth chapter, the performance of silicon carbide devices is compared with that of silicon devices using a typical motor control application as an example to provide the reader with a perspective on the benefits of using silicon carbide devices. From an applications perspective, it is demonstrated that the replacement of the silicon P-i-N rectifier with the silicon carbide JBS rectifier provides a significant reduction in the power losses for the system as well as the stress experienced by the silicon IGBT. This approach is already becoming a commercially viable option due to the availability of silicon carbide Schottky rectifier products from several companies.

Throughout the book, experimental results are included whenever pertinent to each chapter. This provides a historical context and a brief summary of the state of the art for silicon carbide devices. Issues that must be addressed before commercialization of silicon carbide structures becomes viable are pointed out in these sections of the book.

I am hopeful that this book will be used for the teaching of courses on solid state devices and that it will make an essential reference for the power device industry. To facilitate this, analytical solutions are provided throughout the book that can be utilized to understand the underlying physics and model the structures. Comparison of the silicon carbide structures with their silicon counterparts is also included whenever pertinent to the discussion.

Ever since my identification of the benefits of utilizing wide band gap semiconductors for the development of superior power devices twenty-five years ago, it has been my mission to resolve issues that would impede their commercialization. I wish to thank the sponsors of the Power Semiconductor Research Center and the Office of Naval Research for supporting this mission during the last fifteen years. This has been essential to providing the resources required to create many breakthroughs in the silicon carbide technology which I hope will enable commercialization of the technology in the near future. I look forward to observing the benefits accrued to society by adopting the silicon carbide technology for conservation of fossil fuel usage resulting in reduced environmental pollution.

Prof. B. Jayant Baliga
2005

References

[1] B.J. Baliga, "Power Semiconductor Devices", PWS Publishing Company, 1996.

[2] S. K. Ghandhi, "Semiconductor Power Devices", John Wiley and Sons, 1977.

[3] B. J. Baliga, "Evolution of MOS-Bipolar Power Semiconductor Technology", Proceedings IEEE, pp. 409-418, April 1988.

[4] B.J. Baliga, "Semiconductors for High Voltage Vertical Channel Field Effect Transistors", J. Applied Physics, Vol. 53, pp. 1759-1764, March 1982.

[5] B.J. Baliga, R. Ehle, J.R. Shealy, and W. Garwacki, "Breakdown Characteristics of Gallium Arsenide", IEEE Electron Device Letters, Vol. EDL-2, pp. 302-304, November 1981.

[6] B. J. Baliga, A.R. Sears, M.M. Barnicle, P.M. Campbell, W. Garwacki, and J.P. Walden, "Gallium Arsenide Schottky Power Rectifiers", IEEE Transactions on Electron Devices, Vol. ED-32, pp. 1130-1134, June 1985.

[7] P.M. Campbell, W. Garwacki, A.R. Sears, P. Menditto, and B.J. Baliga, "Trapezoidal-Groove Schottky-Gate Vertical-Channel GaAs FET", IEEE Electron Device Letters, Vol. EDL-6, pp. 304-306, June 1985.

[8] M. Bhatnagar, P.K. McLarty, and B.J. Baliga, "Silicon-Carbide High-Voltage (400 V) Schottky Barrier Diodes", IEEE Electron Device Letters, Vol. EDL-13, pp.501-503, October 1992.

[9] R.K. Chilukuri and B.J. Baliga, "High Voltage Ni/4H-SiC Schottky Rectifiers", International Symposium on Power Semiconductor Devices and ICs, pp. 161-164, May 1999.

[10] R. Raghunathan and B.J. Baliga, "Temperature dependence of Hole Impact Ionization Coefficients in 4H and 6H-SiC", Solid State Electronics, Vol. 43, pp. 199-211, February 1999.

[11] P.M. Shenoy and B.J. Baliga, "High Voltage Planar 6H-SiC ACCUFET", International Conference on Silicon Carbide, III-Nitrides, and Related Materials, Abstract Tu3b-3, pp. 158-159, August 1997.

[12] R.K. Chilukuri and B.J. Baliga, PSRC Technical Report TR-00-007, May 2000.

[13] B.J. Baliga, See Chapter 7 in "Power Semiconductor Devices", pp. 417-420, PWS Publishing Company, 1996.

Contents

Chapter 1

Introduction

The increasing dependence of modern society on electrical appliances for comfort, transportation, and healthcare has motivated great advances in power generation, power distribution and power management technologies. These advancements owe their allegiance to enhancements in the performance of power devices that regulate the flow of electricity. After the displacement of vacuum tubes by solid state devices in the 1950s, the industry relied upon silicon bipolar devices, such as bipolar power transistors and thyristors. Although the ratings of these devices grew rapidly to serve an ever broader system need, their fundamental limitations in terms of the cumbersome control and protection circuitry led to bulky and costly solutions. The advent of MOS technology for digital electronics enabled the creation of a new class of devices in the 1970s for power switching applications as well. These silicon power MOSFETs have found extensive use in high frequency applications with relatively low operating voltages (under 100 volts). The merger of MOS and bipolar physics enabled creation of yet another class of devices in the 1980s. The most successful innovation in this class of devices has been the Insulated Gate Bipolar transistor (IGBT). The high power density, simple interface, and ruggedness of the IGBT have made it the technology of choice for all medium and high power applications with perhaps the exception of high voltage DC transmission systems.

Power devices are required for systems that operate over a broad spectrum of power levels and frequencies. In Fig. 1.1, the applications for power devices are shown as a function of operating frequency. High power systems, such as HVDC power distribution and locomotive drives, requiring the control of megawatts of power operate at relatively low frequencies. As the operating frequency increases, the power ratings decrease for the devices with typical microwave devices handling about 100 watts. All of these applications are served by silicon devices. Thyristors are favored for the low frequency, high power applications,

IGBTs for the medium frequency and power applications, and power MOSFETs for the high frequency applications.

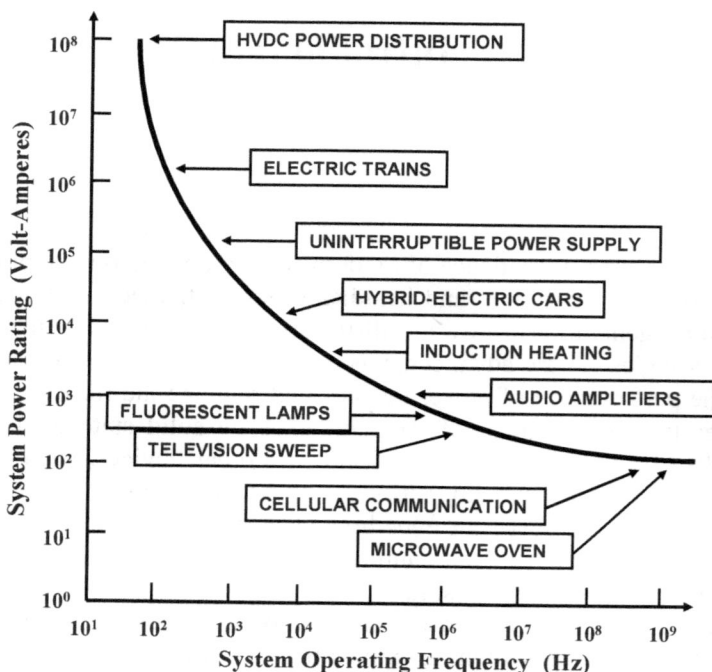

Fig. 1.1 Applications for Power Devices.

Another approach to classification of applications for power devices is in terms of their current and voltage handling requirements as shown in Fig. 1.2. On the high power end of the chart, thyristors are available that can individually handle over 6000 volts and 2000 amperes enabling the control of over 10 megawatts of power by a single monolithic device. These devices are suitable for the HVDC power transmission and locomotive drive (traction) applications. For the broad range of systems that require operating voltages between 300 volts and 3000 volts with significant current handling capability, the IGBT has been found to be the optimum solution. When the current requirements fall below 1 ampere, it is feasible to integrate multiple devices on a single monolithic chip to provide greater functionality for systems such as telecommunications and display drives. However, when the current exceeds a few amperes, it is more cost effective to use discrete power

MOSFETs with appropriate control ICs to serve applications such as automotive electronics and switch mode power supplies.

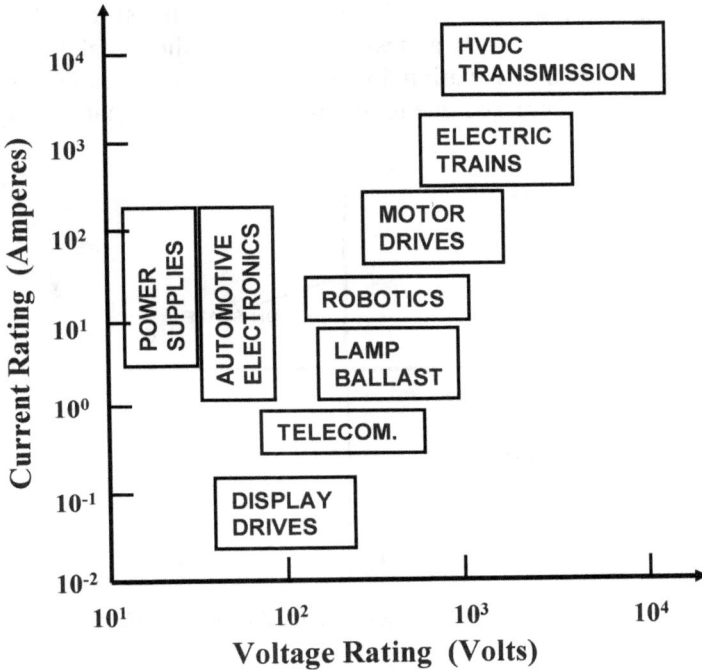

Fig. 1.2 System Ratings for Power Devices.

1.1 Ideal and Typical Power Device Characteristics

Although silicon power devices have served the industry for well over five decades, they cannot be considered to have ideal device characteristics. An ideal power rectifier should exhibit the *i-v* characteristics shown in Fig. 1.3. In the forward conduction mode, the first quadrant of operation the figure, it should be able to carry any amount of current with zero on-state voltage drop. In the reverse blocking mode, the third quadrant of operation in the figure, it should be able to hold off any value of voltage with zero leakage current. Further, the ideal rectifier should be able to switch between the on-state and the off-state with zero switching time. Actual silicon power rectifiers exhibit the *i-v* characteristics illustrated in Fig. 1.4. They have a finite voltage drop (V_{ON}) when carrying current on the on-state leading to 'conduction' power loss. They also have a finite leakage current (I_{OFF}) when blocking voltage in the off-state creating power loss. In addition, the doping concentration and thickness of the drift region of the silicon device must be carefully chosen with a design target for the breakdown voltage (BV)[1]. The power dissipation in power devices increases when their voltage rating is increased due to an increase in the on-state voltage drop.

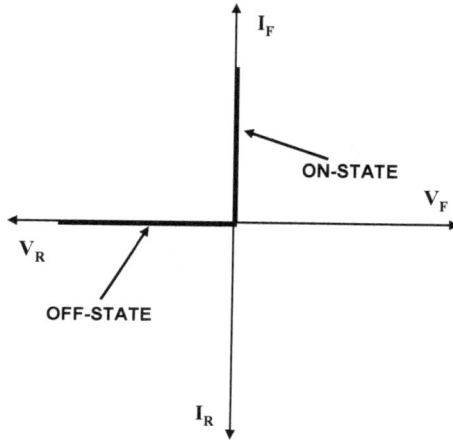

Fig. 1.3 Characteristics of an Ideal Power Rectifier.

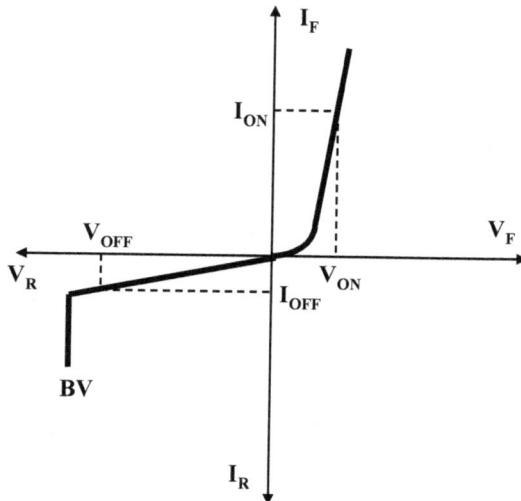

Fig. 1.4 Characteristics of a Typical Power Rectifier.

The *i-v* characteristics of an ideal power switch are illustrated in Fig. 1.5. As in the case of the ideal rectifier, the ideal transistor conducts current in the on-state with zero voltage drop and blocks voltage in the off-state with zero leakage current. In addition, the ideal device can operate with a high current and voltage in the active region with the forward current in this mode controlled by the applied gate bias. The spacing between the characteristics in the active region is uniform for an ideal transistor indicating a gain that is independent of the forward current and voltage.

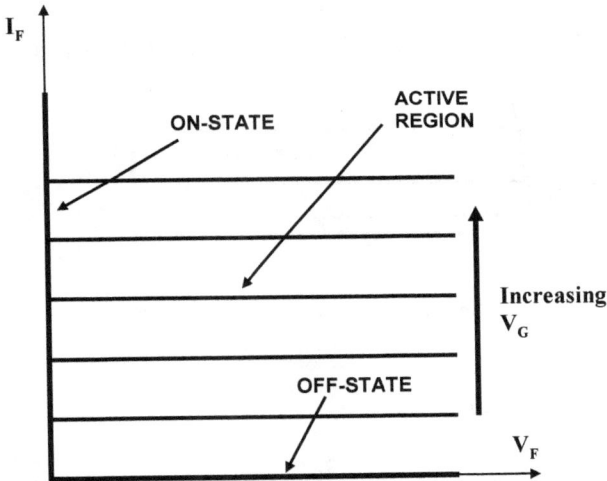

Fig. 1.5 Characteristics of an Ideal Transistor.

The *i-v* characteristics of a typical power switch are illustrated in Fig. 1.6. This device exhibits a finite resistance when carrying current in the on-state as well as a finite leakage current while operating in the off-state (not shown in the figure because its value is much lower than the on-state current levels). The breakdown voltage of a typical transistor is also finite as indicated in the figure with 'BV'. The typical transistor can operate with a high current and voltage in the active region. This current is controlled by the base current for a bipolar transistor while it is determined by a gate voltage for a MOSFET or IGBT (as indicated in the figure). It is preferable to have gate voltage controlled characteristics because the drive circuit can be integrated to reduce its cost. The spacing between the characteristics in the active region is non-uniform for a typical transistor with a square-law behavior for devices operating with channel pinch-off in the current saturation mode[1]. Recently, devices operating under a new super-linear mode have been proposed and demonstrated for wireless base-station applications[2].

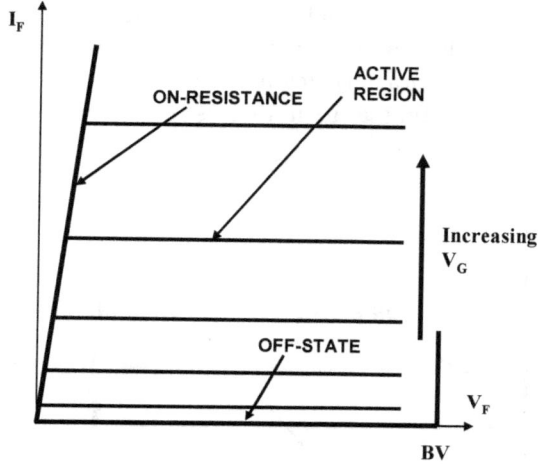

Fig. 1.6 Characteristics of a Typical Transistor.

1.2 Unipolar Power Devices

Bipolar power devices operate with the injection of minority carriers during on-state current flow. These carriers must be removed when the switching the device from the on-state to the off-state. This is accomplished by either charge removal via the gate drive current or via the electron-hole recombination process. These processes introduce significant power losses that degrade the power management efficiency. It is therefore preferable to utilize unipolar current conduction in a power device. The commonly used unipolar power diode structure is the Schottky rectifier that utilizes a metal-semiconductor barrier to produce current rectification. The high voltage Schottky rectifier structure also contains a drift region, as show in Fig. 1.7, which is designed to support the reverse blocking voltage. The resistance of the drift region increases rapidly with increasing blocking voltage capability as discussed later in this chapter. Silicon Schottky rectifiers are commercially available with blocking voltages of up to 100 volts. Beyond this value, the on-state voltage drop of silicon Schottky rectifiers becomes too large for practical applications[1]. Silicon P-i-N rectifiers are favored for designs with larger breakdown voltages due to their lower on-state voltage drop despite their slower switching properties. As shown later in the book, silicon carbide Schottky rectifiers have much lower drift region resistance enabling design of very high voltage devices with low on-state voltage drop.

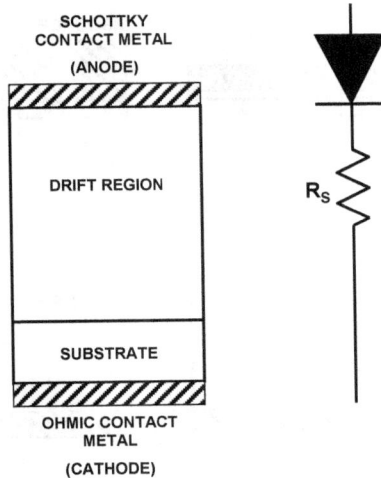

Fig. 1.7 The Power Schottky Rectifier Structure and its Equivalent Circuit.

The most commonly used unipolar power transistor is the silicon power Metal-Oxide-Semiconductor Field-Effect-Transistor or MOSFET. Although other structures, such as JFETs or SITs have been explored[3], they have not been popular for power electronic applications because of their normally-on behavior. The commercially available silicon power MOSFETs are based upon the structures shown in Fig. 1.8. The D-MOSFET was first commercially introduced in the 1970's and contains a 'planar-gate' structure. The P-base region and the N^+ source regions are self-aligned to the edge of the Polysilicon gate electrode by using ion-implantation of boron and phosphorus with their respective drive-in thermal cycles. The n-type channel is defined by the difference in the lateral extension of the junctions under the gate electrode. The device supports positive voltage applied to the drain across the P-base/N-drift region junction. The voltage blocking capability is determined by the doping and thickness of the drift region. Although low voltage (< 100 V) silicon power MOSFET have low on-resistances, the drift region resistance increases rapidly with increasing blocking voltage limiting the performance of silicon power MOSFETs to below 200 volts. It is common-place to use the Insulated Gate Bipolar Transistors (IGBT) for higher voltage designs.

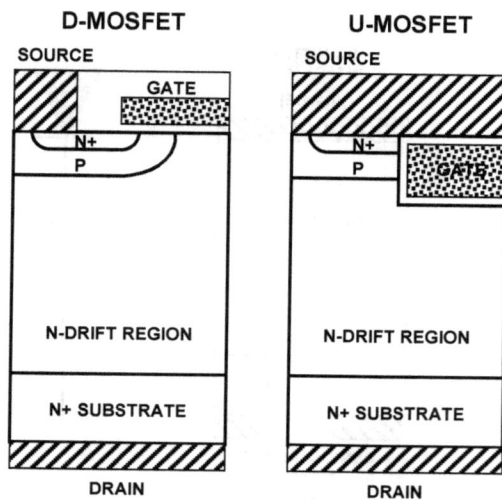

Fig. 1.8 Silicon Power MOSFET Structures.

The silicon U-MOSFET structure became commercially available in the 1990s. It has a gate structure embedded within a trench etched into the silicon surface. The N-type channel is formed on the side-wall of the trench at the surface of the P-base region. The channel length is determined by the difference in vertical extension of the P-base and N^+ source regions as controlled by the ion-implant energies and drive times for the dopants. The silicon U-MOSFET structure was developed to reduce the on-state resistance by elimination of the JFET component within the D-MOSFET structure[1].

1.3 Bipolar Power Devices

The commonly available silicon power bipolar devices are the bipolar transistor and the gate turn-off thyristor or GTO. These devices were originally developed in the 1950's and widely used for power switching applications until the 1970s when the availability of silicon power

MOSFETs and IGBTs supplanted them. The structures of the bipolar transistor and GTO are shown in Fig. 1.9. In both devices, injection of minority carriers into the drift region modulates its conductivity reducing the on-state voltage drop. However, this charge must be subsequently removed during switching resulting in high turn-off power losses. These devices require a large control (base or gate) current which must be implemented with discrete components leading to an expensive bulky system design.

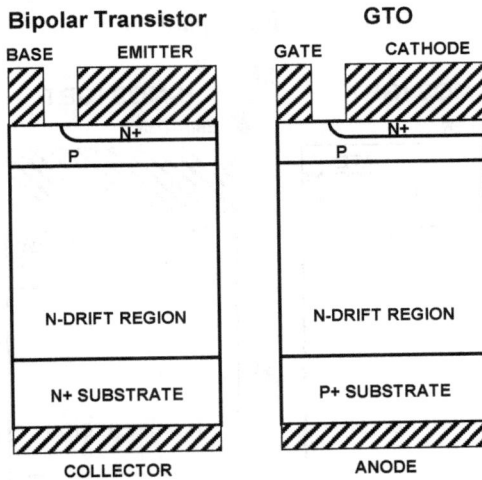

Fig. 1.9 Silicon Bipolar Device Structures.

Several groups have worked on the development of bipolar transistors[4,5,6,7] and GTOs[8,9] using silicon carbide. The large junction potential for silicon carbide results in a relatively high on-state voltage drop for these devices when compared with commercially available silicon devices. For this reason, the focus of this book is on unipolar silicon carbide structures.

1.4 MOS-Bipolar Power Devices

The most widely used silicon high voltage (>300 volt) device for power switching applications is the *Insulated Gate Bipolar Transistor* (IGBT) which was developed in the 1980s by combining the physics of operation of bipolar transistors and MOSFETs[10]. The structure of the IGBT is deceptively similar to that for the power MOSFET as shown in Fig. 1.10 if viewed as a mere replacement of the N^+ substrate with a P^+ substrate. This substitution creates a four-layer parasitic thyristor which can latch up resulting in destructive failure due to loss of gate control. Fortunately the parasitic thyristor can be defeated by the addition of the P^+ region within the cell[1]. The benefit of the P^+ substrate is the injection of minority carriers into the N-drift region resulting in greatly reducing its resistance. This has enabled the development of high voltage IGBT products with high current carrying capability.

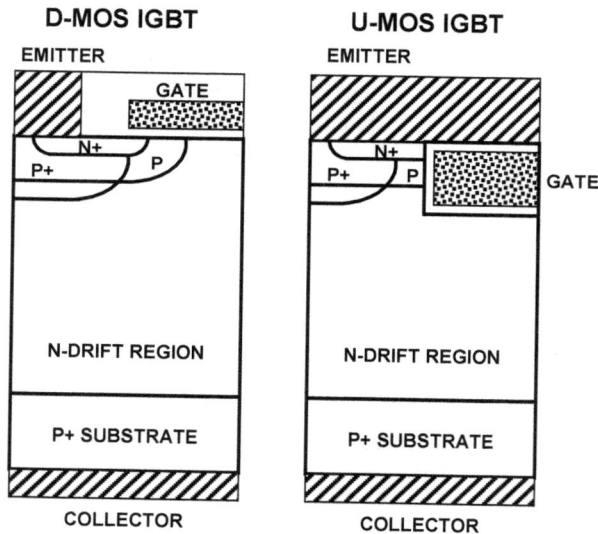

Fig. 1.10 IGBT Device Structures.

The development of the IGBT in silicon carbide has been analyzed and attempted by a few research groups[11,12,13]. The large junction potential and high resistance of the P^+ substrate for silicon carbide results in a relatively high on-state voltage drop for these devices when compared with commercially available silicon devices. Their switching speed is also compromised by the injected stored charge in the on-state. For these reasons, silicon carbide based IGBTs are not discussed in this book.

1.5 Ideal Drift Region for Unipolar Power Devices

The unipolar power devices discussed above all contain a drift region which is designed to support the blocking voltage. The properties (doping concentration and thickness) of the *ideal drift region* can be analyzed by assuming an abrupt junction profile with high doping concentration on one side and a low uniform doping concentration on the other side, while neglecting any junction curvature effects by assuming a parallel-plane configuration. The resistance of the ideal drift region can then be related to the basic properties of the semiconductor material[14].

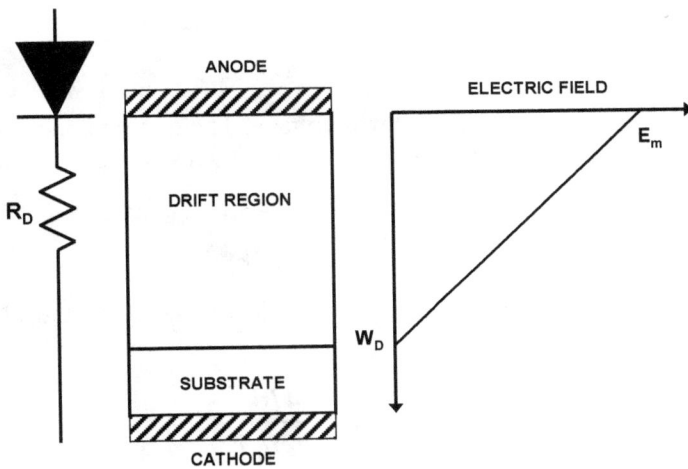

Fig. 1.11 The *Ideal Drift Region* and its Electric Field Distribution.

The solution of Poisson's equation leads to a triangular electric field distribution within a uniformly doped drift region[1] with the slope of the field profile being determined by the doping concentration. The maximum voltage that can be supported by the drift region is determined by the maximum electric field (E_m) reaching the critical electric field (E_c) for breakdown for the semiconductor material. The critical electric field for breakdown and the doping concentration then determine the maximum depletion width (W_D).

The specific resistance (resistance per unit area) of the ideal drift region is given by:

$$R_{on.sp} = \left(\frac{W_D}{q \mu_n N_D} \right)$$

[**1.1**]

Since this resistance was initially considered to be the lowest value achievable with silicon devices, it has historically been referred to as the *ideal specific on-resistance of the drift region*. More recent introduction of the charge-coupling concept, described later in this chapter, has enabled reducing the drift region resistance of silicon devices to below the values predicted by this equation. The depletion width under breakdown conditions is given by:

$$W_D = \frac{2BV}{E_C}$$

[**1.2**]

where BV is the desired breakdown voltage. The doping concentration in the drift region required to obtain this breakdown voltage is given by:

$$N_D = \frac{\varepsilon_S . E_C^2}{2.q.BV}$$

[**1.3**]

Combining these relationships, the specific resistance of the ideal drift region is obtained:

$$R_{on-ideal} = \frac{4BV^2}{\varepsilon_S \mu_n E_C^3}$$

[**1.4**]

The denominator of this equation ($\varepsilon_S . \mu_n . E_C^3$) is commonly referred to as *Baliga's Figure of Merit for Power Devices*. It is an indicator of the impact of the semiconductor material properties on the resistance of the drift region. The dependence of the drift region resistance on the mobility (assumed to be for electrons here because in general they have higher mobility values than for holes) of the carriers favors semiconductors such as Gallium Arsenide. However, the much stronger (cubic) dependence of the on-resistance on the critical electric field for breakdown favors wide band gap semiconductors such as silicon carbide. The critical electric field for breakdown is determined by the impact ionization coefficients for holes and electrons in semiconductors as discussed in the next chapter.

The change in the specific on-resistance for the drift region with critical electric field and mobility is shown in Fig. 1.12 for the case of a breakdown voltage of 1000 volts. The location of the properties for silicon, gallium arsenide, and silicon carbide are shown in the figure by the points. The improvement in drift region resistance for GaAs in comparison with silicon is largely due to its much greater mobility for electrons. The improvement in drift region resistance for SiC in comparison with silicon is largely due to its much larger critical electric field for' breakdown. Based upon these considerations, excellent high voltage Schottky rectifiers were developed from GaAs in the 1980s[15]. Much greater improvement in performance is clearly indicated for silicon carbide devices which is the subject of this book.

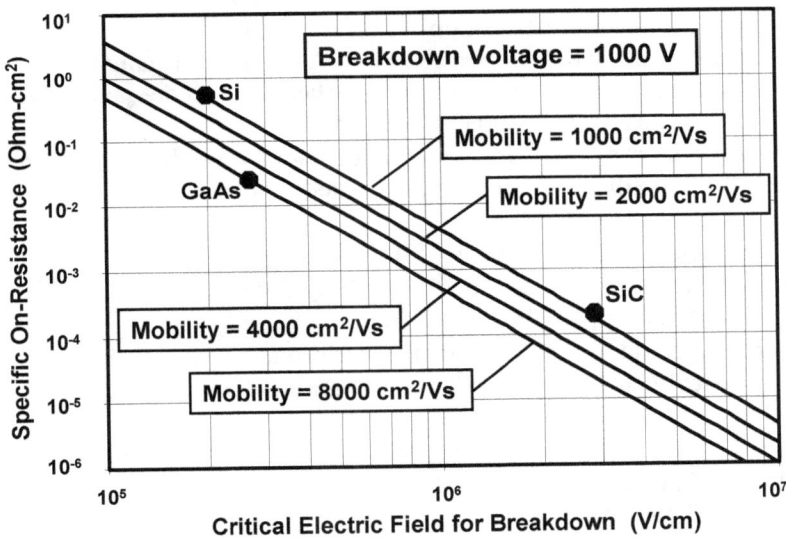

Fig. 1.12 Specific On-Resistance of the *Ideal Drift Region.*

1.6 Summary

The motivation for the development of silicon carbide unipolar devices has been reviewed in this chapter. Although excellent unipolar silicon Schottky rectifiers and power MOSFETs are commercially available with breakdown voltages below 200 volts, the resistance of their drift region increases rapidly at higher breakdown voltages producing significant power losses in applications. This problem can be overcome using silicon carbide based unipolar devices.

References

[1] B.J. Baliga, "Power Semiconductor Devices", PWS Publishing Company, 1996.

[2] B. J. Baliga, "Silicon RF Power Devices", World Scientific Press, 2005.

[3] B. J. Baliga, "Modern Power Devices", John Wiley and Sons, 1987.

[4] E. Danielsson, et al, "Extrinsic base design of SiC bipolar transistors", Silicon Carbide and Related Materials, pp. 1117-1120, 2003.

[5] I. Perez-Wurfll, et al, "Analysis of power dissipation and high temperature operation in 4H-SiC Bipolar Junction Transistors", Silicon Carbide and Related Materials, pp. 1121-1124, 2003.

[6] J. Zhang, et al, "High Power (500V-70A) and High gain(44-47) 4H-SiC Bipolar Junction Transistors", Silicon Carbide and Related Materials, pp. 1149-1152, 2003.

[7] A. Agarwal, et al, "SiC BJT Technology for Power Switching and Rf Applications", Silicon Carbide and Related Materials 2003, pp. 1141-1144, 2004.

[8] P. Brosselard, et al, "Influence of different peripheral protections on the breakover voltage of a 4H-SiC GTO thyristor", Silicon Carbide and Related Materials, pp. 1129-1132, 2003.

[9] A.K. Agarwal, et al, "Dynamic Performance of 3.1 kV 4H-SiC Asymmetrical GTO Thyristors", Silicon Carbide and Related Materials, pp. 1349-1352, 2003.

[10] B. J. Baliga, "Evolution of MOS-Bipolar Power Semiconductor Technology", Proceedings of the IEEE, pp. 409-418, 1988.

[11] T.P. Chow, N. Ramaungul, and M. Ghezzo, "Wide Bandgap Semiconductor Power devices", Materials Research Society Symposium, Vol. 483, pp. 89-102, 1998.

[12] R. Singh, "Silicon Carbide Bipolar power Devices – Potentials and Limits", Materials Research Society Symposium, Vol. 640, pp. H4.2.1-12, 2001.

[13] J. Wang, et al, "Comparison of 5kV 4H-SiC N-channel and P-channel IGBTs", Silicon Carbide and Related Materials, pp. 1411-1414, 2000.

[14] B. J. Baliga, "Semiconductors for High Voltage Vertical Channel Field Effect Transistors", J. Applied Physics, Vol. 53, pp. 1759-1764, 1982.

[15] B. J. Baliga, et al, "Gallium Arsenide Schottky Power Rectifiers", IEEE Transactions on Electron Devices, Vol. ED-32, pp. 1130-1134, 1985.

Chapter 2

Material Properties and Technology

In the previous chapter, it was demonstrated the resistance of the drift region in unipolar vertical power devices is strongly dependent upon the fundamental properties of the semiconductor material. In this chapter, the measured properties for silicon carbide are reviewed and compared with those for silicon. These properties are then used to obtain other parameters (such as the built-in potential) which are relevant to the analysis of the performance of devices discussed in the rest of the book. The discussion is constrained to the 4H poly-type of silicon carbide because its properties are superior to those reported for the 6H poly-type. The parameters given in this chapter were used during two-dimensional numerical simulations to determine the characteristics of various devices discussed in this book. Data on the 6H poly-type is sometimes included to demonstrate the reasons that the 4H poly-type is preferred for power device structures. The 4H-SiC material is commercially available from several sources for the development and manufacturing of power devices because of its superior properties. Manufacturers are continually reducing the micropipe density in this material while increasing the diameter of commercially available wafers.

The second portion of this chapter provides a review of the technology developed for the fabrication of silicon carbide power devices. One of the advantages of silicon carbide is that, unlike gallium arsenide, its processing is compatible with silicon devices without the problem of contamination of production lines. It is advantageous to implement any available silicon process technologies, such as ion-implantation or reactive ion etching, to the fabrication of silicon carbide devices. This approach has been generally successful with the exception of the need to anneal the implants at much higher temperatures. Empirical studies performed on the oxidation and passivation of the silicon carbide surface, as well those on the formation of metal-contacts, are also reviewed in this portion of the chapter.

15

2.1 Fundamental Properties

This section provides a summary of the fundamental properties of silicon carbide[1,2,3] and compares them with those for silicon[4]. Only the properties for the 4H poly-type have been included here as discussed above. The basic properties of relevance to power devices are the energy band gap, the impact ionization coefficients, the dielectric constant, the thermal conductivity, the electron affinity, and the carrier mobility. Since the intrinsic carrier concentration and the built-in potential can be extracted by using this information, their values have also been computed and compared with those for silicon in this section.

Properties	Silicon	4H-SiC
Energy Band Gap (eV)	1.10	3.26
Relative Dielectric Constant	11.7	9.7
Thermal Conductivity (W/cm K)	1.5	3.7
Electron Affinity (eV)	4.05	3.7
Density of States Conduction Band (cm^{-3})	2.80×10^{19}	1.23×10^{19}
Density of States Valence Band (cm^{-3})	1.04×10^{19}	4.58×10^{18}

Fig. 2.1 Fundamental Material Properties.

2.1.1 Energy Band Gap

The energy band gap of silicon carbide is much larger than that for silicon allowing its classification as a wide band gap semiconductor. The band gap for 4H-SiC is 3.26 eV, which is 3 times larger than that for silicon. The band gap for 6H-SiC of 3.0 eV is slightly smaller. Since a larger band gap results in a smaller generation of carriers in the depletion regions of devices, it is favorable for reducing the leakage current of devices which utilize P-N junctions to support voltages. The larger band gap is also favorable for producing Metal-Semiconductor contacts with larger Schottky barrier heights. This enables reduction of the leakage current in Schottky rectifiers and MESFETs.

The intrinsic carrier concentration is determined by the thermal generation of electron hole pairs across the energy band gap of a semiconductor. Its value can be calculated by using the energy band gap (E_G) and the density of states in the conduction (N_C) and valence (N_V) bands:

$$n_i = \sqrt{n.p} = \sqrt{N_C.N_V}\, e^{-(E_G/2kT)} \qquad [2.1]$$

where k is Boltzmann's constant (1.38 x 10^{-23} J/°K) and T is the absolute temperature. For silicon, the intrinsic carrier concentration is given by:

$$n_i = 3.87x10^{16} T^{3/2} e^{-(7.02x10^3)/T} \qquad [2.2]$$

while that for 4H-SiC, it is given by:

$$n_i = 1.70x10^{16} T^{3/2} e^{-(2.08x10^4)/T} \qquad [2.3]$$

Using these equations, the intrinsic carrier concentration can be calculated as a function of temperature. The results are plotted in Fig. 2.2 and Fig. 2.3. The traditional method for making this plot is shown in Fig. 2.2 by using an inverse temperature scale for the x-axis. The additional Fig. 2.3 is provided here because it is easier to relate the value of the intrinsic carrier concentration to the temperature with it plotted on the x-axis. It is obvious that the intrinsic carrier concentration for silicon carbide is far smaller than for silicon due to the large difference in band gap energy. At room temperature (300°K), the intrinsic carrier concentration for silicon is 1.4 x 10^{10} cm^{-3} while that for 4H-SiC is only 6.7 x 10^{-11} cm^{-3}. This demonstrates that the bulk generation current is negligible for the determination of the leakage current in silicon carbide

devices. Surface generation currents may be much larger due to the
presence of states within the band gap.

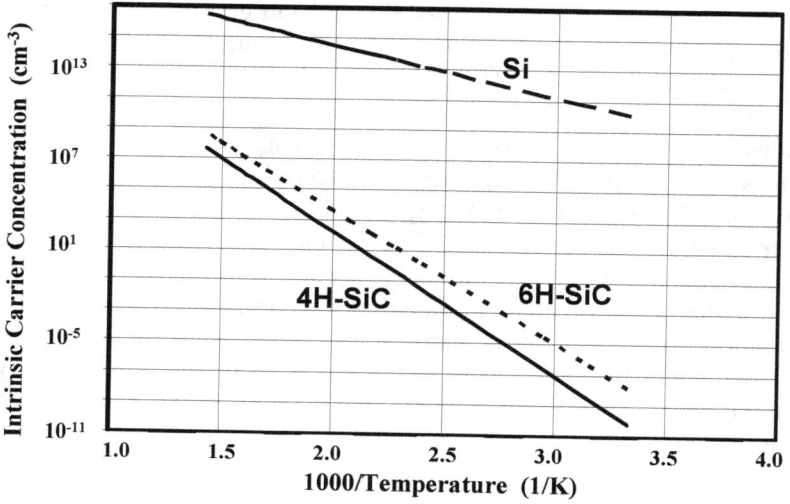

Fig. 2.2 Intrinsic Carrier Concentration.

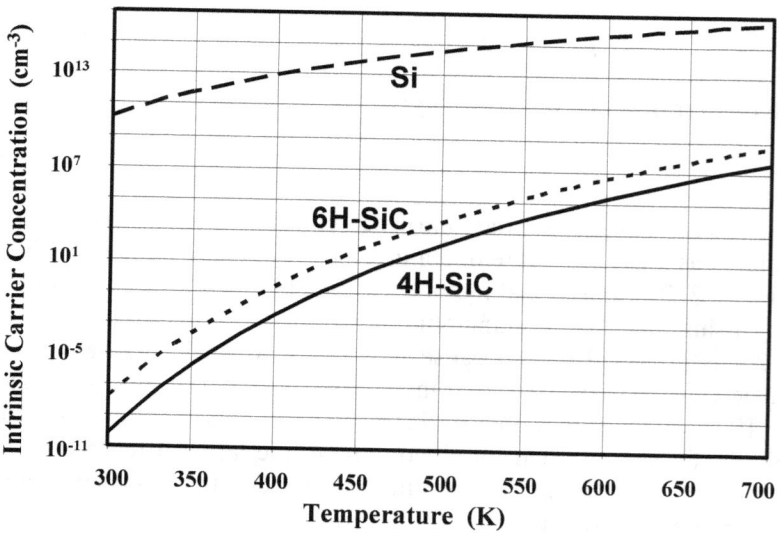

Fig. 2.3 Intrinsic Carrier Concentration.

For silicon, the intrinsic carrier concentration becomes equal to a typical doping concentration of 1 x 10^{15} cm^{-3} at a relatively low temperature of 540 °K or 267 °C. In contrast, the intrinsic carrier concentration for 4H-SiC is only 3.9 x 10^7 cm^{-3} even at 700 °K or 427 °C. The development of mesoplasmas has been associated with the intrinsic carrier concentration becoming comparable to the doping concentration[5]. Mesoplasmas create current filaments that have very high current density leading to destructive failure in semiconductors. This is obviously much less likely in silicon carbide.

The built-in potential of P-N junctions can play an important role in determining the operation and design of power semiconductor devices. As an example, the built-in potential determines the zero-bias depletion width which is important for calculation of the on-resistance in power DMOSFETs. It has a strong impact on the on-state voltage drop in JBS rectifiers, and is an important parameter used for the design of normally-off accumulation mode MOSFETs. These structures are discussed in detail later in the book. The built-in voltage is given by:

$$V_{bi} = \frac{kT}{q} \ln\left(\frac{N_A^- . N_D^+}{n_i^2} \right)$$ [2.4]

where N_A^- and N_D^+ are the ionized impurity concentrations on the two sides of an abrupt P-N junction. For silicon, their values are equal to the doping concentration because of the small dopant ionization energy levels. This does not apply to silicon carbide because of the much larger dopant ionization levels as discussed later in this chapter.

The calculated built-in potential is shown in Fig. 2.4 as a function of temperature for a typical operating range for power devices. In making the plots, the product ($N_A^- . N_D^+$) was assumed to be 10^{35} cm^{-6}. This would be applicable for a typical doping concentration of 1 x 10^{19} cm^{-3} on the heavily doped side of the junction and 1 x 10^{16} cm^{-3} on the lightly doped side of the junction. The built-in potential for silicon carbide is much larger than that for silicon due to the far smaller values for the intrinsic carrier concentration. This can be a disadvantage due to the larger zero bias depletion width for silicon carbide as shown later in the chapter. As an example, the larger zero bias depletion width consumes space within the D-MOSFET cell structure increasing the on-resistance by constricting the area through which current can flow. In contrast, the larger built-in potential for SiC and its associated larger zero bias depletion width can be taken advantage of in making innovative device structures, such as the ACCUFET, that are tailored to the unique properties of SiC.

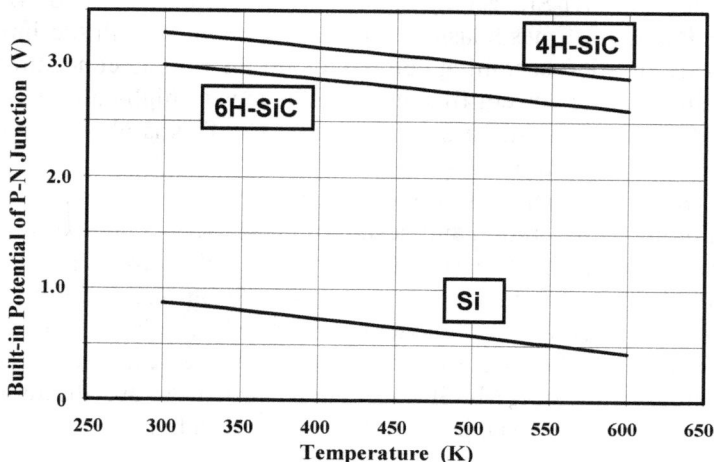

Fig. 2.4 Built-in Potential for a P-N Junction in Si and SiC.

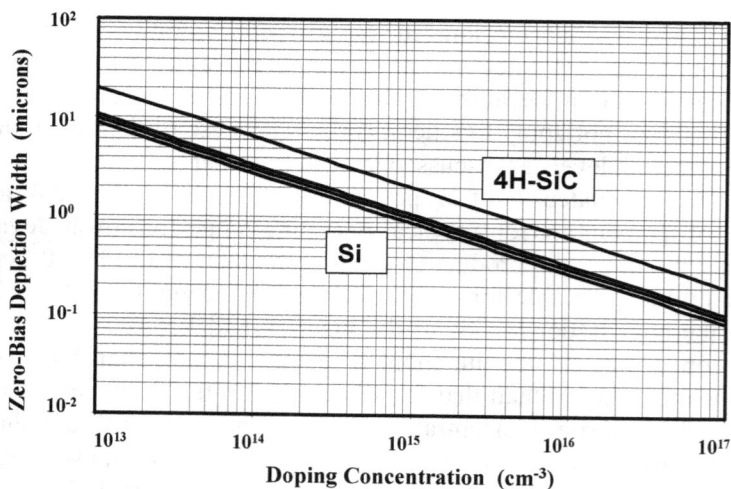

Fig. 2.5 Zero-Bias Depletion Width in Si and SiC.

The calculated zero bias depletion width is shown in Fig. 2.5 as a function of the doping concentration on the lightly doped side of the abrupt P-N junction. For the case of silicon, three lines are given for 300°K, 400°K and 500°K with largest depletion width corresponding to the lowest temperature. For silicon carbide, the lines for these temperatures are very close to each other. For both silicon and silicon carbide, it can be seen that the zero bias depletion width can be substantial in size at low doping concentrations making it important to take this into account during device design and analysis. The zero bias depletion width for SiC is much (about 2x) larger than for silicon for the same doping concentration. However, it is worth noting that the doping concentration in SiC is also much greater for a given breakdown voltage than for silicon as discussed later in the chapter.

2.1.2 Impact Ionization Coefficients

The main advantage of wide band gap semiconductors for power device applications is the greater breakdown voltage capability of these materials. The higher breakdown voltage for these materials is associated with the reduced impact ionization coefficients at any given electric field when compared with silicon. The impact ionization coefficients for semiconductors is dictated by Chynoweth's Law[6,7]:

$$\alpha = a.e^{-b/E} \qquad\qquad [2.5]$$

where E is the electric field component in the direction of current flow. The parameters a and b are constants that depend upon the semiconductor material and the temperature. For silicon, the impact ionization rate is much larger for electrons than for holes. However, for silicon carbide the opposite has been found to hold true. For silicon, the impact ionization rates have been measured as a function of electric field and temperature[4,8,9]. The data can be modeled by using $a = 7$ x 10^5 per cm and $b = 1.23$ x 10^6 V/cm at room temperature[10].

The impact ionization coefficients for silicon carbide have been recently measured as a function of temperature by using an electron beam excitation method[11]. This method allowed extraction of impact ionization rates in defect free regions of the material by isolating diodes containing defects with an EBIC (Electron Beam Induced Current) image. This was important because substantially enhanced impact ionization rates were discovered when the measurements were conducted at defect sites[12]. For defect free material, the extracted values for the impact ionization coefficient parameters for holes in 4H-SiC were found to be:

$$a(4H - SiC) = 6.46x10^6 - 1.07x10^4 T \qquad [\ 2.6]$$

with

$$b(4H - SiC) = 1.75x10^7 \qquad [\ 2.7]$$

while for 6H-SiC:

$$a(6H - SiC) = 4.6x10^6 - 7.4x10^3 T \qquad [\ 2.8]$$

with

$$b(6H - SiC) = 1.5x10^7 \qquad [\ 2.9]$$

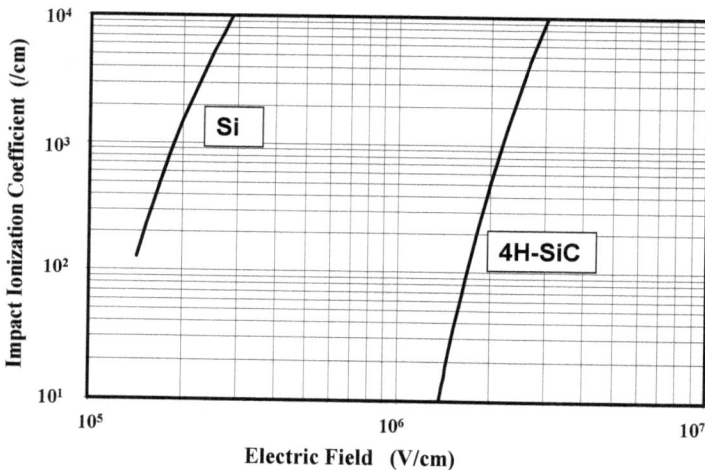

Fig. 2.6 Impact Ionization Coefficients in 4H-SiC and Silicon.

　　　The impact ionization coefficients for 4H-SiC can be compared with those for silicon in Fig. 2.6. This behavior was incorporated in the analytical models used for two-dimensional numerical simulations of device structures discussed later in the book. It can be seen that the onset of significant generation of carriers by impact ionization occurs at much larger electric fields in 4H-SiC when compared with silicon. As a consequence, breakdown in 4H-SiC devices occurs when the electric fields are in the range of 2-3 x 10^6 V/cm – an order of magnitude larger than that for silicon. This implies a much larger critical electric field E_c

for breakdown for 4H-SiC resulting in a big increase in the *Baliga's Figure of Merit*. The critical electric field for breakdown is discussed in more detail in the next chapter.

2.1.3 Electron Mobility

The mobility for electrons is larger than that for holes in silicon carbide. It is therefore favorable to make unipolar devices using N-type drift regions rather than P-type drift regions. The conductivity of the drift region is given by:

$$\rho_n = \frac{1}{q\mu_n N_D} \qquad\qquad [\,2.10\,]$$

where μ_n by the mobility for electrons which is a function of doping concentration (N_D) and temperature (T).

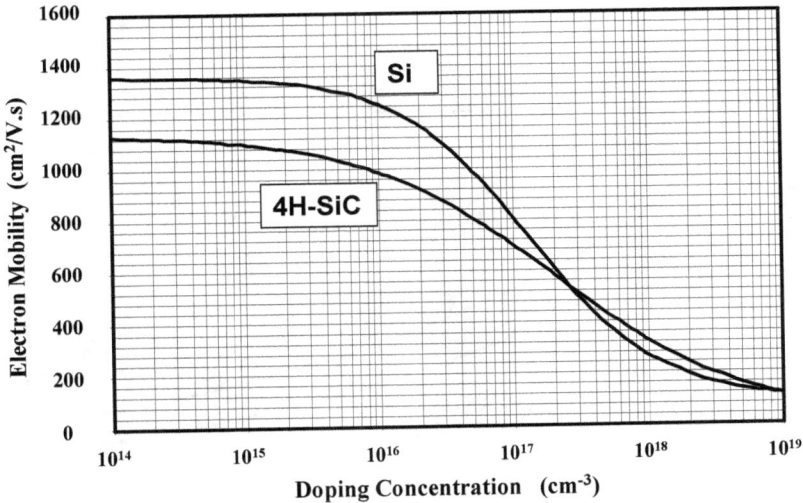

Fig. 2.7 Bulk Mobility for Electrons in 4H-SiC and Silicon.

For silicon, the measured data[13] for mobility of electrons at room temperature as a function of the doping concentration can be modeled by:

$$\mu_n(Si) = \frac{5.10x10^{18} + 92N_D^{0.91}}{3.75x10^{15} + N_D^{0.91}} \qquad [2.11]$$

For silicon carbide, the mobility of electrons at room temperature as a function of the doping concentration can be modeled by[3]:

$$\mu_n(4H-SiC) = \frac{4.05x10^{13} + 20N_D^{0.61}}{3.55x10^{10} + N_D^{0.61}} \qquad [2.12]$$

This behavior has also been theoretically modeled taking into account acoustic and polar optical phonon scattering as well as intervalley scattering[14]. The electron mobility at room temperature is plotted in Fig. 2.7 as a function of doping concentration for silicon and silicon carbide. In both cases, the mobility decreases with increasing doping due to enhanced Coulombic scattering of electrons by the ionized donors. This behavior was incorporated in the analytical models used for two-dimensional numerical simulations of device structures discussed later in the book.

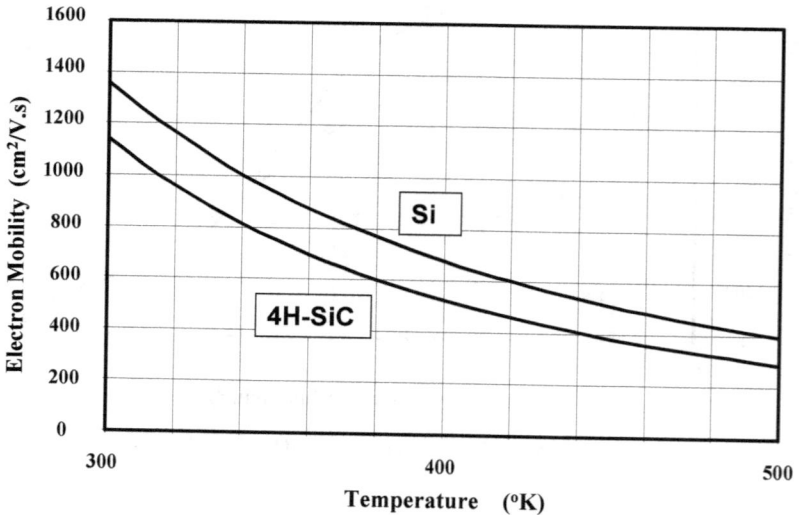

Fig. 2.8 Bulk Mobility for Electrons in 4H-SiC and Silicon.

The mobility decreases in semiconductors with increasing temperature due to enhanced phonon scattering. This holds true for silicon and 4H-SiC as shown in Fig. 2.8. For silicon, the temperature dependence of the mobility at low doping concentrations can be modeled by[15]:

$$\mu_n(Si) = 1360\left(\frac{T}{300}\right)^{-2.42}$$ [2.13]

For 4H-SiC, the temperature dependence of the mobility at low doping concentrations can be modeled by[3]:

$$\mu_n(4H - SiC) = 1140\left(\frac{T}{300}\right)^{-2.70}$$ [2.14]

This behavior was also incorporated in the analytical models used for two-dimensional numerical simulations of device structures discussed later in the book. The variation of the mobility with temperature is shown in the above figure only up to 500 °K (227 °C). Although there is considerable interest in the performance of silicon carbide devices at much higher temperatures due to its wide band gap structure, the operation of devices above 500 °K is problematic due to degradation of the ohmic contacts and the surface passivation.

2.2 Other Properties Relevant to Power Devices

In this section, other properties for 4H-SiC are provided, which are required for the analysis of power device structures. These properties are compared with those for silicon whenever it is pertinent.

2.2.1 Donor and Acceptor Ionization Energy Levels

In silicon, the donor and acceptor ionization energy is small (50 meV) allowing the assumption that all the donors and acceptor impurities in the lattice are completely ionized at room temperature as well as at higher temperatures. This is not a valid assumption for silicon carbide due to the relatively larger ionization energies for donors and acceptors. The most commonly used donor in 4H-SiC is nitrogen which occupies Carbon sites in the lattice. Its ionization energy is reported to be about 100 meV[1]. To create p-type 4H-SiC, the most commonly used acceptor impurity is Aluminum with an ionization energy of about 200 meV[1]. At above room temperature, the ionized donor concentration is over ninety percent allowing the approximation that the electron concentration is equal to the donor concentration when calculating the resistivity of drift regions.

2.2.2 Recombination Lifetimes

In silicon, high voltage devices are designed to operate with the injection of minority carriers into the drift region to reduce its resistivity by the conductivity modulation phenomenon[10]. The recombination of holes and electrons plays an important role in determining the switching speed of these structures. The recombination process can occur from band-to-band but is usually assisted by the presence of deep levels in the band gap. The deep levels can also produce leakage current during reverse blocking in power devices by the generation of carriers in the depletion region.

As already pointed out, for silicon carbide, it is appropriate to only consider unipolar devices for power switching applications. This is fortunate because the lifetime values reported for this material is relatively low (10^{-9} to 10^{-7} seconds)[3]. This is consistent with diffusion lengths of between 1 and 2 microns measured by using the EBIC technique[16]. It is likely that these low lifetime and diffusion lengths are associated with surface recombination. The performance of bipolar devices made from 4H-SiC, such as high voltage P-i-N rectifiers, indicates that higher values of bulk lifetime are applicable[17]. However, the large on-state voltage drop and slow switching speed of these devices makes them much less attractive for applications than unipolar devices.

2.2.3 Metal-Semiconductor Contacts

Contacts to power devices are made by using metal films deposited on the surface of its semiconductor layers. A common requirement is to make ohmic contacts to n-type and p-type regions. This can be achieved by using a metal-semiconductor contact with low barrier height and a high doping concentration in the semiconductor to promote tunneling current across the contact. For metal-semiconductor contacts with high doping level in the semiconductor, the contact resistance determined by the tunneling process is dependent upon the barrier height and the doping level[18]:

$$R_c = \exp\left[\frac{2\sqrt{\varepsilon_S m^*}}{h}\left(\frac{\phi_{bn}}{\sqrt{N_D}}\right)\right] \qquad [2.15]$$

In order to take full advantage of the low specific on-resistance of the drift region in silicon carbide devices, it is important to obtain a specific contact resistance that is several orders of magnitude smaller than that of the drift region. This is also necessary because the contact areas are often only a small fraction (less than 10 percent) of the active area of most power device structures. Typically, specific contact resistances of less than 1 x 10^{-5} ohm-cm^2 are desirable to n-type regions.

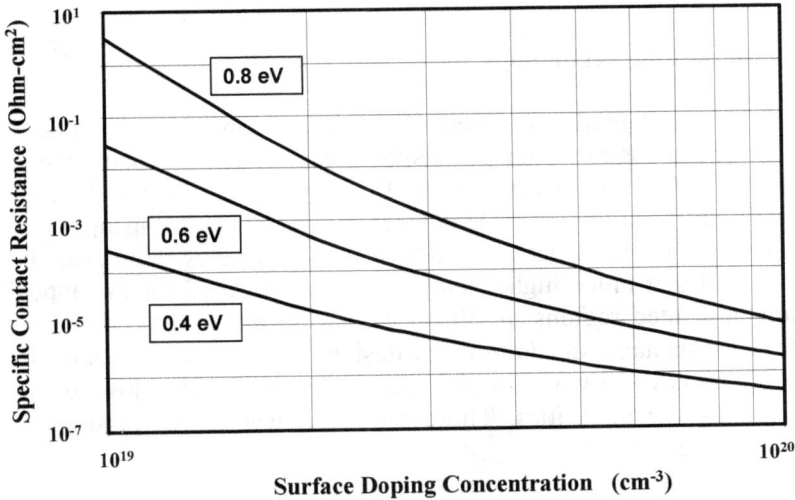

Fig. 2.9 Specific Resistance at Metal-Semiconductor Contacts.

The specific contact resistance calculated using the above formula is plotted in Fig. 2.9 as a function of the doping concentration using the barrier height as a parameter. Unfortunately, the barrier heights of metal contacts to silicon carbide tend to be large due to its wide band gap. From the figure, it can be concluded that a specific contact resistance of 1 x 10^{-5} ohm-cm^2 can be obtained for a doping concentration of 5 x 10^{19} cm^{-3} if the barrier height is 0.6 eV. Such high surface doping concentrations can be achieved for N-type 4H-SiC by using hot-implantation of nitrogen, phosphorus or arsenic followed by appropriate high temperature annealing[19]. Ohmic contacts with specific resistances of less than 10^{-6} Ohm-cm^2 have been reported to ion-implanted layers by using nickel and titanium[20].

Metal-semiconductor contacts can also be used to make Schottky barrier rectifiers with high breakdown voltages by using silicon carbide. In this case, it is advantageous to have a relatively large barrier height to reduce the leakage current. The commonly used metals for formation of Schottky barriers to 4H-SiC are Titanium and Nickel. The barrier heights measured for these contacts range from 1.10 to 1.25 eV for Titanium and 1.30 to 1.60 eV for Nickel[21,22,23]. The use of these metals to fabricate high voltage device structures is discussed in subsequent chapters of the book.

2.3 Fabrication Technology

Due to the substantial infrastructure available to fabricate silicon devices, it is advantageous to manufacture silicon carbide devices using the same technology platform. During the last fifteen years of research, it has been established that silicon carbide devices can be fabricated using the same equipment used for silicon devices in most instances. However, it has been found that much higher temperatures are needed for the annealing of ion implanted regions in silicon carbide to activate the dopants and remove the damage. In addition, the design of device structures in silicon carbide requires giving special consideration to the low diffusion coefficients for impurities. These unique issues are discussed in this section of the chapter.

2.3.1 Diffusion Coefficients and Solubility of Dopants

In silicon power devices, it is common-place to drive the dopants after ion-implantation to increase the junction depth. The widely used double-diffused (DMOS) process for silicon power MOSFETs utilizes the difference in the junction depth of the P-base and N^+ source regions to define the channel length without relying upon high resolution lithography to achieve sub-micron dimensions[10]. In silicon carbide, the diffusion coefficients are very small even at relatively high temperatures[1]. For nitrogen - the commonly used N-type dopant, the effective diffusion coefficient is reported as 5×10^{-12} cm^2/s even at 2450 °C. For aluminum - the commonly used P-type dopant, the effective diffusion coefficient is reported as 3×10^{-14} to 6×10^{-12} cm^2/s even at 1800 - 2000 °C. Due to the on-set of dissociation at these temperatures with attendant generation of defects, it is not practical to drive dopants in silicon carbide after ion implantation. The design of power device structures in silicon carbide must take this limitation in process technology into account.

The solid solubility of dopants in silicon carbide has been reported to be comparable to that for dopants in silicon[1]. For nitrogen and aluminum, the solubility is in excess of 1×10^{20} cm^{-3}. Ion implanted nitrogen[24] and aluminum[25] doped layers have been formed with impurity concentrations ranging from 1×10^{19} cm^{-3} to 1×10^{20} cm^{-3}. These values are sufficient for the fabrication of most high voltage power device structures.

2.3.2 Ion Implantation and Annealing

A significant body of literature has developed on the ion implantation of impurities into silicon carbide and the subsequent annealing process to activate the dopants and remove lattice damage. The reader should refer to the proceeding of the conferences on 'Silicon Carbide and Related Materials' for an abundance of information on this topic.

Although room temperature ion implantation can be successful, it has been found that hot-implantation produces a higher degree of impurity activation and a smaller number of defects. The hot-implants are usually performed in the 500 °C range followed by anneals performed at between 1200 and 1600 °C. At above 1600 °C, degradation of the silicon carbide surface is observed due to sublimation or evaporation[19,25]. The surface degradation can be mitigated by using a graphite boat to host the implanted wafer with another silicon carbide wafer placed on top (proximity anneals)[26].

2.3.3 Gate Oxide Formation

Silicon bipolar power devices, such as the bipolar power transistor and the gate turn-off thyristor (GTO), were supplanted with MOS-gated devices in the 1980s. The silicon power MOSFET replaced the bipolar transistor for lower voltage (< 200 V) applications while the IGBT replaced bipolar transistors and GTOs for high voltage (>200V) applications. The main advantage of these MOS-gated device architectures was voltage controlled operation which greatly simplified the control circuit making it amenable to integration[10]. This feature must be extended to silicon carbide power devices to make them attractive from an applications perspective.

The issues that must be considered for the gate dielectric in MOS-gated silicon carbide devices are the quality of the oxide-semiconductor interface and the ability of the oxide to withstand higher electric fields than within silicon devices[27]. The quality of the oxide-semiconductor interface determines not only the channel mobility but impacts the threshold voltage of the devices. The electric field in the oxide is related to the electric field in the semiconductor by Gauss's Law:

$$E_{ox} = \frac{\varepsilon_{semi}}{\varepsilon_{ox}} E_{semi} \approx 3 E_{semi}$$

[2.16]

In silicon, the maximum electric field at breakdown is in the 3×10^5 V/cm range. Consequently, the maximum electric field in the oxide remains well below its breakdown field strength of 10^7 V/cm. In contrast, the maximum electric field for breakdown in silicon carbide is in the 3×10^6 V/cm range. Consequently, the electric field in the oxide can approach its breakdown strength and easily exceed a field of 3×10^6 V/cm, which is considered to be the threshold for reliable operation. One approach to overcome this problem is to use gate dielectric material with a larger permittivity[28,29]. In these references, it has been theoretically shown that the specific on-resistance can be reduced by an order of magnitude by using high dielectric constant gate material. A permittivity of about 15 (versus 3.85 for silicon dioxide) was found to be adequate for allowing full use of the high electric field strength for breakdown in silicon carbide without problems with unacceptably high oxide electric fields. One example of such a dielectric is zirconium oxide[30].

A second approach utilizes improved device structural architecture to screen the gate dielectric from the high electric fields within the silicon carbide. This is discussed in detail in the chapters on shielded planar and shielded trench gate MOSFETs in this book. Such devices can then be made using silicon dioxide whose properties are well understood. In this case, the silicon dioxide can be either grown by the thermal oxidation of silicon dioxide or by the formation of the oxide using chemical vapor deposition processes. Many studies on both of these techniques are available in the literature.

The thermal oxidation of SiC has been reviewed in the literature and the growth rate of the oxide has been compared with that on silicon surfaces[31]. Some selected data is shown in Fig. 2.10 to illustrate the much lower growth rate of thermal oxide on SiC when compared with silicon for both dry and wet oxidation conditions. The data shown in this figure are for 6H-SiC but similar values are applicable to the 4H-SiC polytype[32]. It is obvious that a much higher temperature and longer time duration is required for the growth of oxides on silicon carbide. This oxide has been shown to be essentially silicon dioxide with no carbon incorporated in the film. The small amount of aluminum incorporated into the oxide, when grown on p-type SiC, has been found to have no adverse effect[33,34] on the MOS properties.

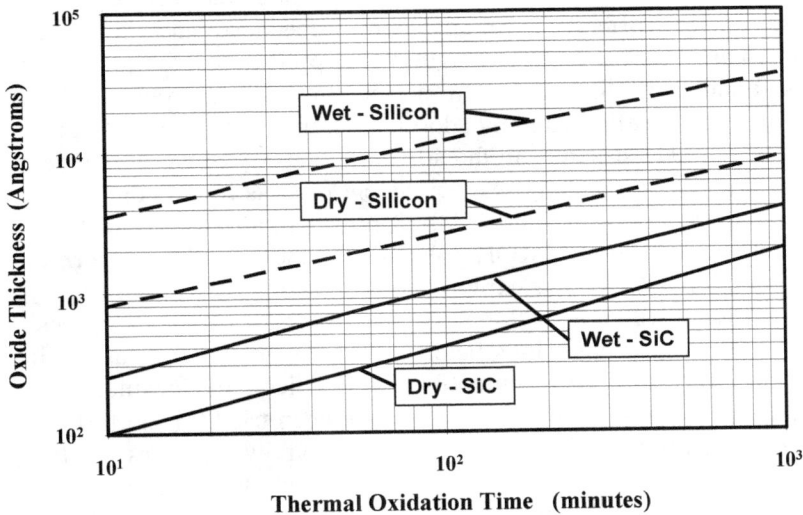

Fig. 2.10 Comparison of Thermal Oxidation of SiC with silicon at 1200 °C.

The gate dielectric produced with thermal oxidation of 6H-SiC has been found to be satisfactory for making MOSFETs with reasonable inversion layer mobility[35]. However, this method did not produce sufficiently high quality interfaces on 4H-SiC to make MOSFETs. It was discovered at PSRC that a low temperature oxide subjected to a wet nitrogen anneal enabled fabrication of n-channel inversion layer MOSFETs in 4H-SiC with an effective inversion layer mobility of 165 cm^2/Vs[36,37]. Although the gate oxide for these MOSFETs was unusually large (9000 angstroms), high performance MOSFETs with similar mobility values, were subsequently fabricated at PSRC using a gate oxide thickness of 850 angstroms[38]. At PSRC, an extensive evaluation of process conditions was performed to determine their impact on the inversion layer mobility. This work has been validated by other groups[39].

In addition, inversion layer mobility in the range of 50 cm^2/V-s has been reported on thermally grown oxides subjected to annealing in nitric oxide[31,40]. These experiments indicate that power MOSFETs with low specific on-resistance can be developed from 4H-SiC.

2.3.4 Reactive Ion Etching of Trenches

The first silicon power MOSFETs were manufactured using the Double-diffused or DMOS process to form a planar gate architecture[41]. Although these devices offered excellent input impedance and fast switching performance for low voltage (< 100 volt) applications, their on-resistance was found to be limited by the presence of a JFET region. The JFET region could be eliminated by adopting a trench gate or UMOS process[10]. The trench gate structure in the silicon power MOSFET is fabricated by using reactive ion etching. An equivalent process for silicon carbide was not developed until the mid 1990s. Until then, alternative methods such as etching with molten KOH solutions or the use of amorphization of selective SiC regions was explored[42].

Reactive ion etching is the most convenient process for formation of trenches in silicon carbide enabling the fabrication of devices with processes compatible with silicon device manufacturing technology. Many gas compositions have been tried for the formation of trenches in silicon carbide. For application to power devices, the trenches should have a nearly vertical profile with a slightly rounded bottom to reduce electric field enhancement. The trench surface must be smooth and free of damage in order to reduce interface states and obtain high mobility along the inversion channel formed on the trench sidewalls. Reactive ion etching using a SF$_6$/O$_2$ mixture has been found to produce trenches that meet these requirements[43]. Other reactive ion etching chemistries using fluorinated gases[44] have also been reported to produce trenches with vertical profiles and smooth surfaces.

2.4 Summary

The properties of silicon carbide relevant to power devices have been reviewed in this chapter. An improved knowledge of the impact ionization coefficients has allowed projection of very high performance for SiC based unipolar devices. The status of process technology for the fabrication of these unipolar devices has also been summarized here.

References

[1] G. L. Harris, "Properties of Silicon Carbide", IEE Inspec, 1995

[2] M. Ruff, H. Mitlehner, and R. Helbig, "SiC Devices: Physics and Numerical Simulations", IEEE Transactions on Electron Devices, Vol. ED41, pp. 1040-1954, 1994.

[3] N.G. Wright, et al, "Electrothermal Simulation of 4H-SiC Power Devices", Silicon Carbide, III-Nitrides, and Related Materials - 1997, Material Science Forum, Vol. 264, pp. 917-920, 1998.

[4] S.M. Sze, "Physics of Semiconductor Devices", John Wiley and Sons, 1981

[5] S. K. Ghandhi, "Semiconductor Power Devices", John Wiley and Sons, 1977.

[6] A.G. Chynoweth, "Ionization Rates for Electrons and Holes in Silicon, Physical Review, Vol. 109, pp.1537-1545, 1958.

[7] A. G. Chynoweth, "Uniform Silicon P-N Junctions II. Ionization rates for Electrons", J. Applied Physics, Vol. 31, pp 1161-1165, 1960.

[8] C. R. Crowell and S. M. Sze, "Temperature dependence of Avalanche Multiplication in Semiconductors", Applied Physics Letters, Vol. 9, pp 242-244, 1966.

[9] R. Van Overstraeten and H. De Man, "Measurement of the Ionization Rates in Diffused Silicon P-N Junctions", Solid State Electronics, Vol. 13, pp. 583-590, 1970.

[10] B. J. Baliga, "Power Semiconductor Devices", PWS Publishing Company, 1996.

[11] R. Raghunathan and B. J. Baliga, "Temperature dependence of Hole Impact Ionization Coefficients in 4H and 6H SiC", Solid State Electronics, Vol. 43, pp. 199-211, 1999.

[12] R. Raghunathan and B. J. Baliga, "Role of Defects in producing Negative Temperature Dependence of Breakdown Voltage in SiC", Applied Physics Letters, Vol. 72, pp. 3196-3198, 1998.

[13] C. Jacobini, et al, "A Review of some Charge Transport Properties of Silicon", Solid State Electronics, Vol. 20, pp. 77-89, 1977.

[14] H. Iwata and K. M. Itoh, "Theoretical Calculation of the Electron Hall Mobility in n-type 4H- and 6H-SiC", Silicon Carbide and Related Materials – 1999, Materials Science Forum, Vol. 338-342, pp. 879-884, 2000.

[15] C. Canali, et al, "Electron Drift Velocity in Silicon", Phys. Rev. Vol. B12, pp. 2265-2284, 1975

[16] R. Raghunathan and B.J. Baliga, "EBIC Measurements of Diffusion Lengths in Silicon Carbide", 38th Electronic Materials Conference, Abstr. I-6, 1996.

[17] R. Singh, "Silicon Carbide Bipolar Power Devices – Potentials and Limits", Material Research Society Proceedings, Vol. 640, pp. H4.2.1-H4-2-12, 2001.

[18] S.M. Sze, "Physics of Semiconductor Devices", page 304, John Wiley and Sons, 1981.

[19] S. Imai, et al, "Hot-Implantation of Phosphorus Ions into 4H-SiC", Silicon Carbide and Related Materials – 1999, Materials Science Forum, Vol. 338-342, pp. 861-864, 2000.

[20] S. Tanimoto, et al, "Ohmic Contact Structure and Fabrication Process Applicable to Practical SiC Devices", Silicon Carbide and Related Materials – 2001, Materials Science Forum, Vol. 389-393, pp. 879-884, 2002.

[21] A. Kestle, et al., "A UHV Study of Ni/SiC Schottky Barrier and Ohmic Contact Formation", Silicon Carbide and Related Materials – 1999, Materials Science Forum, Vol. 338-342, pp. 1025-1028, 2000.

[22] K.V. Vassilevski, et al., "4H-SiC Schottky Diodes with high On/Off Current Ratio", Silicon Carbide and Related Materials – 2001, Materials Science Forum, Vol. 389-393, pp. 1145-1148, 2002.

[23] R. Raghunathan, D. Alok, and B.J. Baliga, "High Voltage 4H-SiC Schottky Barrier Diodes", IEEE Electron Device letters, Vol. EDL-16, pp. 226-227, 1995.

[24] S. Blanque, et al., "Room Temperature Implantation and Activation Kinetics of Nitrogen and Phosphorus in 4H-SiC crystals", Silicon Carbide and Related Materials – 2003, Materials Science Forum, Vol. 457-460, pp. 893-896, 2004.

[25] Y. Negoro, et al., "Low Sheet Resistance of High Dose Aluminum Implanted 4H-SiC using (1120) Face", Silicon Carbide and Related Materials – 2003, Materials Science Forum, Vol. 457-460, pp. 913-916, 2004.

[26] R.K. Chilukuri, P. Ananthanarayanan, V. Nagapudi, and B.J. Baliga, "High Voltage P-N Junction Diodes in Silicon Carbide using Field Plate Edge Termination", Material Research Society Proceedings, Vol. 572, pp. 81-86, 1999.

[27] B. J. Baliga, "Critical Nature of Oxide/Interface Quality for SiC Power Devices", Microelectronics Engineering, Vol. 28, pp. 177-184, 1995.

[28] S. Sridevan, P.K. McLarty, and B.J. Baliga, "Silicon Carbide Switching Devices having Nearly Ideal Breakdown Voltage Capability and Ultra-Low On-State Resistance", U.S. Patent # 5,742,076, Issued April 21, 1998.

[29] S. Sridevan, P.K. McLarty, and B.J. Baliga, "Analysis of Gate Dielectrics for SiC Power UMOSFETs", IEEE Int. Symposium on Power Devices and ICs, pp. 153-156, 1997.

[30] V.V. Afanas'ev, et al, "Oxidation of Silicon Carbide: Problems and Solutions", Silicon Carbide and Related Materials – 2001, Materials Science Forum, Vol. 389-393, pp. 961-966, 2002.

[31] J. A. Cooper, "Silicon Carbide MOSFETs", in 'Wide Energy Bandgap Electronic Devices', Edited by F. Ren and J. C. Zolper, World Scientific Press, 2003.

[32] A. Golz, et al, "Oxidation Kinetics of 3C, 4H, and 6H silicon carbide", Silicon Carbide and Related Materials – 1995, Institute of Physics Conference Series, Vol. 142, pp. 633-636, 1996.

[33] S. Sridevan, P.K. McLarty, and B.J. Baliga, "On the Presence of Aluminum in Thermally Grown Oxides on 6H-SiC", IEEE Electron Device Letters, Vol. 17, pp. 136-138, 1996.

[34] G. Gudjonsson, et al, "High Field Effect Mobility in n-Channel Si Face 4H-SiC MOSFETs with Gate Oxide Grown on Aluminum Ion-Implanted Material", IEEE Electron Device Letters, Vol. 26, pp. 96-98, 2005.

[35] L.A. Lipkin and J.W. Palmour, "Improved Oxidation Procedures for reduced SiO$_2$/SiC Defects". J. Electronic Materials, Vol. 25, pp. 909-915, 1996.

[36] S. Sridevan and B.J. Baliga, "Lateral n-Channel Inversion Mode 4H-SiC MOSFETs", Silicon Carbide and Related Materials – 1997, Materials Science Forum, Vol. 264-268, pp. 997-1000, 1998.

[37] S. Sridevan and B.J. Baliga, "Lateral n-Channel Inversion Mode 4H-SiC MOSFETs", IEEE Electron Device Letters, Vol. 19, pp. 228-230, 1998.

[38] S. Sridevan and B.J. Baliga, "Phonon Scattering limited Mobility in SiC Inversion Layers", PSRC Technical Report, TR-98-03.

[39] D. Alok, E. Arnold, and R. Egloff, "Process dependence of Inversion Layer Mobility in 4H-SiC Devices", Silicon Carbide and Related Materials – 1999, Materials Science Forum, Vol. 338-342, pp. 1077-1080, 2000.

[40] J.R. Williams, et al, "Passivation of the 4H-SiC/SiO$_2$ Interface with Nitric Oxide", Silicon Carbide and Related Materials – 2001, Materials Science Forum, Vol. 389-393, pp. 967-972, 2002.

[41] D.A. Grant and J. Gowar, "Power MOSFETs", John Wiley and Sons, 1989.

[42] D. Alok and B.J. Baliga, "A Novel Method for Etching Trenches in Silicon Carbide", J. Electronic Materials, Vol. 24, pp. 311-314, 1995.

[43] M. Kothandaraman, D. Alok, and B.J. Baliga, "Reactive Ion Etching of Trenches in 6H-SiC", J. Electronic Materials, Vol. 25, pp. 875-878, 1996.

[44] V. Saxena and A.J. Steckl, "Fast and Anisotropic Reactive Ion Etching of 4H and 6H SiC in NF$_3$", Silicon Carbide and Related Materials – 1997, Materials Science Forum, Vol. 264-268, pp. 829-832, 1998.

Chapter 3

Breakdown Voltage

The main advantage of a wide band gap semiconductor for power device applications stems from the very low resistance of the drift region even when it is designed to support larger voltages. The largest voltage that can be supported by a drift region is determined by the onset of impact ionization with increasing electric field within the region. In the previous chapter, it was shown that the onset of impact ionization occurs at much larger electric fields in silicon carbide when compared with silicon. In this chapter, the design of the drift region is discussed and it is demonstrated that its specific resistance is much smaller than for silicon devices designed to support the same voltage. In addition to the basic parallel plane junction, the influence of the edge termination design is analyzed. The electric field in power devices is invariably enhanced at its edges. Consequently, the edge termination design limits the maximum voltage that can be supported by the device. The results of two-dimensional numerical simulations are used to elucidate the physics of operation within the structures.

3.1 One-Dimensional Abrupt Junction

The analysis of a one-dimensional abrupt junction can be used to understand the design of the drift region within power devices[1]. The case of a P^+/N junction is illustrated in Fig. 3.1 where the P^+ side is assumed to be very highly doped so that the electric field supported within it can be neglected. When this junction is reverse-biased by the application of a positive bias to the N-region, a depletion region is formed in the N-region together with the generation of a strong electric field within it that supports the voltage. The Poisson's equation for the N-region is then given by:

$$\frac{d^2V}{dx^2} = -\frac{dE}{dx} = -\frac{Q(x)}{\varepsilon_S} = -\frac{qN_D}{\varepsilon_S} \qquad [3.1]$$

where $Q(x)$ is the charge within the depletion region due to the presence of ionized donors, ε_S is the dielectric constant for the semiconductor, q is the electron charge, and N_D is the donor concentration in the uniformly doped N-region.

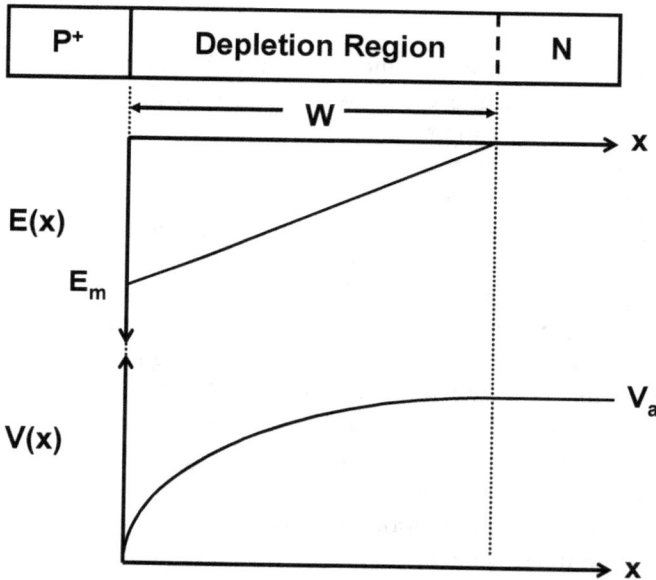

Fig. 3.1 Electric Field and Potential Distribution for an Abrupt Parallel-Plane P+N Junction.

Integration of the above equation with the boundary condition that the electric field must go to zero at the edge of the depletion region (i.e. at x=W) provides the electric field distribution:

$$E(x) = -\frac{qN_D}{\varepsilon_S}(W - x) \qquad [3.2]$$

The electric field has a maximum value of E_m at the P+/N junction (x=0) and decreases linearly to zero at x=W. Integration of the electric field distribution through the depletion region provides the potential distribution:

$$V(x) = -\frac{qN_D}{\varepsilon_S}(Wx - \frac{x^2}{2})$$ [3.3]

This equation is obtained by using the boundary condition that the potential is zero at x=0 within the P^+ region. The potential varies quadratically as illustrated in the figure. The thickness of the depletion region (W) is related to the applied reverse bias (V_a):

$$W = \sqrt{\frac{2\varepsilon_S V_a}{qN_D}}$$ [3.4]

Using these equations, the maximum electric field at the junction can be obtained:

$$E_m = \sqrt{\frac{2qN_D V_a}{\varepsilon_S}}$$ [3.5]

When the applied bias increases, the maximum electric field approaches values at which significant impact ionization begins to occur. The breakdown voltage is determined by the ionization integral becoming equal to unity:

$$\int_0^W \alpha.dx = 1$$ [3.6]

where α is the impact ionization coefficient discussed in chapter 2. In order to obtain a closed form solution for the breakdown voltage, it is convenient to use a power law behavior for the impact ionization coefficient in place of Chynoweth's law. For silicon, this is usually performed using the Fulop's power law formula:

$$\alpha_F(Si) = 1.8x10^{-35} E^7$$ [3.7]

A fit to the 4H-SiC data provided in chapter 2 gives the equivalent Baliga's power law formula for the impact ionization coefficient:

$$\alpha_B(4H - SiC) = 3.9x10^{-42} E^7$$ [3.8]

The impact ionization coefficient obtained by using these equations are compared with those obtained using the Chynoweth formula in Fig. 3.2.

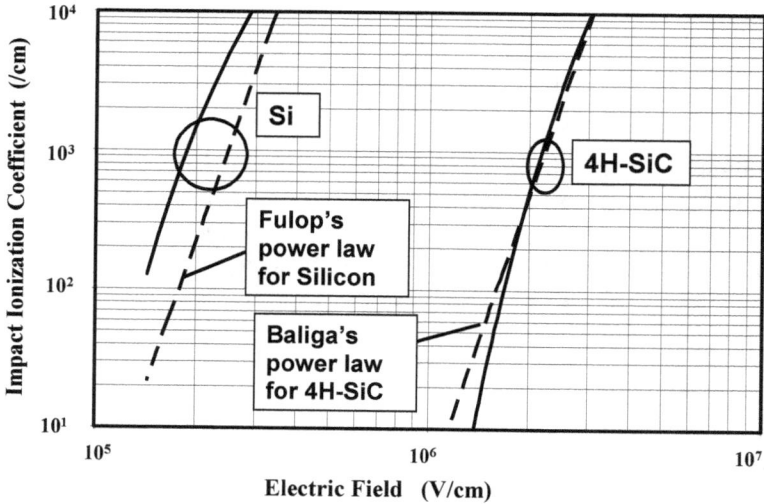

Fig. 3.2 Impact Ionization Coefficients for Si and 4H-SiC using a Power Law.

Using the power law equation, analytical solutions for the breakdown voltage and the corresponding maximum depletion layer width can be derived:

$$BV_{PP}(4H - SiC) = 3.0x10^{15} N_D^{-3/4}$$ [3.9]

and

$$W_{PP}(4H - SiC) = 1.82x10^{11} N_D^{-7/8}$$ [3.10]

The breakdown voltage is plotted in Fig. 3.3 as a function of the doping concentration on the lightly doped side of the junction. It is obvious that it is possible to support a much larger voltage in 4H-SiC when compared with silicon for any given doping concentration. The ratio of the breakdown voltage in 4H-SiC to that in silicon for the same doping concentration is found to be 56.2. It is also obvious from this figure that for a given breakdown voltage, it is possible to use a much higher doping concentration in the drift region for 4H-SiC devices when compared with silicon devices. The ratio of the doping concentration in the drift region for a 4H-SiC device to that for a silicon device with the same breakdown voltage is found to be 200.

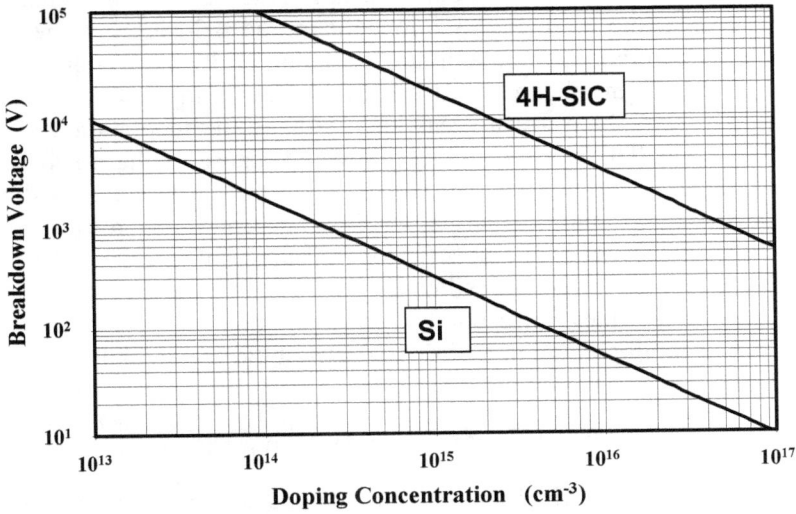

Fig. 3.3 Breakdown Voltage for Abrupt Parallel-Plane Junctions in Si and 4H-SiC.

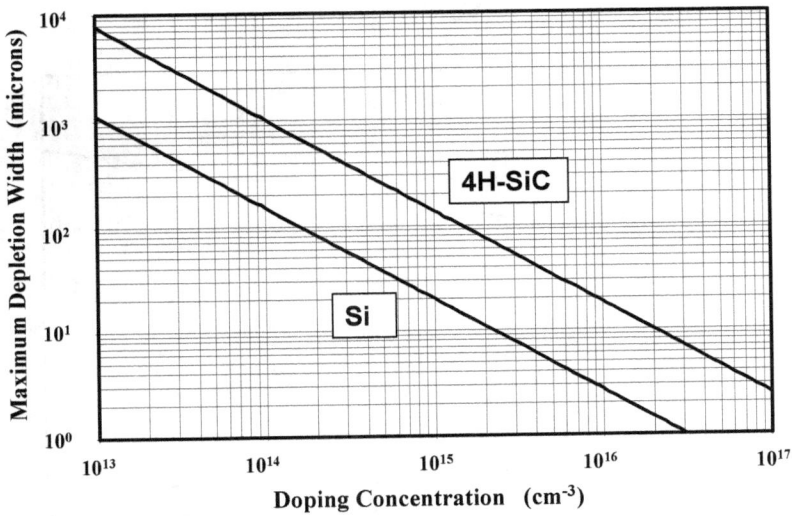

Fig. 3.4 Maximum Depletion Width at Breakdown in Si and 4H-SiC.

The maximum depletion width reached at the onset of breakdown is shown in Fig. 3.4 for silicon and 4H-SiC. For the same doping concentration, the maximum depletion width in 4H-SiC is 6.8x larger than that in silicon because it can sustain a much larger electric field. However, for a given breakdown voltage, the depletion width in 4H-SiC is smaller than for a silicon device because of the much larger doping concentration in the drift region. This smaller depletion width, in conjunction with the far larger doping concentration, results in an enormous reduction in the specific on-resistance of the drift region in 4H-SiC when compared with silicon.

These values correspond to the maximum electric field at the junction becoming equal to a critical electric field for breakdown, which is given by:

$$E_C(4H - SiC) = 3.3x10^4 N_D^{1/8} \qquad [3.11]$$

The critical electric field for 4H-SiC can be compared with that for silicon using Fig. 3.5. For the same doping concentration, the critical electric field in 4H-SiC is 8.2x larger than in silicon. The larger critical electric field in 4H-SiC results in a much larger *Baliga's Figure of Merit*.

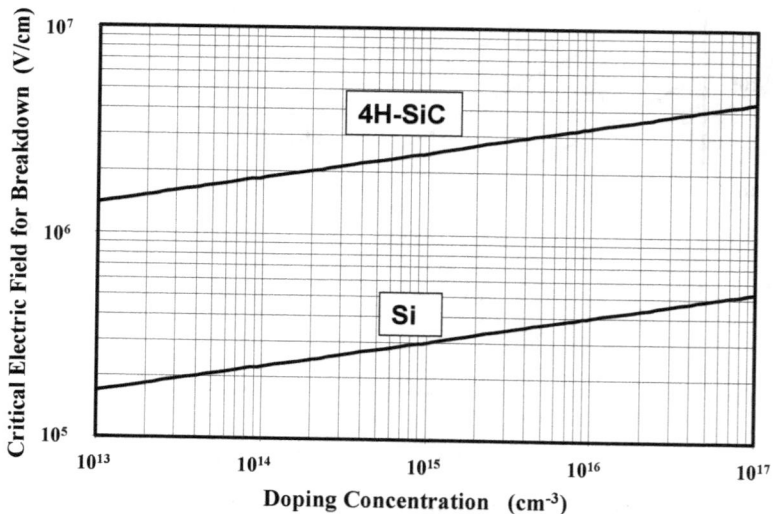

Fig. 3.5 Critical Electric Field for Breakdown in Si and 4H-SiC.

The specific on-resistance of the drift region is related to the breakdown voltage by equation [1.4] which is repeated here for discussion:

$$R_{on,sp} = \frac{4BV^2}{\varepsilon_S \mu_n E_C^3}$$ [3.12]

A complete modeling of the specific on-resistance requires taking into account the variation of the critical electric field and mobility on the doping concentration, which varies as the breakdown voltage is changed. It is possible to do this by computing the doping concentration for achieving a given breakdown voltage and then using the equations for the depletion width and mobility as a function of doping concentration to obtain the specific on-resistance:

$$R_{on,sp} = \frac{W_{PP}}{q \mu_n N_D}$$ [3.13]

The specific on-resistance projected for the drift region in 4H-SiC devices by using the above method is compared with that for silicon devices in Fig. 3.6. The 4H-SiC values are about 2000 times smaller than for silicon devices for the same breakdown voltage.

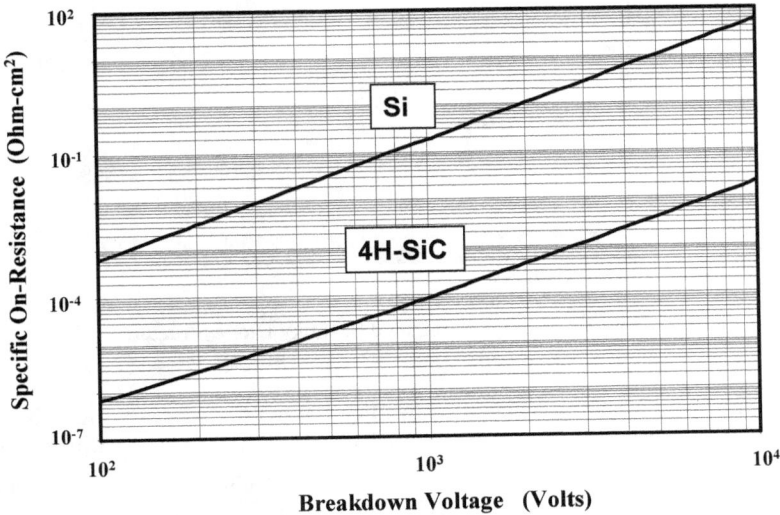

Fig. 3.6 Specific On-Resistance of Drift Regions in 4H-SiC and Silicon.

3.2 Schottky Diode Edge Terminations

The silicon Schottky diode is commonly used for power supply designs operating at low voltages. It offers a much lower on-state voltage drop albeit at a larger leakage current and reduced maximum operating temperature capability. The resistance of the drift region becomes prohibitive for silicon Schottky rectifiers at breakdown voltages above 200 volts[1]. However, the much lower drift region resistance in silicon carbide enables scaling the breakdown voltage to over 3000 volts while obtaining reasonable on-state voltage drops. This is discussed in detail in the next chapter. The realization of the full potential for a silicon carbide Schottky rectifier requires attainment of close to the ideal breakdown voltage, i.e. the parallel-plane breakdown voltage. This requires proper edge termination designs that are suitable for implementation in silicon carbide as discussed in this section.

3.2.1 Planar Schottky Diode Edge Termination

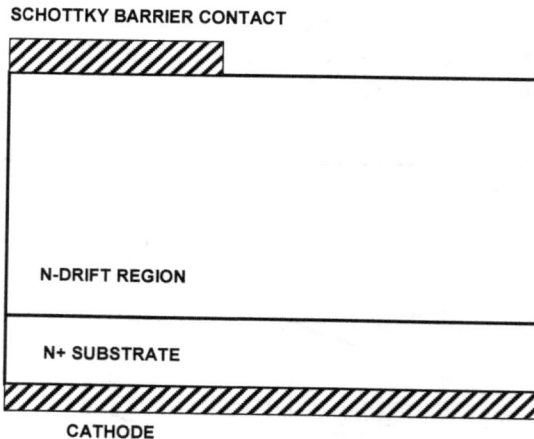

Fig. 3.7 The Un-terminated Schottky Diode Structure.

The first high voltage vertical Schottky barrier rectifiers were fabricated at PSRC with the remarkably simple process of evaporating metal through a 'dot' mask onto a silicon carbide surface after appropriate cleaning[2]. This structure, illustrated in Fig. 3.7, contains a sharp edge at the periphery of the diode leading to electric field enhancement and reduction of the breakdown voltage. To illustrate this phenomenon, two-dimensional numerical simulations were performed on a Schottky diode with a drift region with doping concentration of 1×10^{16} cm^{-3}. A metal work function of 4.5 eV was used to form the Schottky barrier. The diode was found to exhibit a breakdown voltage of 650 volts. A three dimensional view of the electric field for the un-terminated Schottky diode is shown in Fig. 3.8 at a bias of 500 volts. It can be seen that there is a sharp increase in the electric field at the edge of the metal leading to the reduced breakdown voltage.

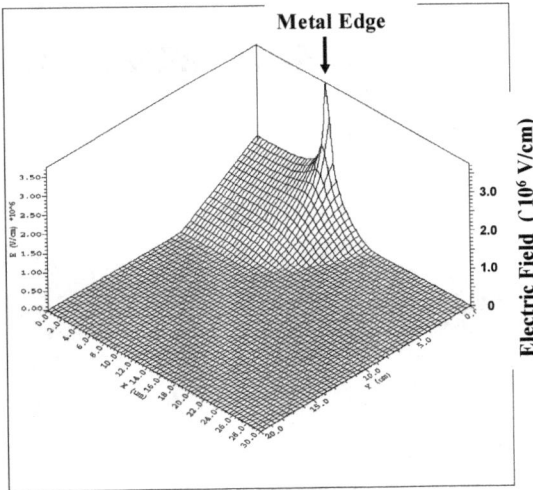

Fig. 3.8 Electric Field distribution at the 4H-SiC un-terminated Schottky diode for a Reverse Bias of 500 volts.

3.2.2 Planar Schottky Diode with Field Plate Edge Termination

The breakdown voltage for the planar Schottky diode can be significantly improved by using a field plate at its periphery to reduce the electric field at the metal edge. This termination is illustrated in Fig. 3.9 with oxide as the dielectric. Field plates have also been reported using silicon nitride as the dielectric[3]. To understand the benefits of the field plate structure, the results of two-dimensional numerical simulations are discussed here with variations to the field plate length and the oxide thickness.

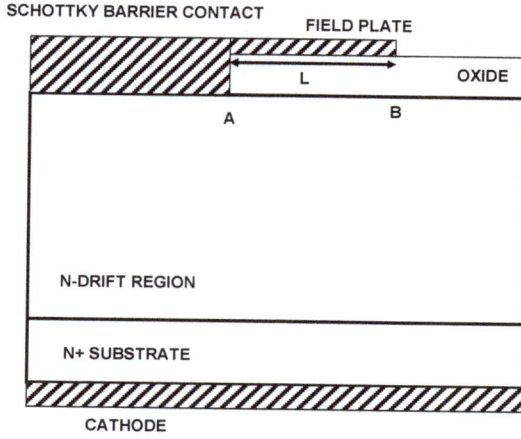

Fig. 3.9 Schottky Diode Structure with Field Plate Termination.

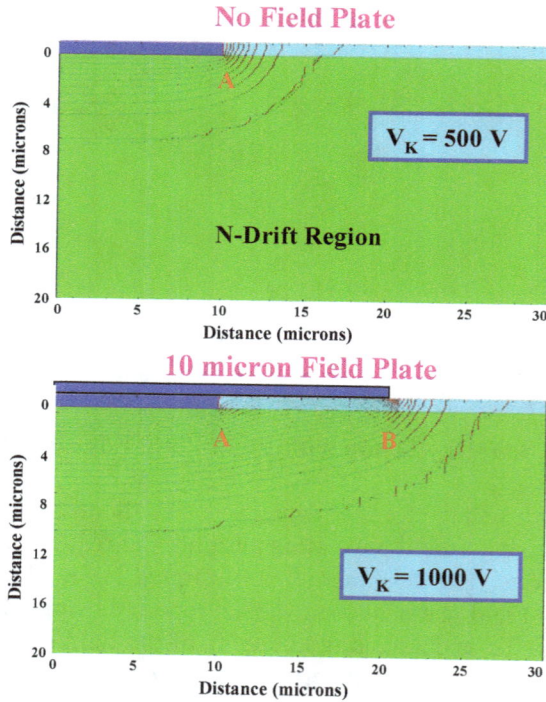

Fig. 3.10 Potential Contours for Schottky Diode with Field Plate.

The potential contours for the case of a field plate edge termination using a length of 10 microns and an oxide thickness of 1 micron are shown in Fig. 3.10 at a reverse bias of 500 volts. The potentials contours for the un-terminated diode are also shown in this figure for comparison at a reverse bias of 1000 volts. It can be seen that the depletion region spreads along the surface in the presence of the field plate reducing the potential crowding at point A. Consequently, the structure with the field plate had a breakdown voltage of 1300 volts compared with 650 volts for the un-terminated diode. The reduction in electric field at the edge (point A) of the Schottky diode can be examined in the three-dimensional view of the electric field provided in Fig. 3.11. The ratio of the peak electric field at point A to that in the middle of the diode is reduced to 2.3x in comparison with a ratio of 2.7x for the un-terminated diode.

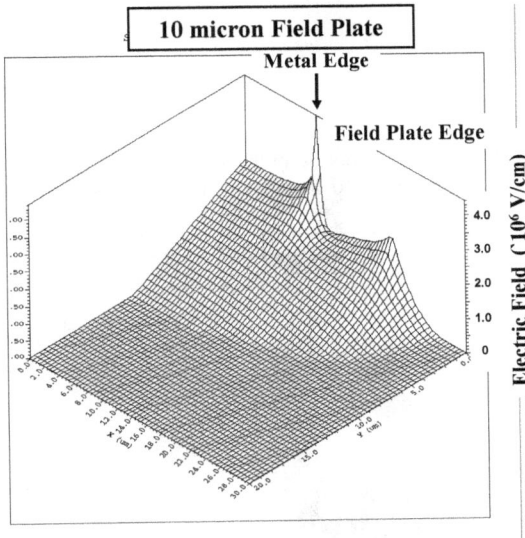

Fig. 3.11 Electric Field distribution in the 4H-SiC Schottky diode with Field Plate for a Reverse Bias of 1000 volts.

A further reduction of the electric field at the metal edge (point A) can be obtained by reducing the oxide thickness. The electric field distribution for the case of an oxide thickness of 0.6 microns is shown in Fig. 3.12. The peak electric field at the metal corner is now 2x of that in the middle. This allowed the breakdown voltage to increase to 1430 volts. A further reduction in oxide thickness leads to an even lower electric field at the metal corner as shown in Fig. 3.13. Unfortunately, the electric field at the field plate edge (point B) now becomes larger than at point A leading to a reduction of the breakdown voltage to 840 volts.

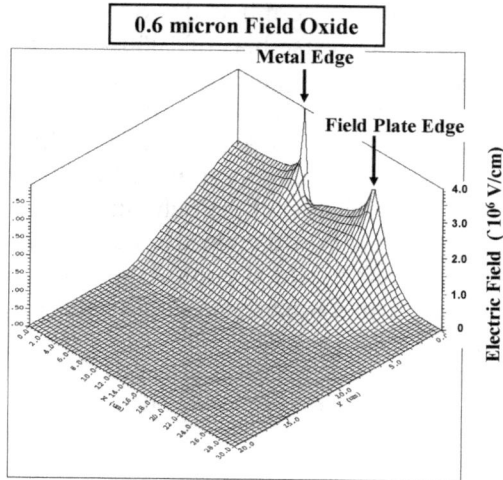

Fig. 3.12 Electric Field distribution in the 4H-SiC Schottky diode with Field Plate for a Reverse Bias of 1000 volts.

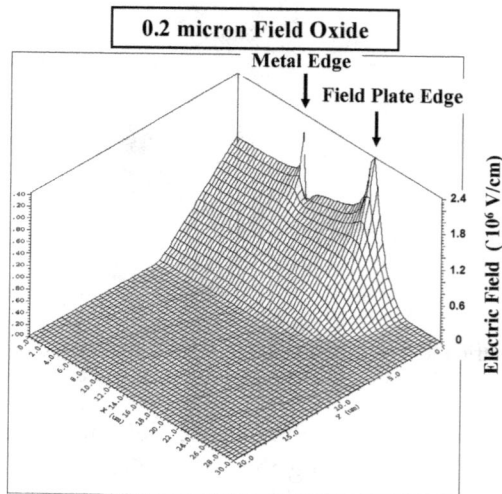

Fig. 3.13 Electric Field distribution in the 4H-SiC Schottky diode with Field Plate for a Reverse Bias of 500 volts.

The variation of the breakdown voltage with field oxide thickness is plotted in Fig. 3.14. It is obvious that there is an optimum oxide thickness at which the breakdown voltage reaches a maximum value. It is also important to use a sufficient length (L) for the field plate as shown in Fig. 3.15. A length of 10 microns is adequate for the simulated case.

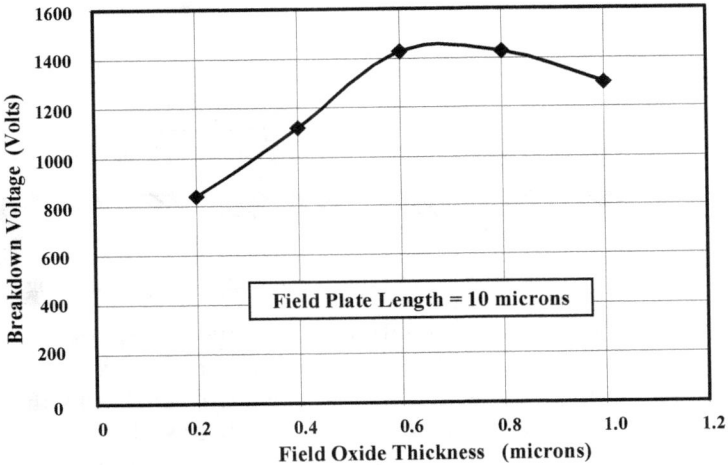

Fig. 3.14 Schottky Diode in 4H-SiC with Field Plate.

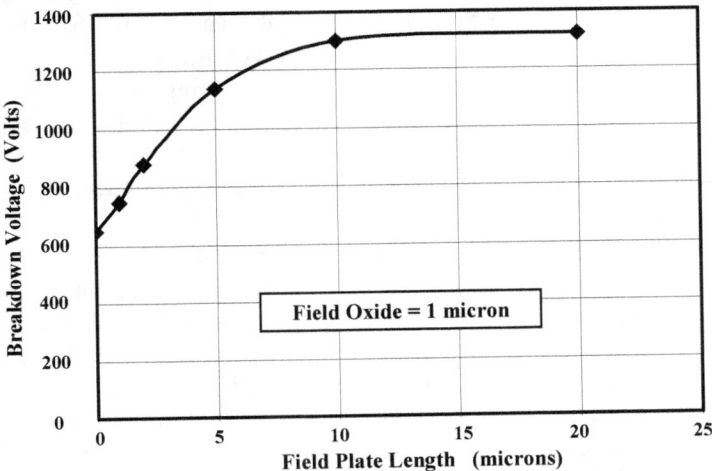

Fig. 3.15 Schottky Diode in 4H-SiC with Field Plate.

3.2.3 Schottky Diode with Floating Metal Rings

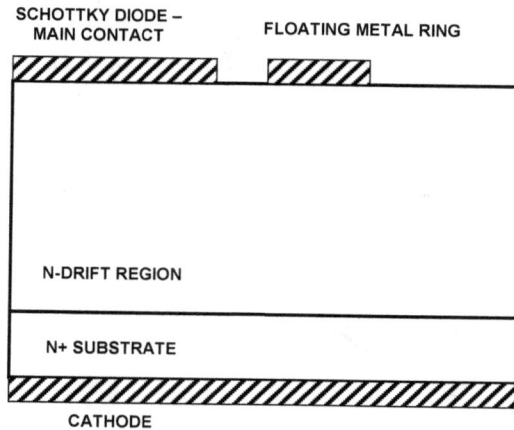

Fig. 3.16 Schottky Diode Edge Termination with Floating Metal Ring.

A floating guard ring is commonly used for silicon devices to enhance their breakdown voltage[1]. This concept of using a ring around the main junction with a floating potential can be extended to Schottky diodes. In this case, it is convenient to use the Schottky barrier metal to simultaneously form floating metal rings around the main diode. This structure is illustrated in Fig. 3.16. The optimum spacing for the floating metal ring is governed by the same considerations as for the floating field ring (see later section in this chapter). In the case of 6H-SiC Schottky diodes, it was found that the breakdown voltage increased with ring spacing up to a distance of 5 microns. The resulting breakdown voltage was twice that for the un-terminated Schottky barrier diode[4].

3.2.4 Schottky Diode Edge Termination with Argon Implant

Since the breakdown voltage of the planar Schottky diode is limited by electric field enhancement at the edge of the metal, its breakdown voltage can be increased by spreading the potential along the surface. One effective method for achieving this is by creating a highly resistive layer along the surface as illustrated in Fig. 3.17. This was first demonstrated at PSRC by using Argon ion implantation around Schottky diodes in 6H-SiC to produce nearly ideal breakdown voltages[5] followed by its successful application to 4H-SiC[6]. A remarkable increase in the breakdown voltage from 300 V to 1000 V was achieved with this technique.

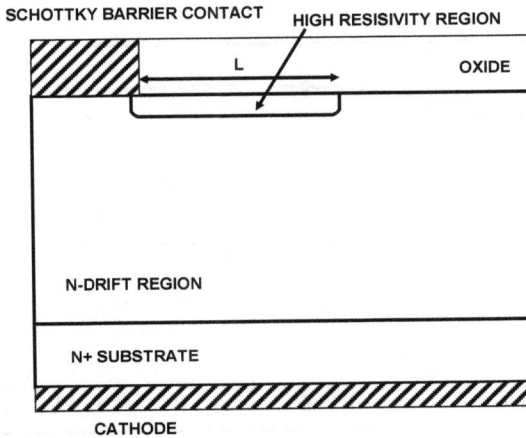

Fig. 3.17 Schottky Diode Edge Termination with High Resistivity Extension.

The principle behind the approach is to utilize ion implantation to create damage in the silicon carbide lattice to produce deep levels in its band gap. This moves the Fermi level close to the center of the band gap within the implanted zone producing a high resistivity layer due to the large band gap of the semiconductor[7]. The presence of the deep level traps has been confirmed using Deep Level Transient Spectroscopy (DLTS) measurements. The damage can be created by using any implantation species (including dopants, such as Aluminum and Boron, without sufficient annealing to remove all the deep level defects) as subsequently reported by other groups[8].

The design of the argon implanted edge termination has been studied to determine its optimum length[9]. It was found that a length of 100 microns is adequate to achieve a breakdown voltage of 800 V for a 6H-SiC diode fabricated using an epitaxial layer doping concentration of 2×10^{16} cm^{-3} as shown in Fig. 3.18. An argon implant dose of 1×10^{15} cm^{-2} was used at the periphery of the Schottky diodes. The length of the implanted zone is about ten times the width of the depletion region at breakdown. The leakage current was found to increase linearly with the length of the implanted zone. It is therefore important to use a length just sufficient to achieve the parallel plane breakdown voltage. It has been found that the leakage current can be reduced by two-orders of magnitude with post implantation annealing at 600 °C[10].

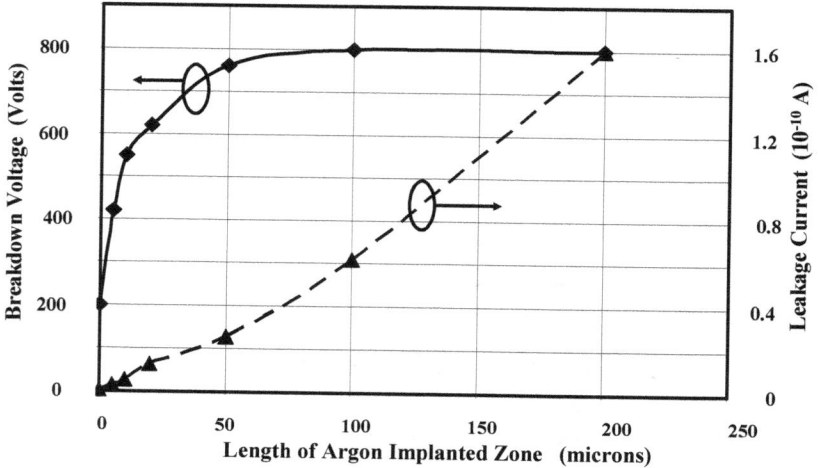

Fig. 3.18 Design of the High Resistivity Extension for the Planar Junction Edge Termination.

3.2.5 Schottky Diode Edge Termination with RESP Region

The breakdown voltage of a Schottky diode can also be improved by spreading the potential along the surface using a 'resistive Schottky barrier field plate' or RESP termination[4]. The RESP layer is a high resistivity film formed on the silicon carbide surface with a rectifying interface. In the case of silicon carbide devices, this has been achieved by depositing a very thin layer of Titanium (50 to 80 angstroms), which is subsequently oxidized to increase its resistivity in a controlled manner. It has been demonstrated that the oxidation can be performed at 300 °C in air without damaging the quality of the main Schottky barrier contact. Prior to oxidation of the titanium layer, the breakdown voltage of the structure was only 150 volts, as in the case of un-terminated diodes. The breakdown voltage was found to increase to 500 volts when the sheet resistance of the RESP layer reached 1 MΩ/square. Modeling of the RESP termination using a distributed resistor/diode network[4] indicates that larger breakdown voltages would be achieved when the sheet resistance reaches 1 GΩ/square. Beyond this, the breakdown voltage decreases because the RESP layer begins to behave like a dielectric.

3.3 P-N Junction Edge Terminations

The P-N junction is commonly used for the formation of a variety of power device structures. Although the emphasis in this book is on unipolar devices due to their high performance characteristics when fabricated from 4H-SiC, these devices often utilize P-N junctions to control the electric field distribution within the active region. The P-N junction can then also be used at the edges of the devices to reduce electric field crowding. Device edge terminations that utilize P-N junctions are discussed in this section.

3.3.1 Planar Junction Edge Termination

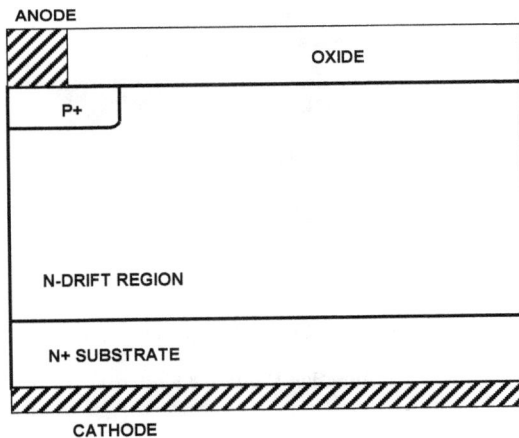

Fig. 3.19 Ion-Implanted Planar Junction Edge Termination in 4H-SiC.

One of the simplest edge terminations used for silicon power devices is the cylindrical junction. The breakdown voltage of this type of termination is well below that for the parallel-plane junction due to electric field enhancement at the corner of the cylindrical junction[1]. This behavior has been analytically modeled[11]. In silicon carbide, the dopants do not move during the ion implant annealing cycles despite the relatively high temperature because of the extremely low diffusion coefficients (see chapter 2). The ion-implanted junctions must therefore be modeled with very small characteristic diffusion lengths that are representative of the ion implant straggle. This type of planar edge termination is illustrated in Fig. 3.19 with an oxide passivation[3]. This structure was analyzed by performing two-dimensional numerical simulations using a drift region doping concentration of 1 x 10^{16} cm^{-3} and thickness of 30 microns. The depth of the P$^+$ region was varied to examine its impact on the breakdown voltage. Three-dimensional views of the doping profile for a shallow and deep junction used during the simulations are provided below in Fig. 3.20 and 3.21.

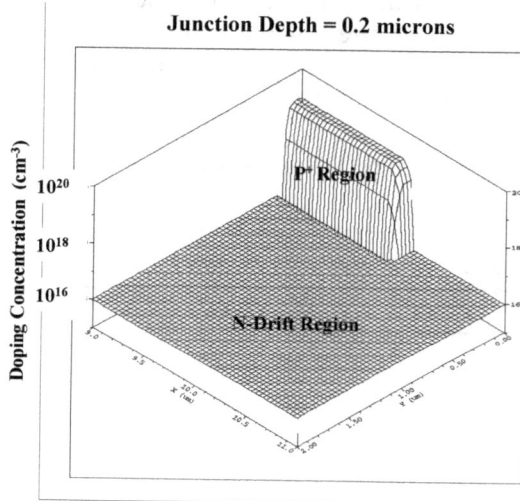

Fig. 3.20 Doping Profile for the Shallow Planar Cylindrical Junction.

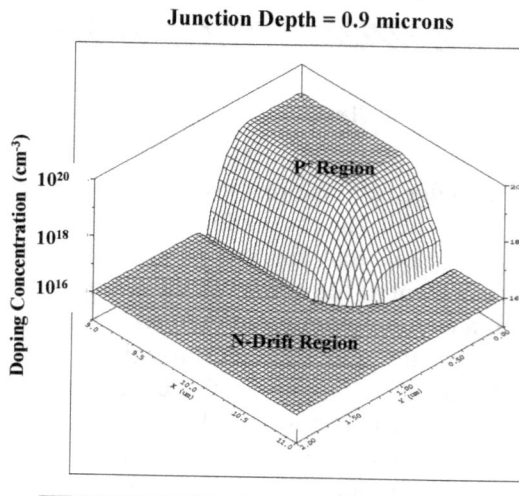

Fig. 3.21 Doping Profile for the Deep Planar Cylindrical Junction.

Junction Depth = 0.2 microns

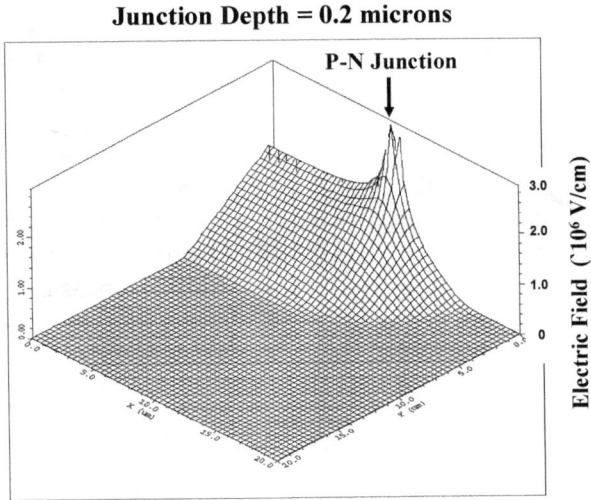

Fig. 3.22 Electric Field distribution at the Shallow Planar Junction in 4H-SiC
for a Reverse Bias of 500 volts.

Junction Depth = 0.9 microns

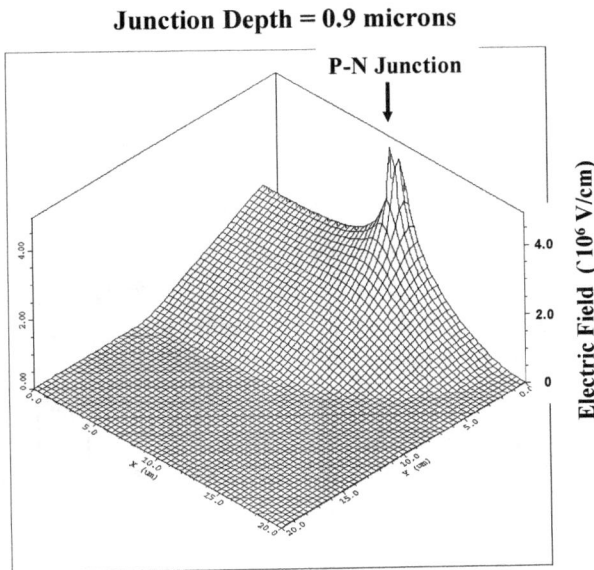

Fig. 3.23 Electric Field distribution at the Deep Planar Junction in 4H-SiC for
a Reverse Bias of 950 volts.

The breakdown voltage of the structure was found to increase with increasing junction depth. For a shallow junction depth of 0.2 microns, the breakdown occurred at 400 volts. This is due to a significant enhancement in the electric field at the corner of the junction as shown in Fig. 3.22. This field enhancement was found to be reduced as shown in Fig. 3.23 when the junction depth was increased to 0.9 microns leading to an increase in the breakdown voltage to 900 volts. These values are well below the breakdown voltage (3000 volts) for the ideal parallel plane junction. They are consistent with the measured breakdown voltages for nitrogen implanted N[+]/P diodes formed in 4H-SiC with oxide surface passivation[3] and boron implanted 6H-SiC diodes[12].

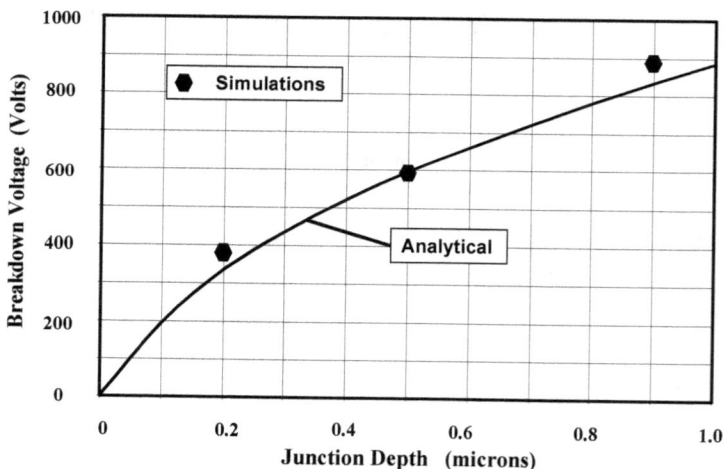

Fig. 3.24 Breakdown Voltage of Cylindrical Planar Junctions in 4H-SiC.

The breakdown voltage of cylindrical junctions can be predicted by using the following analytical relationship[11]:

$$\frac{BV_{CYL}}{BV_{PP}} = \frac{1}{2}\left[\left(\frac{r_J}{W_{PP}}\right)^2 + 2\left(\frac{r_J}{W_{PP}}\right)^{6/7}\right].\ln\left[1 + 2\left(\frac{W_{PP}}{r_J}\right)^{8/7}\right] - \left(\frac{r_J}{W_{PP}}\right)^{6/7}$$

[3.14]

where r_J is the radius of curvature of the junction. The breakdown voltage calculated using this equation is shown in Fig. 3.24 by the solid line for the case of 4H-SiC with a doping concentration of 1×10^{16} cm^{-3}. The breakdown voltage obtained by the two dimensional numerical simulations are in reasonable agreement with the analytical formula as shown by the symbols.

3.3.2 Planar Junction Edge Termination with Field Plate

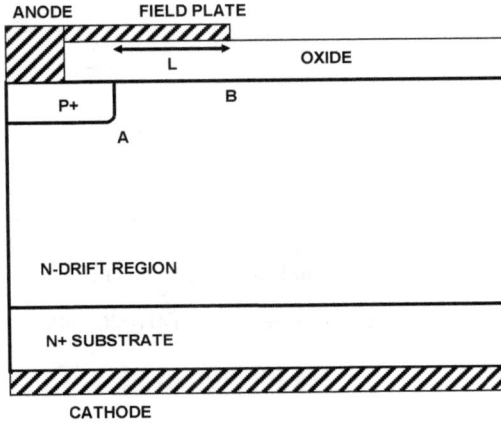

Fig. 3.25 Planar Junction Edge Termination with Field Plate.

In the previous section, it was demonstrated that the breakdown voltage of a planar junction is limited by the enhanced electric field at the corner of the P-N junction (location A in Fig. 3.25 above). This electric field enhancement can be ameliorated by forming a field plate by extending the anode metal over the edge of the junction on top of the oxide. For silicon devices, the improvement in the breakdown voltage depends upon the length (L) of the field plate and the thickness of the oxide underneath the field plate[1]. Although this also true in principle for the case of silicon carbide structures, an additional consideration is the much higher electric field within the oxide.

Two-dimensional numerical simulations of the planar junction with field plate were performed using an N-type drift region with doping concentration of 1×10^{16} cm^{-3}. The junction depth was chosen as 0.9 microns. The increase in the breakdown voltage is shown in Fig. 3.26 with increasing length of the field plate. In these simulations, the oxide thickness was kept at 1 micron. The breakdown voltage can be increased two-fold with sufficient field plate length. A field plate length of 15 microns is adequate for this particular N-region doping concentration.

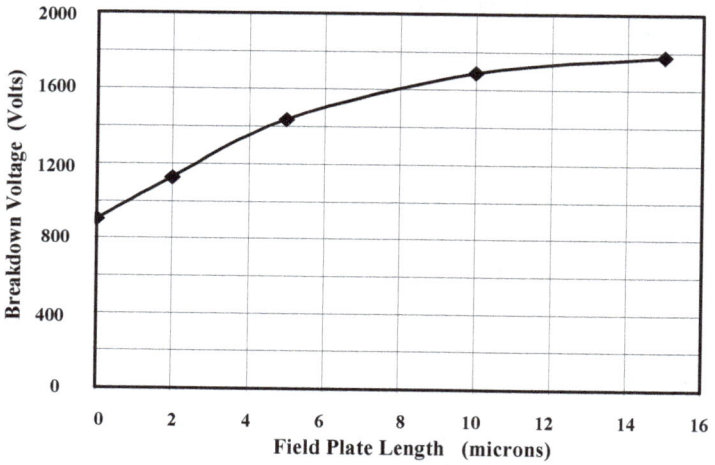

Fig. 3.26 Cylindrical Planar Junctions in 4H-SiC with Field Plate.

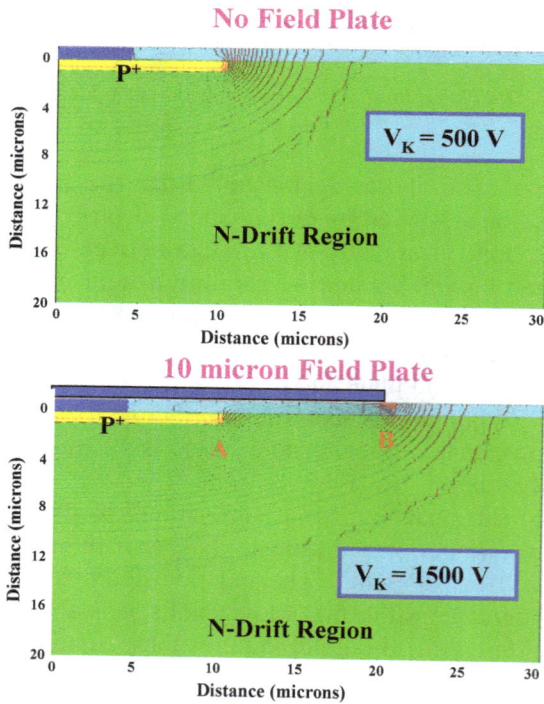

Fig. 3.27 Potential Contours for Cylindrical Planar Junctions with Field Plate.

The reason for the increase in the breakdown voltage with the addition of the field plate can be understood by examining the potential contours shown in Fig. 3.27. In the case of the planar junction shown at the top, potential crowding is observed at the junction (location A in Fig. 3.25). The addition of the field plate reduces the potential crowding at the junction by spreading the depletion region along the surface as shown in the lower portion of Fig. 3.27. However, potential crowding is observed at the edge of the field plate (location B in Fig. 3.25).

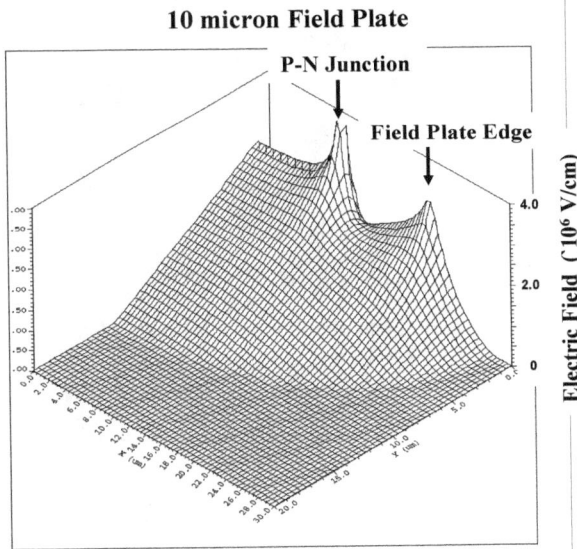

Fig. 3.28 Electric Field distribution at the Field Plate in 4H-SiC for a Reverse Bias of 1500 volts.

The impact of the potential crowding at the edge of the field plate can be seen in the three-dimensional electric field distribution plot in Fig. 3.28. Although there is an electric field peak located at its edge (point B) inside the semiconductor, this peak is smaller than that at the junction. This implies that the breakdown will be initiated at the junction. The electric field at the edge of the field plate can become greater than that at the junction if the oxide thickness is reduced. As an example, the electric field for the case of 4000 angstrom oxide is shown in Fig. 3.29. It can be seen that the electric field at the edge of the field plate exceeds that at the junction for this smaller oxide thickness.

4000 Angstrom Field Oxide

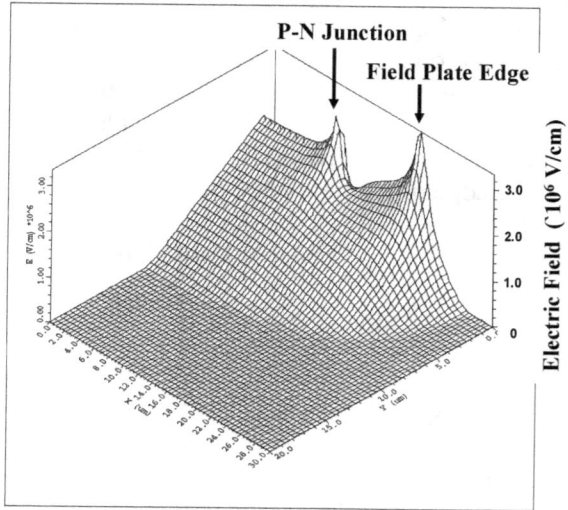

Fig. 3.29 Electric Field distribution at the Field Plate in 4H-SiC for a
Reverse Bias of 1000 volts.

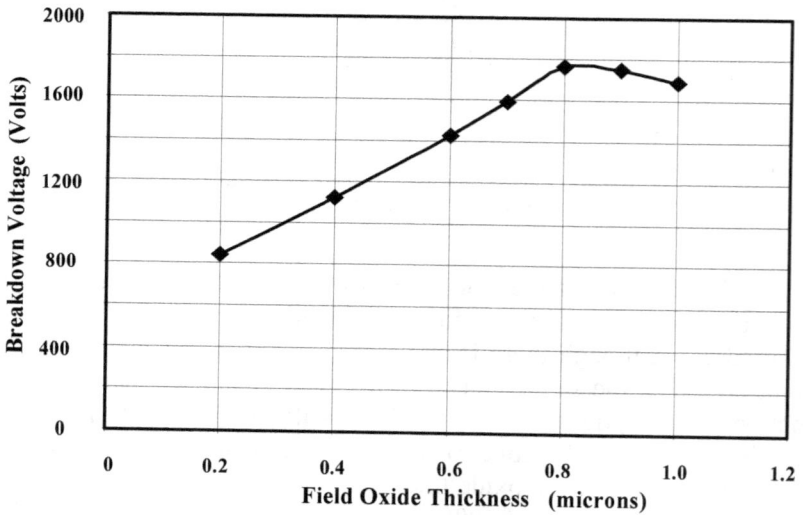

Fig. 3.30 Cylindrical Planar Junctions in 4H-SiC with Field Plate.

The impact of changing the field oxide thickness on the breakdown voltage is shown in Fig. 3.30. It can be seen that the breakdown voltage reaches a maximum value at a field oxide thickness of 0.8 microns. However, the change in breakdown voltage is small for larger oxide thickness values. These results indicate that a simple edge termination for 4H-SiC devices can be constructed using a P-N junction with a field plate extending over an oxide at the edges. The main advantage of this approach is that no additional masking or processing steps are required during device fabrication to prepare the edge termination. However, the breakdown voltage achieved with this method is well below that for the ideal parallel plane junction.

3.3.3 Planar Junction Edge Termination with Floating Rings

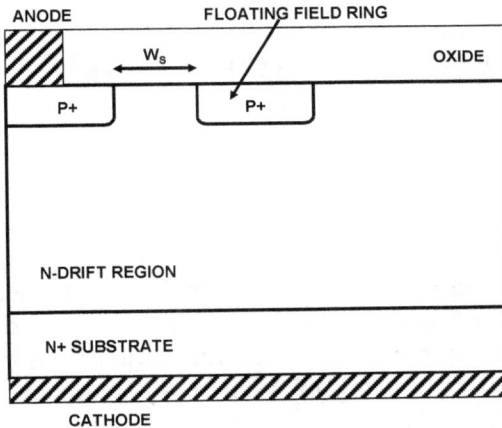

Fig. 3.31 Planar Junction Edge Termination with Floating Field Ring.

The planar junction edge termination with a single floating field ring is shown in Fig. 3.31. This method for edge termination is commonly used in silicon devices, particularly at lower breakdown voltages, to improve the operating voltages. One of the advantages of this approach is that the floating field ring can be fabricated simultaneously with the main junction without adding process steps. The breakdown voltage of the termination depends upon the spacing (W_S) of the floating field ring from the main junction[1]. If the spacing is too large, the electric field at the edge of the main junction remains high leading to a breakdown voltage similar to the planar junction without the field ring. If the spacing is too small, a high electric field develops at the outer edge of the floating field ring leading to a breakdown voltage equal to that for a planar junction without the field ring. An optimum spacing is required to achieve an increase in the breakdown voltage. The breakdown voltage of a planar junction with a single floating field ring can be calculated using an analytical method[13]:

$$\frac{BV_{FFR} - BV_{CYL}}{BV_{PP}} = \left[0.5\left(\frac{r_J}{W_{PP}}\right)^2 - 0.96\left(\frac{r_J}{W_{PP}}\right)^{6/7} \right]$$

$$+ 1.92\left(\frac{r_J}{W_{PP}}\right)^{6/7} . \ln\left[1.386\left(\frac{W_{PP}}{r_J}\right)^{4/7} \right]$$

[3.15]

with the optimum floating field ring spacing given by:

$$\frac{W_S}{W_{PP}} = \sqrt{\frac{BV_{FFR}}{BV_{PP}}} - \sqrt{\frac{BV_{FFR}}{BV_{PP}} - \frac{BV_{CY}}{BV_{PP}}}$$

[3.16]

For the case of a drift region doping concentration of 1×10^{16} cm^{-3} and a junction depth of 0.9 microns, these equations predict an increase in the breakdown voltage from about 800 V to 1200 V for an optimum field ring spacing of 4.5 microns. Such a modest increase in the breakdown voltage has been experimentally observed for junctions fabricated in 4H-SiC with field ring spacing of 4 microns[3,14]. An optimum field ring spacing of 5 microns has also been reported for diodes with a breakdown voltage of 1600 volts[15], consistent with the predictions of the analytical model.

3.3.4 Planar Junction Edge Termination with P-Extension

For silicon devices, the breakdown voltage has been shown to be greatly improved by using a 'Junction Termination Extension'[16]. This structure, illustrated in Fig. 3.32, contains a P-type region formed at the periphery of a P$^+$/N junction. This lightly doped P-type region is usually formed by using ion implantation to precisely control the dopant charge within the layer. If the doping concentration in the P-type region is too high, the breakdown occurs at its edge (point B) at a lower breakdown voltage than the main junction due to its smaller radius of curvature. If the doping concentration is too low in the P-type region, it becomes completely depleted at low reverse bias voltages resulting in breakdown at the main junction (point A) at the same voltage as the un-terminated junction.

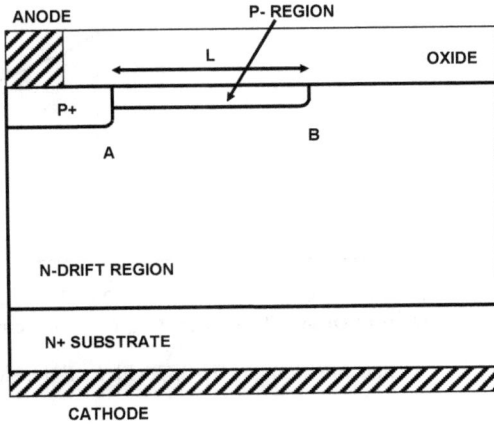

Fig. 3.32 Planar Junction Edge Termination with P-Extension.

The optimum charge in the P-type region is:

$$Q = \varepsilon_S E_C \qquad [3.17]$$

which amounts to 2.58×10^{-6} C/cm^2 for a critical electric field of 3×10^6 V/cm for 4H-SiC. This is equivalent to a dopant dose of 1.6×10^{13} cm^{-2}, which is about 10 times larger than that used in silicon devices[1]. A more precise optimization of the dose can be performed by two-dimensional numerical simulations of the structure.

The results of the simulations, performed using an extension length of 10 microns, are summarized in Fig. 3.33 where the variation in breakdown voltage is plotted as a function of the dopant dose in the P-region. As in the case of silicon devices, there is an optimum dose at which the breakdown voltage has a maximum value. At lower doses, the breakdown voltage continues to occur at the main junction due to the peak in the electric field at point A as shown in the electric field distribution plotted in Fig. 3.34. At higher doses, the breakdown voltage occurs at the edge of the extension due to the peak in the electric field at point B as shown in the electric field distribution plotted in Fig. 3.35.

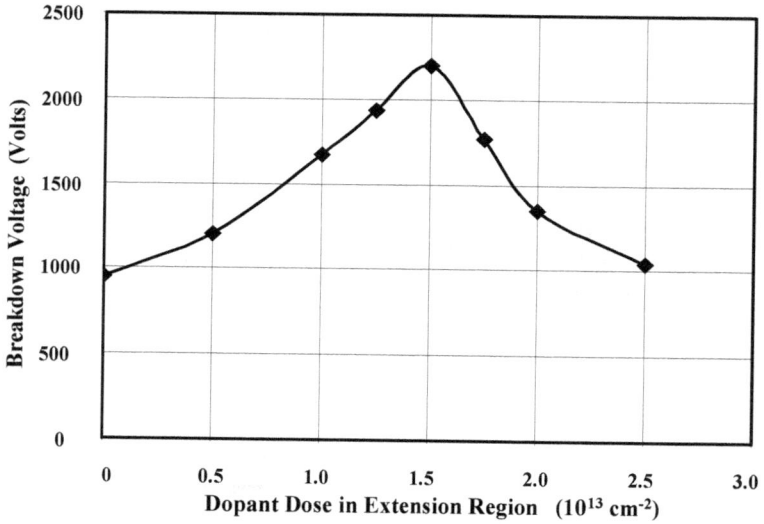

Fig. 3.33 Planar Junctions in 4H-SiC with P- Extension.

Dose = 0.5 x 10^{13} cm^{-3}

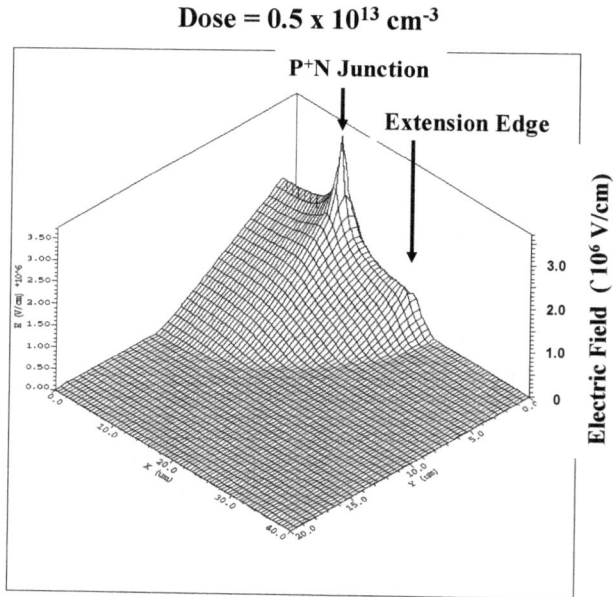

Fig. 3.34 4H-SiC Planar Junction with Extension (Bias = 1000 V).

Dose = 2.5 x 10^{13} cm^{-3}

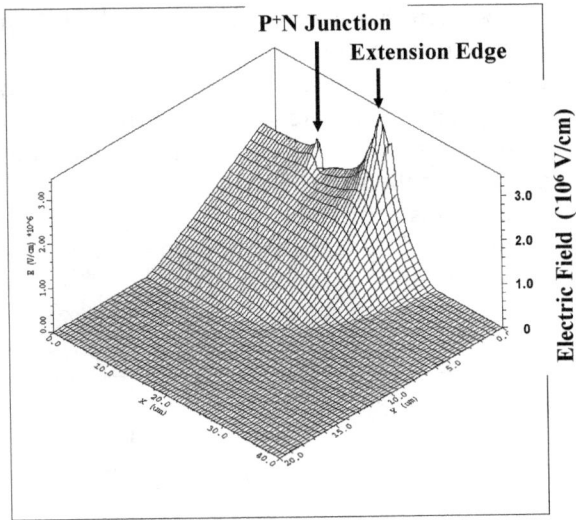

Fig. 3.35 4H-SiC Planar Junction with Extension (Bias = 1000 V).

Dose = 1.5 x 10^{13} cm^{-3}

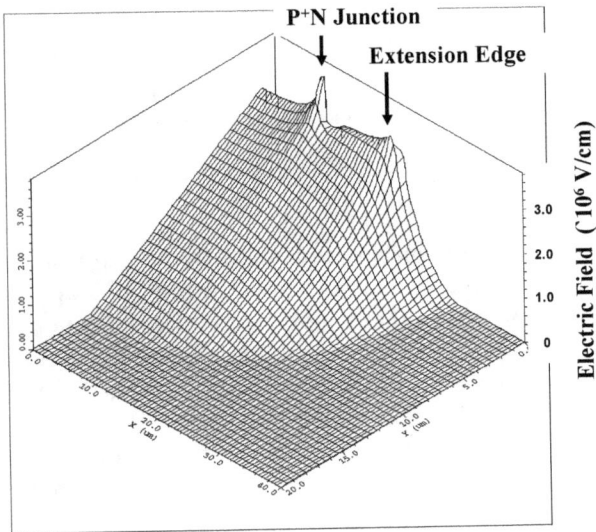

Fig. 3.36 4H-SiC Planar Junction with Extension (Bias = 2000 V).

The electric field distribution at the optimum dose is shown in Fig. 3.36. It can be seen that the electric field at the P$^+$N junction and at the edge of the extension are equal leading to the maxima in the breakdown voltage. In all the cases, the electric field first develops at the edge of the extension and then spreads through the extension region. This can be observed in the electric field profile taken along the surface, shown in Fig. 3.37, for the case of the optimum dose. At a bias of 500 volts, the electric field is confined to point B in Fig. 3.32. As the voltage increases, the electric field extends towards the main P$^+$N junction because the extension region gets depleted.

Dose = 1.5 x 10^{13} cm^{-2}

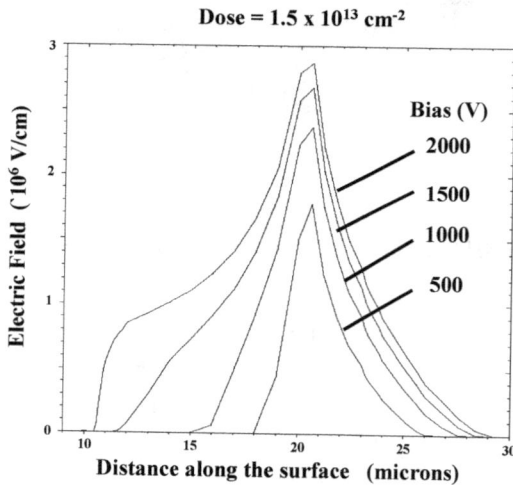

Fig. 3.37 4H-SiC Planar Junction with Extension.

For a comparison between the different dopant dose cases, the electric field along the surface is shown in Fig. 3.38 at a reverse bias of 1000 volts. It can be seen that a high electric field develops at the edge of the extension region if the dose is too high and at the main junction if the dose is too low. At high doses (> 2 x 10^{13} cm^{-2}), the extension region does not get depleted in the region between point A and B in Fig. 3.32 because the charge is too large. Consequently, the electric field continues to escalate with bias at the edge of the extension region leading to breakdown at lower voltages. If the dose is too low, the extension region becomes depleted at low bias values and a high electric field develops at the main junction.

Bias Voltage = 1000 V

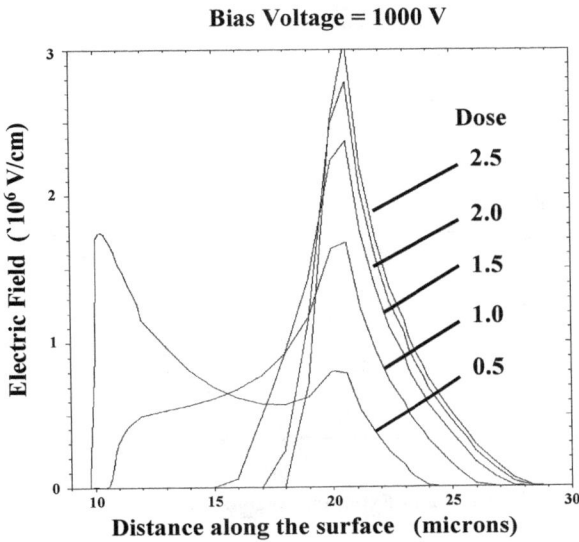

Fig. 3.38 4H-SiC Planar Junction with Extension.

The optimum dose obtained by using the numerical simulations is in excellent agreement with that predicted by using the simple formula in Eq. [3.17]. The breakdown voltage obtained at the optimum dose is twice that for the un-terminated junction for this extension length. The dependence of the breakdown voltage on the extension length is shown in Fig. 3.39 at the optimum dose in the extension. A breakdown voltage close to the parallel plane junction can be obtained by using an extension length of 50 microns.

This method for edge terminations has gained popularity due to its effectiveness in suppressing the edge breakdown for P-N junctions[17,18,19]. In order to implement this technique it was necessary to develop methods for precisely controlling the P-type dopant dose in the extension. Although the dose can be controlled very well by using ion implantation, processes for the annealing and activation of P-type dopants (Boron and Aluminum) needed to be developed. In addition, the performance has been found to be sensitive to surface charge in the passivation as also reported for silicon devices. Enhancements to the breakdown voltage can also be achieved using multiple zones with different dopant dose levels[20] achieved by etching steps in a epitaxially grown P-type layer.

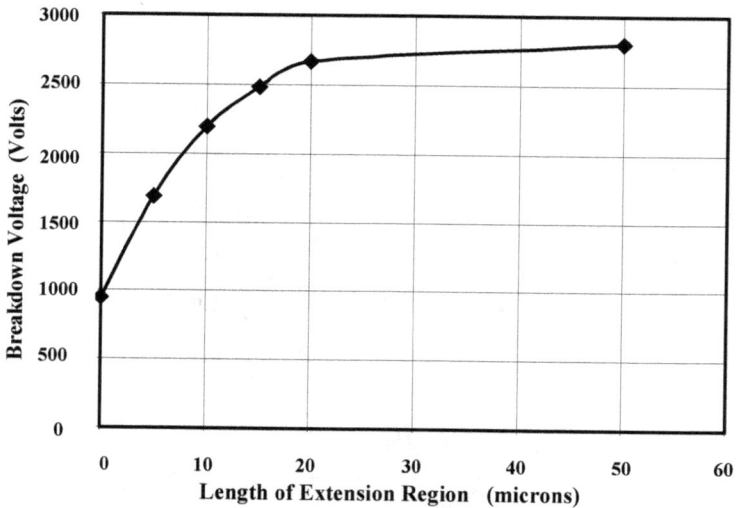

Fig. 3.39 Planar Junctions in 4H-SiC with P- Extension.

3.4 Summary

The design issues pertinent to obtaining a high breakdown voltage in silicon carbide structures have been reviewed in this chapter. As a bench mark, the breakdown voltage of the abrupt parallel plane junction has been analyzed for 4H-SiC and analytical equations developed for the depletion layer width and critical electric field. The breakdown voltage of practical devices is limited by the electric field crowding at their edges. Various edge termination methods have been reviewed to determine their performance relative to the ideal parallel plane junction. The argon implanted high resistance zone and the junction extension using a P- region with a dose of 1.5×10^{13} cm^{-2} have been found to provide the best performance.

References

[1]   B. J. Baliga, "Power Semiconductor Devices", PWS Publishing Company, 1996.

[2]   M. Bhatnagar and B.J. Baliga, "Silicon Carbide High Voltage (400V) Schottky Barrier Diodes", IEEE Electron Device Letters, Vol. ED13, pp. 501-503, 1992.

[3]   R.K. Chilukuri, P. Ananthanarayanan, V. Nagapudi, and B.J. Baliga, "High Voltage P-N Junction Diodes in Silicon Carbide using Field Plate Edge Termination", MRS Symposium Proceedings, Vol. 572, pp. 81-86, 1999.

[4]   M. Bhatnagar, et al, "Edge Terminations for SiC High Voltage Schottky Rectifiers", International Symposium on Power Semiconductor Devices and ICs, pp. 89-94, 1993.

[5]   D. Alok, B.J. Baliga, and P.K. McLarty, "A Simple Edge Termination for Silicon Carbide with Nearly Ideal Breakdown Voltage", IEEE Electron Device Letters, Vol. ED15, pp. 394-395, 1994.

[6]   D. Alok, R. Raghunathan, and B.J. Baliga, "Planar Edge Termination for 4H-SiC Devices", IEEE Transactions on Electron Devices, Vol. 43, pp. 1315-1317, 1996.

[7]   D. Alok, B.J. Baliga, M. Kothandaraman, and P.K. McLarty, "Argon Implanted SiC Device Edge Termination: Modeling, Analysis, and Experimental Results", Institute of Physics Conference Series, Vol. 142, pp. 565-568, 1996.

[8]   A. Itoh, T. Kimoto, and H. Matsunami, "Excellent Reverse Blocking Characteristics of High Voltage 4H-SiC Schottky Rectifiers with Boron Implanted Edge Termination", IEEE Electron Device Letters, Vol. ED17, pp. 139-141, 1996.

[9]   D. Alok and B.J. Baliga, "SiC Device Edge Termination using Finite Area Argon Implantation", IEEE Transactions on Electron Devices, Vol. 44, pp. 1013-1017, 1997.

[10]   A.P. Knights, et al, "The Effect of Annealing on Argon Implanted Edge Terminations for 4H-SiC Schottky Diodes", MRS Symposium Proceedings, Vol. 572, pp. 129-134, 1999.

[11]   B.J. Baliga and S. Ghandhi, "Analytical Solutions for the Breakdown Voltage of Abrupt Cylindrical and Spherical Junctions", Solid State Electronics, Vol. 19, pp.739-744, 1976.

[12]   P.M. Shenoy and B.J. Baliga, "Planar, High Voltage, Boron Implanted 6H-SiC P-N Junction Diodes", Institute of Physics Conference Series, Vol. 142, pp. 717-720, 1996.

[13] B. J. Baliga, "Closed Form Analytical Solutions for the Breakdown Voltage of Planar Junctions Terminated with a Single Floating Field Ring", Solid State Electronics, Vol. 33, pp. 485-488, 1990.

[14] R. Singh and J. W. Palmour, "Planar Terminations in 4H-SiC Schottky Diodes with Low Leakage and High Yields", International Symposium on Power Semiconductor Devices and ICs, pp. 157-160, 1997.

[15] W. Bahng, et al, "Fabrication and Characterization of 4H-SiC pn Diode with Field Limiting Ring", Silicon Carbide and Related Materials – 2003, Materials Science Forum, Vol. 457-460, pp. 1013-1016, 2004.

[16] V.A.K. Temple, "Junction Termination Extension: a New Technique for increasing Avalanche Breakdown Voltage and controlling Surface Electric Fields in P-N Junctions", IEEE International Electron Devices Meeting, Abstract 20.4, pp. 423-426, 1977.

[17] R. Rupp, et al, "Performance and Reliability Issues of SiC Schottky Diodes", Silicon Carbide and Related Materials – 1999, Materials Science Forum, Vol. 338-342, pp. 1167-1170, 2000.

[18] R. Singh, et al, "SiC Power Schottky and PiN Diodes", IEEE Transactions on Electron Devices, Vol. ED49, pp. 665-672, 2002.

[19] H.P. Felsl and G. Wachutka, "Performance of 4H-SiC Schottky Diodes with Al-Doped p-Guard-Ring Junction Termination at Reverse Bias", Silicon Carbide and Related Materials – 2001, Materials Science Forum, Vol. 389-393, pp. 1153-1156, 2002.

[20] X. Li, et al, "Theoretical and Experimental Study of 4H-SiC Junction Edge Termination", Silicon Carbide and Related Materials – 1999, Materials Science Forum, Vol. 338-342, pp. 1375-1378, 2000.

Chapter 4

PiN Rectifiers

As discussed in previous chapters, from a power device perspective, the main advantage of wide band gap semiconductors is the very low resistance of the drift region even when it is designed to support larger voltages. This favors the development of high voltage unipolar devices which have much superior switching speed than bipolar structures. However, the development of high voltage PiN rectifiers has been given significant attention among the SiC research community[1,2]. For this reason, the basic operating principles of the PiN rectifier are reviewed in this chapter. The results of two-dimensional numerical simulations are used to elucidate the physics of operation within these structures. Since the minority carrier lifetime plays an important role in determining the characteristics of these devices, this parameter has been varied to examine its influence on performance.

The silicon PiN rectifiers that are designed to support large voltages rely upon the high level injection of minority carriers into the drift (i) region[3]. This phenomenon greatly reduces the resistance of the thick, very lightly doped drift regions necessary to support high voltages in silicon. In the case of silicon carbide, the drift region doping level is relatively large and its thickness is much smaller than for silicon devices to achieve very high breakdown voltages. The benefits of high level injection are not obvious unless the breakdown voltage approaches 10,000 volts as shown in this section.

4.1 One-Dimensional PiN Structure

The basic one-dimensional PiN rectifier structure is illustrated in Fig. 4.1 together with the electric field profile when reverse biased and the carrier distribution profile when it is forward biased. A punch-through i-region design is favored for PiN rectifiers due to the conductivity modulation of the drift region in the on-state. The breakdown voltage for such regions is provided in reference 3.

When this junction is forward-biased by the application of a negative bias to the N-region, holes and electrons are injected into the drift-region as illustrated in Fig. 4.1. The carrier distribution n(x) can be obtained by solving the continuity equation for the N-region:

$$\frac{dn}{dt} = 0 = -\frac{n}{\tau_{HL}} + D_a \frac{d^2n}{dx^2}$$
[4.1]

where D_a is the ambipolar diffusion coefficient and τ_{HL} is the high level lifetime in the drift region.

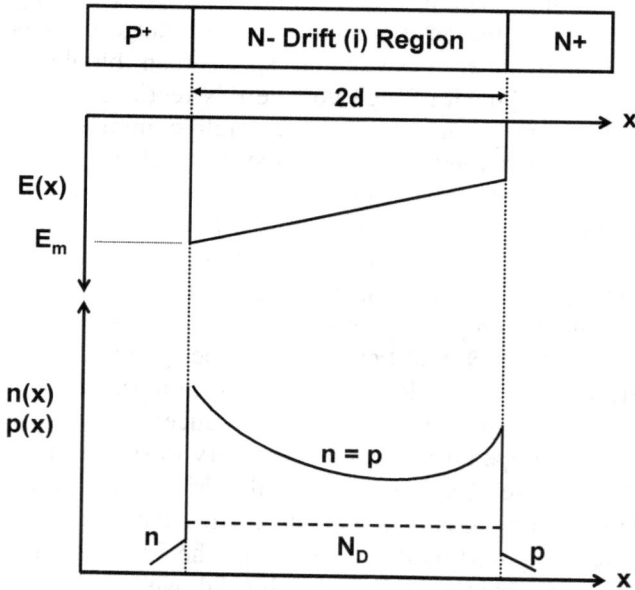

Fig. 4.1 Electric Field and Carrier Distribution for a PiN Rectifier.

The solution for this equation using appropriate boundary conditions[3] yields:

$$n(x) = p(x) = \frac{\tau_{HL}J_F}{2qL_a}\left[\frac{\cosh(x/L_a)}{\sinh(d/L_a)} - \frac{\sinh(x/L_a)}{2\cosh(d/L_a)}\right]$$
[4.2]

The catenary carrier distribution described by this equation is illustrated in the figure. In this expression, the ambipolar diffusion length is given by:

$$L_a = \sqrt{D_a \tau_{HL}} \qquad [4.3]$$

The forward current density J_F can be related to the on-state voltage drop V_F after taking into account voltage drops in the middle (i) region and the two junctions (P$^+$/i and i/N$^+$):

$$J_F = \frac{2qD_a n_i}{d} F\left(\frac{d}{L_a}\right) \exp\left(-\frac{qV_F}{2kT}\right) \qquad [4.4]$$

where $F(d/L_a)$ is a function that reaches a maximum value when $L_a = d$.

For purposes of comparison, numerical simulations were performed on both silicon and 4H-SiC PiN rectifier structures with various breakdown voltages. In each case, the impact of changing the high level lifetime was included in the analysis. The design of the drift region for these rectifiers can be compared in Table 4.1. As discussed earlier, for each breakdown voltage, the doping concentration in the drift region for 4H-SiC is 200 times larger than that for the silicon case.

Breakdown Voltage	Silicon		4H-SiC	
	Doping (cm^{-3})	Thickness (microns)	Doping (cm^{-3})	Thickness (microns)
500 V	5×10^{14}	36	1×10^{17}	2.2
1,000 V	2×10^{14}	81	4×10^{16}	5.3
2,000 V	8×10^{13}	183	1.6×10^{16}	12
3,000 V	5×10^{13}	294	1×10^{16}	20
5,000 V	2×10^{13}	530	4.4×10^{15}	35
10,000 V	9×10^{12}	1200	1.7×10^{15}	80

Table 4.1 Comparison of Drift Region Parameters for Si and 4H-SiC.

Silicon PiN Rectifiers: τ = 1 μsec

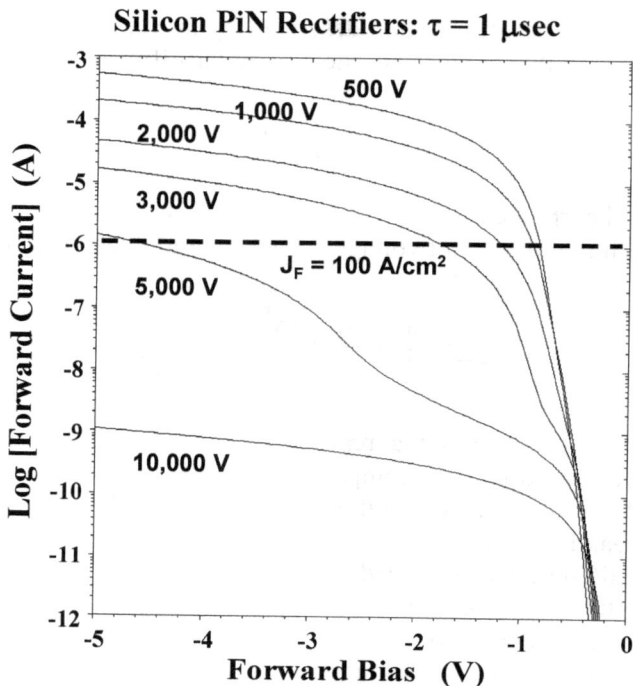

Fig. 4.2 Forward Conduction Characteristics of Silicon PiN Rectifiers.

The forward conduction characteristics for the silicon PiN rectifiers obtained with numerical simulations are shown in Fig. 4.2 for the case of a lifetime of 1 μs (namely, $\tau_{p0} = \tau_{n0} = 1$ μs). Since the area of the simulation structure is 1×10^{-8} cm^{-2}, an on-state current density of 100 A/cm^2 corresponds to a current of 1×10^{-6} A in the figure (as indicated the dashed line). It can be seen that the on-state voltage drop becomes large when the voltage rating becomes greater than 3000 volts for this lifetime value. The on-state voltage drop for the device with breakdown voltage of 10,000 volts can be reduced by increasing the lifetime as shown in Fig. 4.3. This is due to be improved conductivity modulation, which can be observed in Fig. 4.4, where the injected hole concentration is compared with the background doping level in the drift region. At the 1 μs lifetime, very little high level injection occurs due to the short diffusion length. The large resistance of the very lightly doped, thick drift region produces the high on-state voltage drop. At a lifetime of 10 μs, there is weak modulation of the drift region. A lifetime of 100 μs is required to get a low on-state voltage drop in the 10 kV silicon PiN rectifier. This results in a poor switching speed.

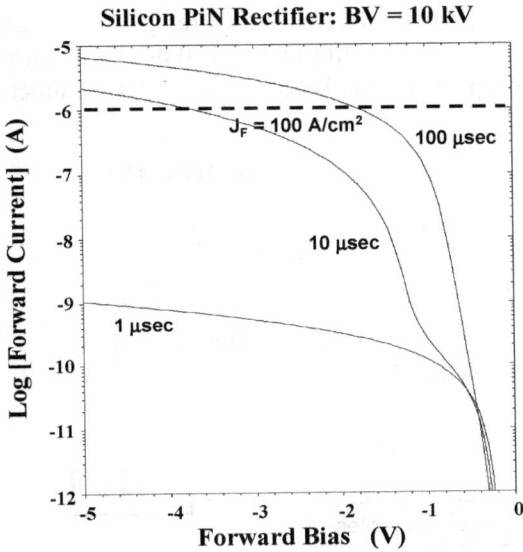

Fig. 4.3 Forward Conduction Characteristics of 10kV Si PiN Rectifiers.

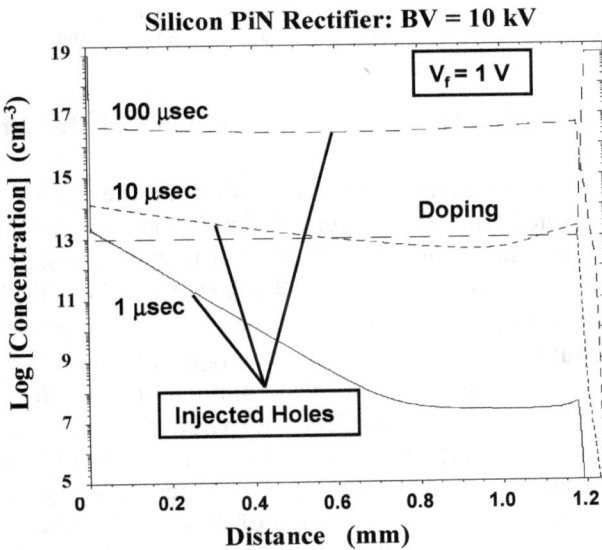

Fig. 4.4 Conductivity Modulation within 10 kV Si PiN Rectifiers.

As shown in Table 4.1, the thickness of the drift region in the 4H-SiC PiN rectifiers is much smaller than for the silicon devices for the same breakdown voltage. This allows good conductivity modulation of the drift region even for the device with breakdown voltage of 10,000 volts as shown in Fig. 4.5 for a lifetime of 100 ns. This plot was done at a forward voltage drop of 4 volts because of the larger junction potential for 4H-SiC

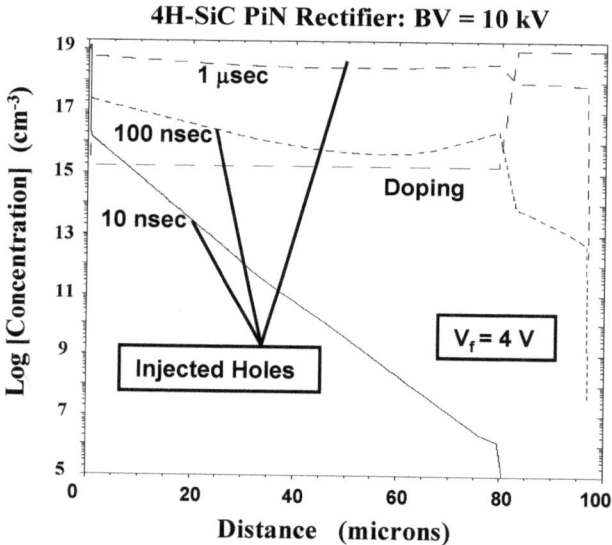

Fig. 4.5 Conductivity Modulation within 10 kV 4H-SiC PiN Rectifier.

The forward conduction characteristics for the 10 kV 4H-SiC PiN rectifier obtained from the numerical simulations are shown in Fig. 4.6 for various lifetime values. It can be seen that the on-state voltage drop is determined by the un-modulated resistance of the drift region when the lifetime is at 5 or 10 nsec. When the lifetime is increased to 100 nsec, the conductivity modulation of the drift region reduces the on-state voltage drop. The results of numerical simulations of the forward characteristics of 4H-SiC PiN rectifiers with various breakdown voltage capabilities are shown in Fig. 4.7 for the case of a lifetime of 5 nsec. The influence of conductivity modulation of the drift region can be observed even for the case of a breakdown voltage of 5000 V. However, the on-state voltage drop for all the devices is relatively high (3 V) when compared to silicon devices.

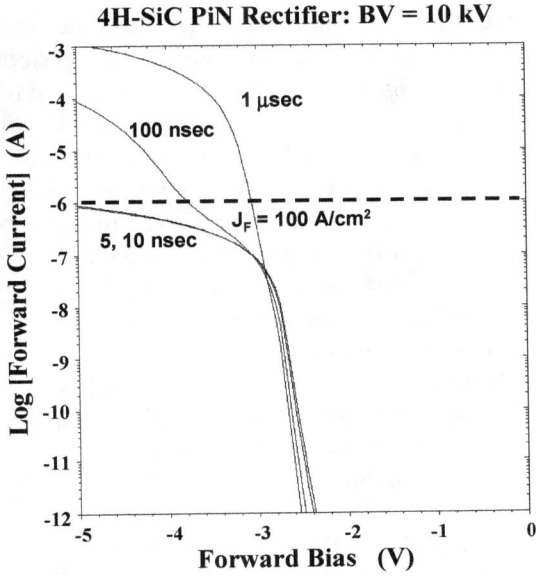

Fig. 4.6 Forward Conduction Characteristics of 10kV 4H-SiC PiN Rectifier.

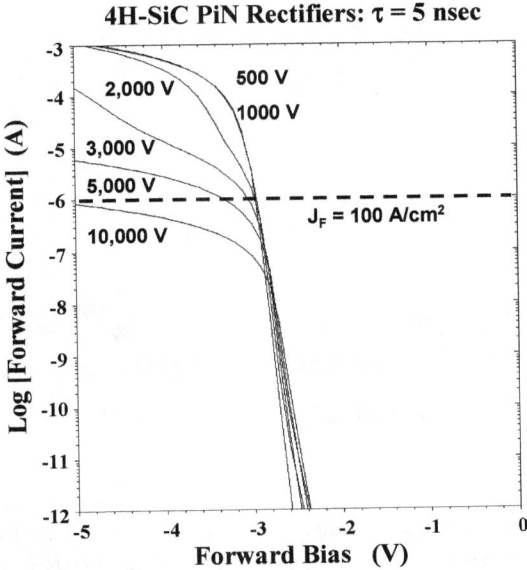

Fig. 4.7 Forward Conduction Characteristics of 4H-SiC PiN Rectifiers.

The 4H-SiC PiN rectifier operates on the same principles as the silicon device. To illustrate this, the on-state carrier distribution within a 3 kV device is shown in Fig. 4.8 at a forward voltage drop of 4 volts. The plot includes both the electron and hole concentrations within the drift region as well as the background doping concentration. High level injection of the mobile carriers is evident leading to strong conductivity modulation even with a lifetime of 10 nsec used in the simulation. The hole and electron carrier profile observes the catenary shape described by Eq. [4.2]. This carrier distribution has been experimentally validated[4]. The behavior of silicon carbide rectifiers can be modeled using the same high level injection physics previous used for silicon devices. However, one advantage of 4H-SiC structures is that the small drift region width required to support high voltages allows reduction of the lifetime to the nanosecond level. This is favorable for achieving fast switching speeds when compared with silicon devices.

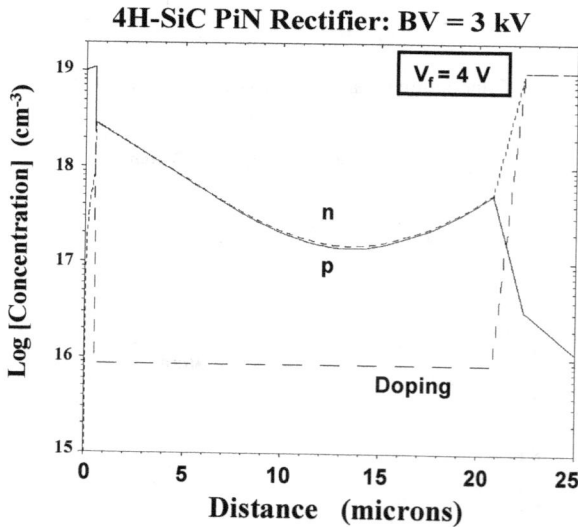

Fig. 4.8 Conductivity Modulation within 3 kV 4H-SiC PiN Rectifier.

The impact of operating temperature on the on-state voltage drop of the 4H-SiC PiN rectifier is shown in Fig. 4.9 for case of a breakdown voltage of 3000 volts. It can be seen that a positive temperature coefficient for on-state voltage drop occurs when the forward current density exceeds 100 A/cm^2. This is favorable for the paralleling of devices and preventing hot spots or current filaments.

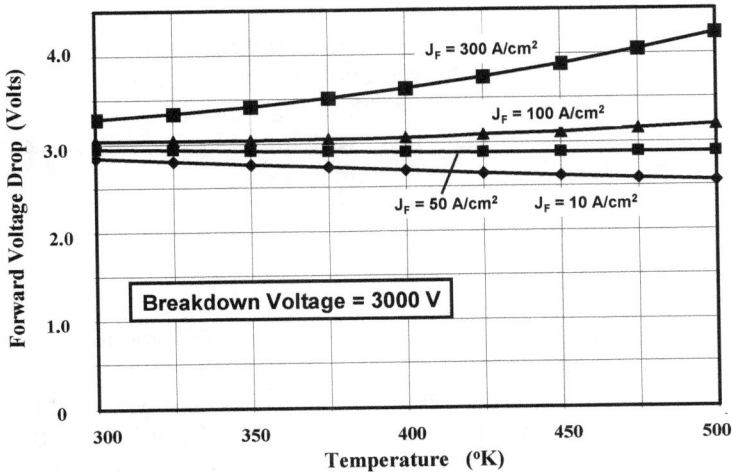

Fig. 4.9 On-state Voltage Drop for a 3 kV 4H-SiC PiN Rectifier.

4.2 Experimental Results

Many groups have worked on the development of high voltage PiN rectifiers[2]. In early studies, the quality of the P$^+$ region was poor due to problems with achieving a high doping concentration. The best results were obtained by epitaxial growth of the anode followed by etching steps to create the edge termination[5]. More recently, the anode has been formed by ion implantation of Al or B or both dopants. The forward characteristic of implanted junction diodes was found to be sensitive to the junction depth and activation process[6]. The on-state voltage drop of a 4H-SiC diode fabricated using 100 micron thick drift region with doping concentration of 1-3 x 10^{14} cm^{-3} has been reported[2] at 7.1 volts for a current density of 100 A/cm^2. These diodes had a breakdown voltage of 8.6 kV. This is significantly worse than that obtained by the numerical simulations despite the reported lifetime in the diodes being about 2 microseconds. Some of this can be attributed to the high contact resistance to the anode.

The on-state voltage drop of diodes formed by co-implantation of aluminum, carbon, and boron to create the anode region[7] has been reported to be about 4.7 volts at room temperature. These diodes, fabricated using a 40 micron thick drift region with doping concentration of 1×10^{15} cm^{-3}, had a breakdown voltage of about 4.5 kV. The on-state voltage drop was found to reduce with increasing temperature at a current density of 100 A/cm^2. This was attributed to a reduction in the contact resistance to the anode region. The lifetime extracted from the reverse recovery measurements was about 21 nanoseconds. These results confirm the conclusion that fast switching PiN rectifiers can be developed from 4H-SiC.

4.3 Summary

The physics of operation of the silicon carbide PiN rectifier has been shown to be similar to that used to describe silicon devices. High level injection in the drift region enables modulating its conductivity to reduce the on-state voltage drop. Due to the much smaller width of the drift region required to obtain a given breakdown voltage when compared with silicon devices, a much smaller lifetime can be used in silicon carbide devices. This is favorable for reducing switching losses. However, the large band gap of silicon carbide results in a large on-state voltage drop in excess of 4 volts. A lower on-state voltage drop can be obtained in a silicon carbide device by using the Schottky barrier contact to achieve rectification as discussed in the next chapter. These Schottky barrier rectifiers are superior to PiN rectifiers for breakdown voltages of at least 5000 volts.

References

[1]   T.P. Chow, N. Ramangul, and M. Ghezzo, "Wide Bandgap Semiconductor Power Devices", MRS Symposium Proceedings, Vol. 483, pp. 89-102, 1998.
[2]   R. Singh, "Silicon Carbide Bipolar Power Devices – Potentials and Limits", MRS Symposium Proceedings, Vol. 640, pp. H4.2.1-H4.2.12, 2001.
[3]   B. J. Baliga, "Power Semiconductor Devices", PWS Publishing Company, 1996.
[4]   O. Tornblad, et al, "Investigation of Excess Carrier distributions in 4H-SiC Power Diodes under Static Conditions and Turn-on", Silicon Carbide and Related Materials – 1997, Material Science Forum, Vol. 264-268, pp. 1053-1056, 1998.
[5]   Y. Sugawara, K. Asano, R. Singh, and J.W. Palmour", "6.2 kV 4H-SiC pin Diode with Low Forward Voltage Drop", Silicon Carbide and Related Materials – 1999, Material Science Forum, Vol. 338-342, pp. 1371-1374, 2000.
[6]   R.K. Chilukuri, P. Ananthanarayanan, V. Nagapudi, and B.J. Baliga, "High Voltage P-N Junction Diodes in Silicon Carbide using Field Plate Edge Termination", MRS Symposium Proceedings, Vol. 572, pp. 81-86, 1999.
[7]   J.B. Fedison, et al, "Al/C/B Co-implanted High Voltage 4H-SiC PiN Junction Rectifiers", Silicon Carbide and Related Materials – 1999, Material Science Forum, Vol. 338-342, pp. 1367-1370, 2000.

Chapter 5

Schottky Rectifiers

The main advantage of wide band gap semiconductors for power device applications is the very low resistance of the drift region even when it is designed to support large voltages. This favors the development of high voltage unipolar devices which have much superior switching speed than bipolar structures. The Schottky rectifier, formed by making a rectifying contact been a metal and the semiconductor drift region, is an attractive unipolar device for power applications. In the case of silicon, the maximum breakdown voltage of Schottky rectifiers has been limited by the increase in the resistance of the drift region[1]. Commercially available devices are generally rated at breakdown voltages of less than 100 volts. Novel silicon structures that utilize the charge-coupling concept have allowed extending the breakdown voltage to the 200 volt range[2,3]. Many applications described in chapter 1 require fast switching rectifiers with low on-state voltage drop that can also support over 500 volts. The much lower resistance of the drift region for silicon carbide enables development of such Schottky rectifiers with very high breakdown voltages. These devices not only offer fast switching speed but also eliminate the large reverse recovery current observed in high voltage silicon rectifiers. This reduces switching losses not only in the rectifier but also in the IGBTs used within the power circuits[4].

This chapter reviews the basic principles of operation of the metal semiconductor contact. Unique issues that relate to the wide band gap of silicon carbide and the presence of defect levels within the band gap must be given special consideration. The much larger electric field in silicon carbide is shown to lead to significant barrier lowering with increasing reverse voltage when compared with silicon devices. Structures designed to reduce the electric field at the Schottky contact are therefore of importance to the development of rectifiers. The results of the analysis of these structures by using two-dimensional numerical simulations are described in the next chapter. The influence of defects on the characteristics of 4H-SiC Schottky rectifiers is discussed in this chapter. Experimental results on relevant structures are provided to define the state of the development effort on Schottky rectifiers.

5.1 Schottky Rectifier Structure: Forward Conduction

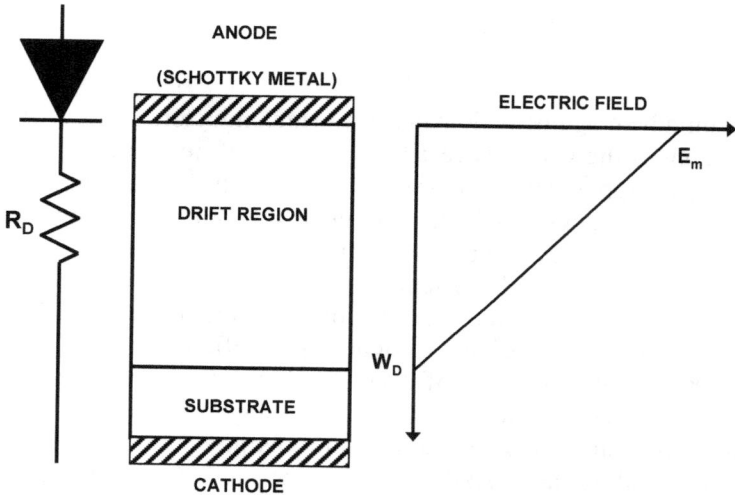

Fig. 5.1 Structure and Electric Field in a Schottky Rectifier.

The basic one-dimensional structure of the metal-semiconductor or Schottky rectifier structure is shown in Fig. 5.1 together with electric field profile under reverse bias operation. The applied voltage is supported by the drift region with a triangular electric field distribution if the drift region doping is uniform. The maximum electric field occurs at the metal contact. The device undergoes breakdown when this field becomes equal to the critical electric field. As discussed in chapter 1, the specific on-resistance of the drift region is given by:

$$R_{on-ideal} = \frac{4BV^2}{\varepsilon_S \mu_n E_C^3}$$

[5.1]

The specific on-resistance of the drift region for 4H-SiC is approximately 2000 times smaller than for silicon devices for the same breakdown voltage as shown earlier in Fig. 3.6. Their values are given by:

$$R_{drift} = R_{on-ideal}(Si) = 5.93x10^{-9} BV^{2.5} \qquad [5.2]$$

and

$$R_{drift} = R_{on-ideal}(4H - SiC) = 2.97x10^{-12} BV^{2.5} \qquad [5.3]$$

In addition, it is important to include the resistance associated with the thick, highly doped N^+ substrate because this is comparable to that for the drift region in some instances. The specific resistance of the N^+ substrate can be determined by taking the product of its resistivity and thickness. For silicon, N^+ substrates with resistivity of 1 mΩ-cm are available. If the thickness of the substrate is 200 microns, the specific resistance contributed by the N^+ substrate is 2 x 10^{-5} Ω-cm^2. For silicon carbide, the available resistivity of the N^+ substrates is substantially larger. For the available substrates with a typical resistivity of 0.02 Ω-cm and thickness of 200 microns, the substrate contribution is 4 x 10^{-4} Ω-cm^2.

The on-state voltage drop for the Schottky rectifier at a forward current density J_F, including the substrate contribution, is given by:

$$V_F = \frac{kT}{q} \ln\left(\frac{J_F}{J_S}\right) + \left(R_{drift} + R_{subs}\right)J_F \qquad [5.4]$$

where J_S is the saturation current density. The saturation current density for the Schottky barrier is related to the Schottky barrier height (ϕ_b):

$$J_S = AT^2 \exp\left(-\frac{q\phi_b}{kT}\right) \qquad [5.5]$$

where A is Richardson's constant. For silicon, the Richardson's constant is 110 A°K^{-2}cm^{-2} while that for 4H-SiC has been calculated[5] to be 146 A°K^{-2}cm^{-2}. In Eq. [5.4], the first term accounts for the voltage drop across the metal-semiconductor contact while the second term accounts for the voltage drop across the series resistance (shown as R_D in Fig. 5.1).

The calculated forward conduction characteristics for silicon Schottky rectifiers are shown in Fig. 5.2 for various breakdown voltages. For this figure, a Schottky barrier height of 0.7 eV was chosen because this is a typical value used in actual devices. It can be seen that the series resistance of the drift region does not adversely impact the on-state voltage drop for the device with breakdown voltage of 50 volts at a nominal on-state current density of 100 A/cm^2. However, this resistance

becomes significant when the breakdown voltage exceeds 100 volts. This
has limited the application of silicon Schottky rectifiers to systems, such
as switch-mode power supply circuits, operating at voltages below 50 V.

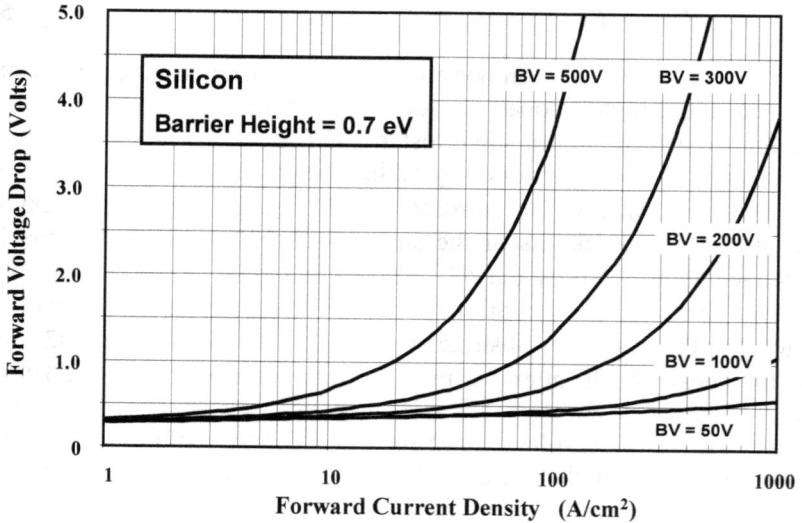

Fig. 5.2 Forward Characteristics of Silicon Schottky Rectifiers.

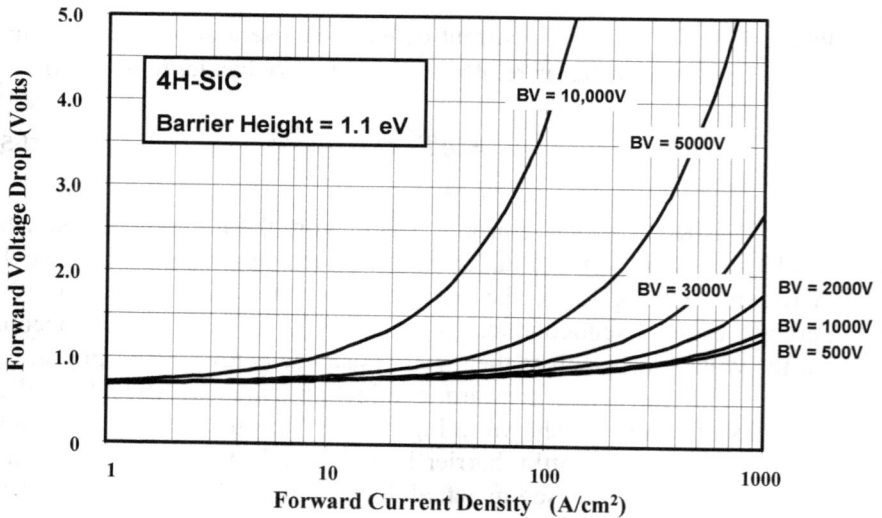

Fig. 5.3 Forward Characteristics of 4H-SiC Schottky Rectifiers.

The significantly smaller resistance of the drift region enables scaling of the breakdown voltage of silicon carbide Schottky rectifiers to much larger voltages typical of medium and high power electronic systems, such as those used for motor control. The forward characteristics of high voltage 4H-SiC Schottky rectifiers are shown in Fig. 5.3 for the case of a Schottky barrier height of 1.1 eV. The N^+ substrate resistance used for these calculations was 4 x 10^{-4} Ω-cm^2. It can be seen that the drift region resistance does not produce a significant increase in on-state voltage drop until the breakdown voltage exceeds 3000 volts. From these results, it can be concluded that silicon carbide Schottky rectifiers are excellent companion diodes for medium and high power electronic systems that utilize Insulated Gate Bipolar Transistors (IGBTs). Their fast switching speed and absence of reverse recovery current can reduce power losses and improve the efficiency in motor control applications[4].

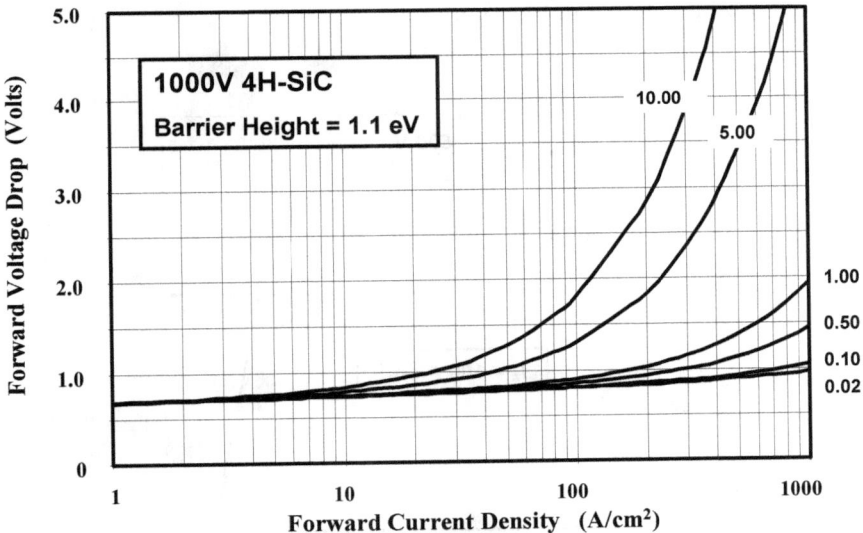

Fig. 5.4 Forward Characteristics of 4H-SiC Schottky Rectifiers: Impact of Substrate and Contact Resistance (in mΩ-cm^2).

The impact of the substrate and contact resistance is shown in Fig. 5.4 for the case of a 4H-SiC rectifier with a breakdown voltage of 1000 volts. The drift region specific on-resistance for this breakdown voltage is 9.4 x 10^{-5} Ω-cm^2. The typical substrate contribution of 4 x 10^{-4} Ω-cm^2 is therefore a dominant resistance for this rectifier. However, the contribution of this resistance to the on-state voltage drop (40 mV) is acceptable when compared with a total on-state voltage drop of 0.9 volts. The ohmic contact resistance to the substrate can substantially add to the total series resistance. It is necessary to reduce this resistance to below 1 mΩ-cm^2 to avoid further increase in the on-state voltage drop.

5.1.1 Forward Conduction: Simulation Results

To confirm the analytically calculated forward characteristics, numerical simulations of the one-dimensional 4H-SiC Schottky rectifier were performed using a metal work function of 4.5 eV (barrier height of 0.8 eV). The drift region parameters were varied in accordance with the parameters given in Table 4.1 to change the breakdown voltage capability of the structures. The forward characteristics obtained from the simulations are shown in Fig. 5.5.

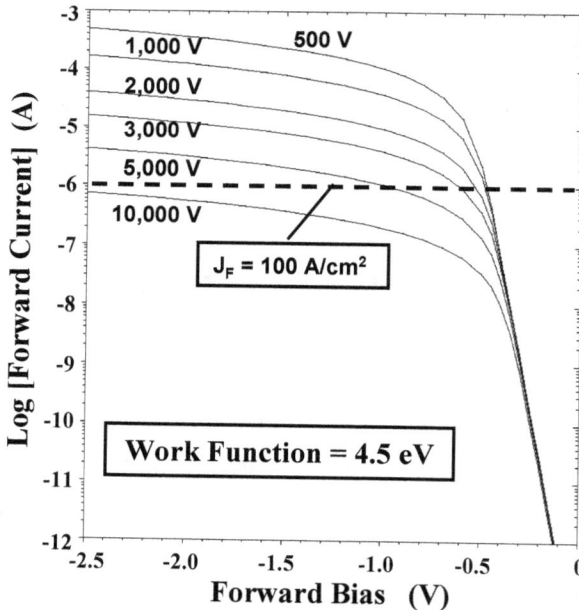

Fig. 5.5 Forward Characteristics of 4H-SiC Schottky Rectifiers.

The results shown in Fig. 5.5 are in agreement with the analytical model when the difference in barrier height is taken into account. The influence of the barrier height on the forward characteristics can be observed from Fig. 5.6 for the case of drift region with breakdown voltage of 3000 volts. When the barrier height is increased to 1.3 eV corresponding to a work function of 5.0 eV, the forward characteristics are close to those for the 3kV structure in Fig. 5.3. Due to the relatively small contribution to the forward voltage drop (at the nominal on-state current density of 100 A/cm^2) from the series resistance of the drift region, the on-state voltage drop is found to strongly depend upon the barrier height.

Breakdown Voltage = 3000 V

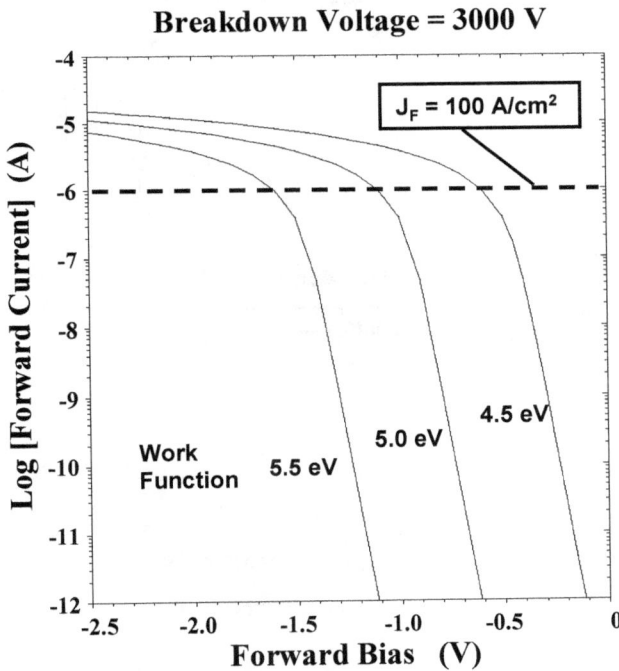

Fig. 5.6 Forward Characteristics of 3 kV 4H-SiC Schottky Rectifiers:

Impact of Schottky Barrier Work-Function.

The impact of changing the ambient temperature on the forward conduction characteristics of the 4H-SiC Schottky rectifier is shown in Fig. 5.7 for the case of a breakdown voltage of 3000 volts. As in the case of silicon devices, the voltage drop associated with the barrier reduces

with increasing temperature leading to a smaller voltage drop at low
current levels. However, the drift region resistance increases with
temperature due to the degradation of the bulk mobility leading to an
increase in the on-state voltage drop at high current levels. For the
nominal operating current density of 100 A/cm^2, the on-state voltage
drop has a positive temperature coefficient indicating stable operation for
the diode.

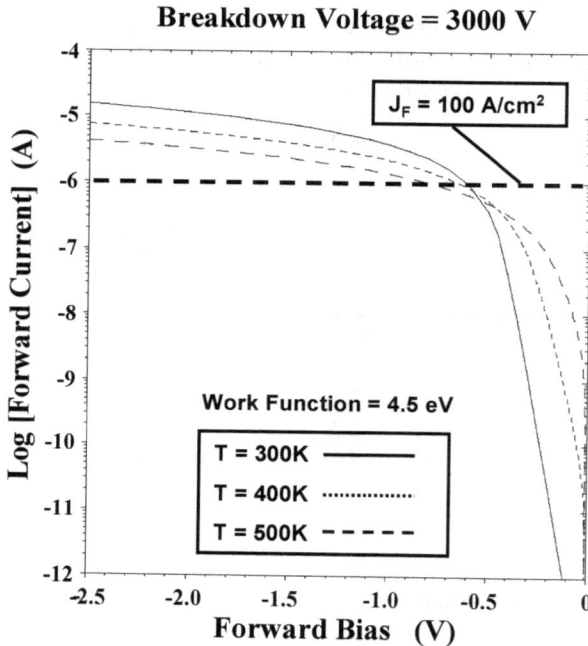

Fig. 5.7 Forward Characteristics of a 3 kV 4H-SiC Schottky Rectifier:

Impact of Operating Temperature.

5.1.2 Forward Conduction: Experimental Results

The first high voltage silicon carbide Schottky rectifiers were developed
at the Power Semiconductor Research Center (PSRC) in 1991 using the
commercially available 6H polytype[6]. The breakdown voltage of these
un-terminated diodes was limited by electric field enhancement at the
edges. The breakdown voltage was subsequently increased to 1000 volts
by using an argon implanted edge termination[7]. The same approach

enabled the development of high voltage Schottky diodes from 4H-SiC with breakdown voltages of up to 1000 volts without edge termination[8]. The breakdown voltage of these diodes was extended to 1330 volts by using the argon implanted edge termination[9].

The 4H-SiC Schottky rectifiers discussed above were fabricated using titanium as the barrier metal due to its low barrier height (1.1 eV). The on-state voltage drop for these 4H-SiC Schottky rectifiers was found to be less than 1.1 volts at a current density of 100 A/cm^2 consistent with the calculated values in Fig. 5.3. Other groups have also reported the successful fabrication of high voltage Schottky barrier rectifiers from 4H-SiC by using various barrier metals. An on-state voltage drop of 1.73V, 1.5V, and 1.12V was reported at an on-state current density of 100 A/cm^2 for diodes with breakdown voltage of 800V fabricated by using gold, nickel and titanium[5], respectively. The on-state voltage drop reduces as the barrier height decreases from 1.80 eV for gold to 1.1 eV for titanium. Schottky rectifiers with a breakdown voltage of 1000 volts were reported with on-state voltage drop of 1.78V and 1.61V, at an on-state current density of 100 A/cm^2, by using nickel and platinum as barrier metals[10]. These relatively larger on-state voltage drops are associated with the larger barrier height of 1.59 eV and 1.39 eV for nickel and platinum, respectively. These demonstrations of low on-state voltage drop in high voltage silicon carbide Schottky rectifiers have encouraged the analysis of manufacturing issues[11] leading to the deployment of commercial products.

5.2 Schottky Rectifier Structure: Reverse Blocking

When a negative bias is applied to the Schottky rectifier, the voltage is supported across the drift region with the maximum electric field located at the metal-semiconductor contact as shown in Fig. 5.1. At a small reverse bias voltage, the leakage current is determined by the saturation current given in Eq. [5.5] as governed by the Schottky barrier height. The saturation current for 4H-SiC calculated by using this expression is shown in Fig. 5.8 for various temperatures. In comparison with silicon, it is possible to obtain a much larger barrier height in silicon carbide metal-semiconductor contacts due to the larger band gap. For this reason, the graph extends to relatively large barrier heights. On the one hand, in order to maintain a low on-state voltage drop, it is preferable to use a low barrier height as discussed in the previous section. On the other hand, as can be seen from Fig. 5.8, it is preferable to use a large barrier height to reduce the leakage current. A good compromise occurs for a barrier height in the range of 1.0 to 1.3 eV. This can be achieved by using titanium as the Schottky contact metal[6,8].

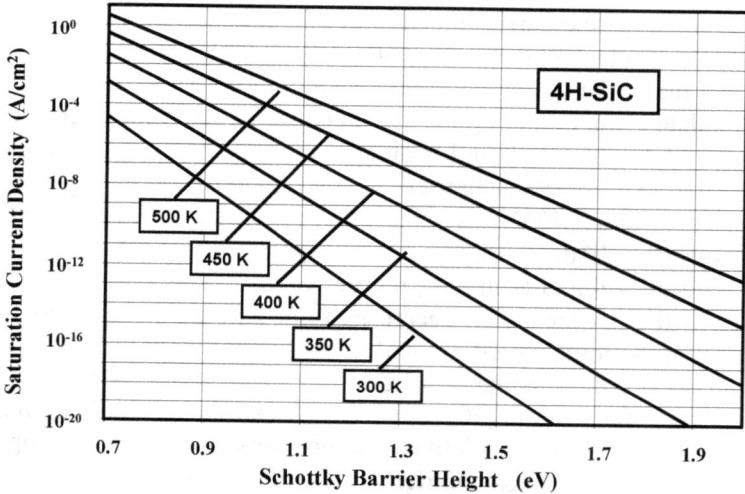

Fig. 5.8 Saturation Current Density for 4H-SiC Schottky Rectifiers.

In silicon Schottky rectifiers, it is well established that the leakage current is exacerbated by the Schottky barrier lowering phenomenon[1]. The barrier lowering is determined by the electric field at the metal-semiconductor interface:

$$\Delta\phi_B = \sqrt{\frac{qE_m}{4\pi\varepsilon_S}} \qquad [5.6]$$

where E_m is the maximum electric field located at the metal-semiconductor interface. For a one-dimensional structure, the maximum electric field is related to the applied reverse bias voltage (V_R) by:

$$E_m = \sqrt{\frac{2qN_D}{\varepsilon_S}(V_R + V_{bi})} \qquad [5.7]$$

In addition, it is necessary to include the effect of pre-avalanche multiplication on the leakage current[12]. The multiplication coefficient

(M) can be determined from the maximum electric field at the metal-semiconductor contact:

$$M = \left\{ 1 - 1.52 \left[1 - \exp\left(-\frac{4.33x10^{-24} E_m^{4.93} W_D}{6} \right) \right] \right\}^{-1} \qquad [5.8]$$

where W_D is the depletion layer width. The leakage current density for a silicon Schottky rectifier with drift region doping concentration of 1 x 10^{16} cm^{-3} is shown in Fig. 5.9. The breakdown voltage for this case is about 50 volts. The impact of including the Schottky barrier lowering and multiplication is shown in the figure to illustrate their relative contributions. The barrier lowering effect increases the leakage current by nearly an order of magnitude. The effect of including the multiplication coefficient is apparent at high voltages when the electric field approaches the critical electric field for breakdown. This behavior is consistent with commercially available silicon devices, which exhibit an order of magnitude increase in leakage current from low reverse bias voltages to the rated voltage (about 80 percent of the breakdown voltage).

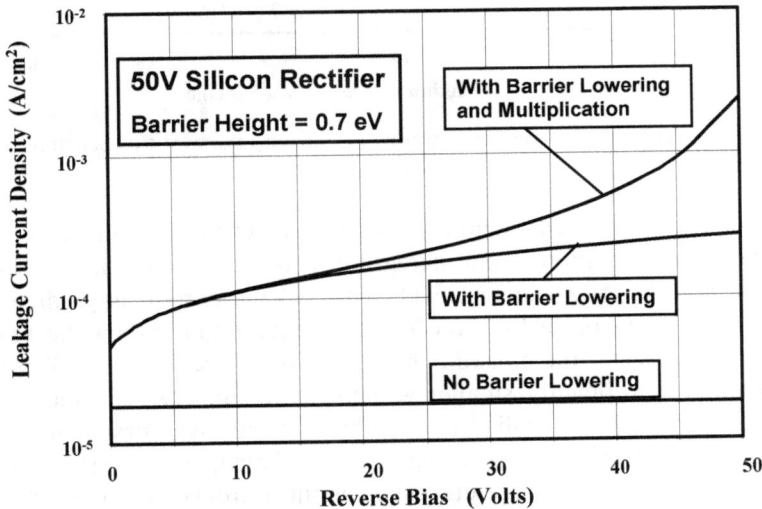

Fig. 5.9 Leakage Current Density for a 50V Silicon Schottky Rectifier.

Since the low specific on-resistance of the drift region in silicon carbide devices is associated with the much larger electric fields in the material before the on-set of impact ionization, the Schottky barrier lowering in silicon carbide rectifiers is significantly larger than in silicon devices. For the case of a drift region doping level of 1×10^{16} cm^{-3} for both silicon and 4H-SiC, the barrier lowering was found to be three times larger in silicon carbide at the breakdown voltage as shown in Fig. 5.10. In preparing this graph, the reverse voltage was normalized to the breakdown voltage.

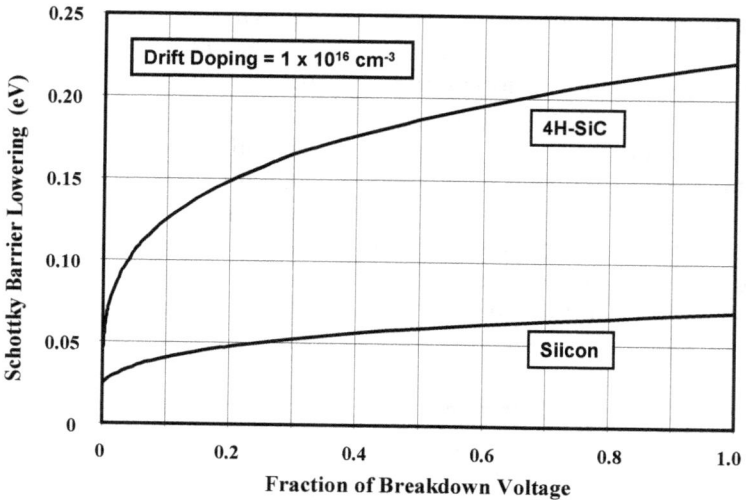

Fig. 5.10 Schottky Barrier Lowering for Silicon versus 4H-SiC Schottky Rectifiers.

The enhanced Schottky barrier lowering in silicon carbide devices leads to a more rapid increase in leakage current with increasing reverse bias as shown in Fig. 5.11. The leakage current is predicted by this model to increase by three orders of magnitude when the reverse voltage approaches the breakdown voltage. In comparison, for silicon the increase is significantly smaller as shown in Fig. 5.9. The increase in leakage current with applied reverse bias voltage was first reported for high voltage 6H-SiC Schottky rectifiers by Bhatnagar et al[6]. It was found that the increase in leakage current is about 6 orders of magnitude with increase in reverse bias voltage. A similar observation was made later for Schottky rectifiers fabricated at PSRC using 4H-SiC[8]. Similar results were subsequently reported by other groups[13,14]. Thus, the observed leakage current in silicon carbide rectifiers cannot be accounted for by using the thermionic emission model with Schottky barrier lowering.

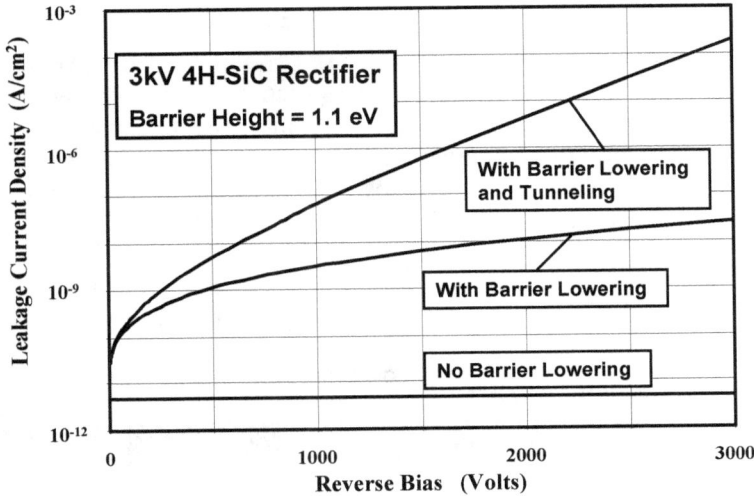

Fig. 5.11 Leakage Current Density for a 3kV 4H-SiC Schottky Rectifier.

In order to explain the more rapid increase in leakage current observed in silicon carbide Schottky rectifiers, it is necessary to include the field emission (or tunneling) component of the leakage current[15]. The thermionic field emission model for the tunneling current leads to a barrier lowering effect proportional to the square of the electric field at the metal-semiconductor interface. When combined with the thermionic emission model, the leakage current density can be written as:

$$ J_S = AT^2 \exp\left(-\frac{q\phi_b}{kT}\right) . \exp\left(\frac{q\Delta\phi_b}{kT}\right) . \exp\left(C_T E_m^2\right) \qquad [5.9] $$

where C_T is a tunneling coefficient. A tunneling coefficient of 8 x 10^{-13} cm²/V² was found to yield an increase in leakage current by six orders of magnitude as shown in Fig. 5.11 consistent with the behavior observed in reference [15] and the experimental results discussed earlier. As can be noted from Fig. 5.11, the inclusion of the tunneling model enhances the leakage current by another three orders of magnitude.

As discussed above, the leakage current in silicon carbide Schottky rectifiers increases much more rapidly with reverse voltage than in silicon devices. Fortunately, larger barrier heights can be utilized in silicon carbide devices to reduce the magnitude of the leakage current density. This enables maintaining an acceptable level of power dissipation in the reverse blocking mode. For example, in the case of the 3kV 4H-SiC Schottky diode discussed above, the reverse power dissipation at room temperature is less than 1 W/cm^2 compared with an on-state power dissipation of 100 W/cm^2. The expected increase in leakage current with temperature must of course be taken into account in order to ensure than the reverse power dissipation remains below the on-state power dissipation for stable operation[1].

5.3 Schottky Rectifier Structure: Impact of Defects

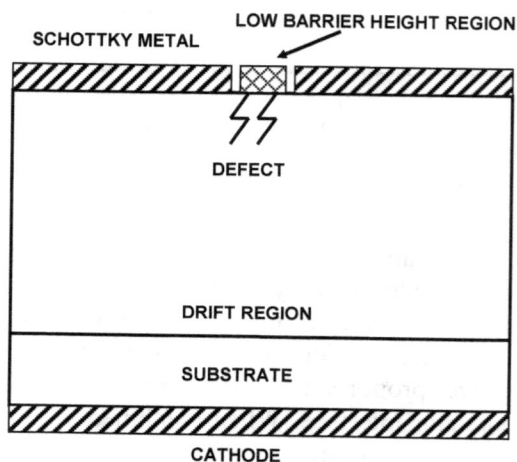

Fig. 5.12 Schottky Rectifier with Localized Low Barrier Region.

The impact of defects or surface inhomogeneities on the forward and reverse characteristics of silicon carbide high voltage Schottky rectifiers was first observed and analyzed at PSRC for diodes made from 6H-SiC[16]. These diodes exhibited a leakage current at low bias voltages (before the on-set of the tunneling currents) that was several orders of magnitude larger than calculated based upon the barrier height extracted from C-V measurements. Meanwhile, the forward characteristics at high current density were consistent with those calculated using this barrier height. This conundrum could be solved by postulating small regions of the contact with lower barrier height when compared with most of the contact layer as illustrated in Fig. 5.12. It was speculated that the lower barrier height originates from the presence of defects in the silicon carbide or the surface preparation process prior to Schottky metal deposition. The influence of inhomogeneous contacts on silicon Schottky rectifiers had been previously postulated by Tung[17].

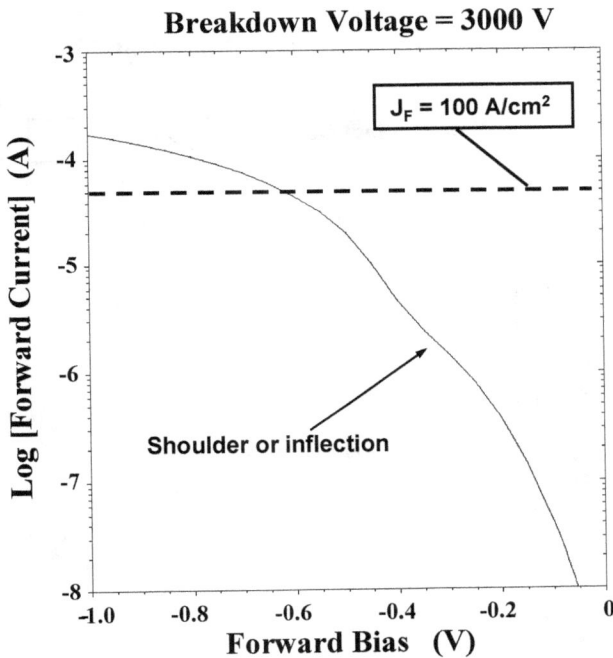

Fig. 5.13 Forward Characteristics of 3 kV 4H-SiC Schottky Rectifiers with 0.02 percent Inhomogeneous Contact Area.

The impact of the inhomogeneous contact depends upon the reduction of the barrier height at the local region as well as its area relative to the entire contact. The results of two dimensional numerical analysis of a Schottky contact with a reduced barrier region operating in parallel with the main contact are discussed here to elucidate the phenomenon. The drift region parameters used for these simulations are the same as those used earlier for the 3kV 4H-SiC Schottky rectifiers. A defect region with an area of 0.02 percent of the total contact area will be considered with a total area of the simulated structure of 5×10^{-7} cm^{-2}.

The work function for the defect area was chosen as 4.0 eV versus 4.5 eV used for the main contact (i.e. a barrier reduction by 0.5 eV). Under these conditions, a shoulder is observed in the forward characteristics, as shown in Fig. 5.13. This shoulder is associated with current flow through the low barrier contact region at smaller applied voltages. This can be confirmed by looking at the currents through the main contact and the low barrier (defect) contact as shown in Fig. 5.14.

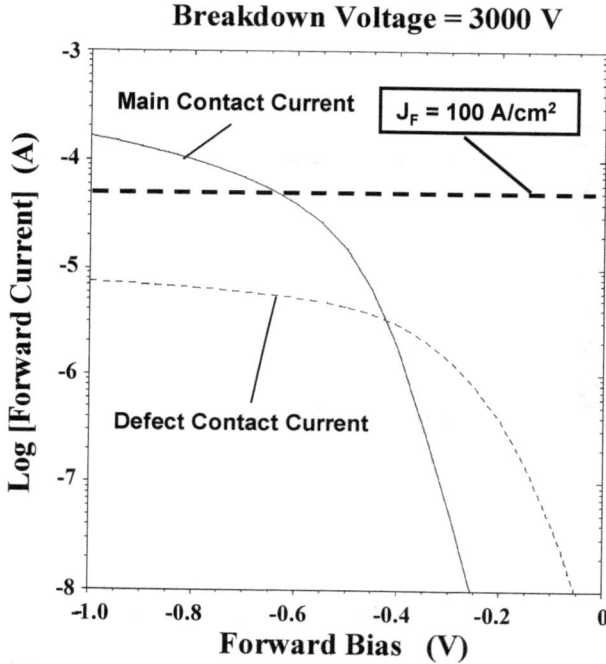

Fig. 5.14 Forward Characteristics of 3 kV 4H-SiC Schottky Rectifiers with 0.02 percent Inhomogeneous Contact Area.

From the forward characteristics shown above, it is apparent that the impact of the low-barrier defect region is not significant on the on-state voltage drop (indicated by the line at a current density of 100 A/cm^2 on the graphs). This is due to the high series resistance for such a small contact within the structure, although this series resistance is not proportional to the area of the defect region because of significant current spreading within the drift region. However, the defect region has a very significant impact on the reverse leakage current as shown in Fig.

5.15. It can be seen that the leakage current contributed from the low barrier (defect) region is four orders of magnitude larger than from the main contact in spite of a difference in area by a factor of 2 x 10^{-4}. Thus, very small defects can have a large influence on the leakage current in the silicon carbide Schottky contacts. The impact of changing the area of the defect region and the amount of barrier height lowering has been reported in reference [16] for 6H-SiC Schottky rectifiers. Similar results have been more recently reported for 4H-SiC Schottky rectifiers[18,19,20]. These diodes display a shoulder in the forward characteristics as shown by the above figures and a much larger leakage current than predicted by the barrier height extracted from C-V measurements. A defect area of about 0.1 percent is indicated in the experimental results. The presence of defects that may be responsible for the low barrier height regions has been confirmed by EBIC measurements[16].

Breakdown Voltage = 3000 V

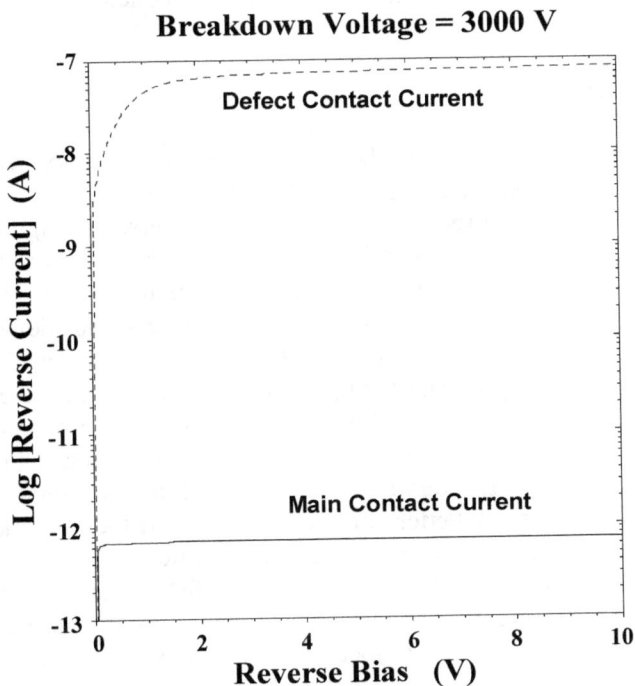

Fig. 5.15 Reverse Characteristics of 3 kV 4H-SiC Schottky Rectifiers with 0.02 percent inhomogeneous contact area.

5.4 Reliability Issues

The low on-state voltage drop of high voltage silicon carbide Schottky rectifiers and their excellent reverse recovery behavior has encouraged the development of commercial products. However, for commercial acceptance, it is necessary to meet various reliability criteria for semiconductor diodes. The results of the following tests on titanium 4H-SiC Schottky rectifiers have been reported:

(a) Thermal cycling up to 400°C
(b) Cycling between -55 and 150°C (1000 times)
(c) High temperature reverse bias testing (150°C, 600V, 1000 hours)
(d) High humidity, high temperature reverse bias testing (85°C, 85% relative humidity, 80V, 1000 hours)

No failures were observed indicating that the reliability of silicon carbide Schottky rectifiers under reverse bias stress is excellent[21].

5.5 Summary

The physics of operation of the silicon carbide Schottky rectifier has been shown to be similar to that used to describe silicon devices in the forward conduction direction. However, the reverse leakage current in silicon carbide devices is significantly enhanced by the thermionic field emission (or tunneling) current at high reverse bias voltages. This is due to the much larger electric field in silicon carbide devices when compared with the silicon counterpart. Fortunately, the availability of much larger Schottky barrier heights in silicon carbide, at the expense of an increased on-state voltage drop, allows reducing the absolute values for the leakage currents to a level where the reverse bias power dissipation is below the forward bias power dissipation enabling stable operation in circuits. A better approach is to shield the Schottky contact from the high electric field developed in the silicon carbide drift region as discussed in the next chapter. It has been found that the presence of small defects can greatly enhance the leakage current in silicon carbide Schottky rectifiers due to a localized barrier lowering phenomenon. However, the defect density in wafers has been reduced to the point at which commercial products are being deployed as companion diodes for IGBTs in motor control applications.

References

[1] B. J. Baliga, "Power Semiconductor Devices", PWS Publishing Company, 1996.

[2] B.J. Baliga, "Schottky Barrier Rectifiers and Methods of Forming the Same", U.S. Patent 5,612,567, March 18, 1997.

[3] B.J. Baliga, "The Future of Power Semiconductor Technology", Proceedings of the IEEE, Vol. 89, pp. 822-832, 2001.

[4] B.J. Baliga, "Power Semiconductor Devices for Variable Frequency Drives", Proceedings of the IEEE, Vol. 82, pp. 1112-1122, 1994.

[5] A. Itoh, T. Kimoto, and H. Matsunami, "High Performance of High Voltage 4H-SiC Schottky Barrier Diodes", IEEE Electron Device Letters, Vol. 16, pp. 280-282, 1995.

[6] M. Bhatnagar, P.K. McLarty, and B.J. Baliga, "Silicon Carbide High Voltage (400V) Schottky Barrier Diodes", IEEE Electron Device Letters, Vol. 13, pp. 501-503, 1992.

[7] D. Alok, B.J. Baliga, and P.K. McLarty, "A Simple Edge Termination for Silicon Carbide with Nearly Ideal Breakdown Voltage", IEEE Electron Device Letters, Vol. ED15, pp. 394-395, 1994.

[8] R. Raghunathan, D. Alok, and B.J. Baliga, "High Voltage 4H-SiC Schottky Barrier Diodes", IEEE Electron Device Letters, Vol. 16, pp. 226-227, 1995.

[9] D. Alok, R. Raghunathan, and B.J. Baliga, "Planar Edge Termination for 4H-SiC Devices", IEEE Transactions on Electron Devices, Vol. 43, pp. 1315-1317, 1996.

[10] V. Saxena, N. Nong, and A.J. Steckl, "High Voltage Ni and Pt SiC Schottky Barrier Diodes utilizing Metal Field Plate Termination", IEEE Transactions on Electron Devices, Vol. 46, pp. 456-464, 1999.

[11] M. Treu, et al, "Challenges and First Results of SiC Schottky Diode Manufacturing using a 3 inch Technology", Silicon Carbide and Related Materials – 2003, Material Science Forum, Vol. 457-460, pp. 981-984, 2004.

[12] S. L. Tu and B. J. Baliga, "On the Reverse Blocking Characteristics of Schottky Power Diodes", IEEE Transactions on Electron Devices, Vol. 39, pp. 2813-2814, 1992.

[13] F. Dahlquist, et al, " A 2.8kV, Forward Drop JBS Diode with Low Leakage", Silicon Carbide and Related Materials – 1999, Material Science Forum, Vol. 338-342, pp. 1179-1182, 2000.

[14]   Y. Sugawara, K. Asano, and R. Saito, "3.6kV 4H-SiC JBS Diodes with Low RonS", Silicon Carbide and Related Materials – 1999, Material Science Forum, Vol. 338-342, pp. 1183-1186, 2000.

[15]   T. Hatakeyama and T. Shinohe, "Reverse Characteristics of a 4H-SiC Schottky Barrier Diode", Silicon Carbide and Related Materials – 2001, Material Science Forum, Vol. 389-393, pp. 1169-1172, 2002.

[16]   M. Bhatnagar, B.J. Baliga, H.R. Kirk, and G.A. Rozgonyi, "Effect of Surface Inhomogeneities on the Electrical Characteristics of SiC Schottky Contacts", IEEE Transactions on Electron Devices, Vol. 43, pp. 150-156, 1996.

[17]   R. T. Tung, "Electron Transport of Inhomogeneous Schottky Barriers", Applied Physics Letters, Vol. 58, pp. 2821-2822, 1991.

[18]   D. Defives, et al, "Barrier Inhomogeneities and Electrical Characteristics of Ti/4H-SiC Schottky Rectifiers", IEEE Transactions on Electron Devices, Vol. 46, pp. 449-455, 1999.

[19]   F. Roccaforte, et al, "Electrical Characterization of Inhomogeneous NiSi/SiC Schottky Contacts", Silicon Carbide and Related Materials – 2003, Material Science Forum, Vol. 457-3460, pp. 869-872, 2004.

[20]   H. Saitoh, T. Kimoto, and H. Matsunami, "Origin of Leakage Current in SiC Schottky Barrier Diodes at High Temperatures", Silicon Carbide and Related Materials – 2003, Material Science Forum, Vol. 457-3460, pp. 997-1000, 2004.

[21]   R. Rupp, et al, "Performance and Reliability Issues of SiC Schottky Diodes", Silicon Carbide and Related Materials – 1999, Material Science Forum, Vol. 338-342, pp. 1167-1170, 2000.

Chapter 6

Shielded Schottky Rectifiers

In the case of silicon Schottky rectifiers, it has been traditional to trade-off the on-state (or conduction) power loss against the reverse blocking power loss by optimizing the Schottky barrier height. As the Schottky barrier height is reduced, the on-state voltage drop decreases producing a smaller conduction power loss. At the same time, the smaller barrier height produces an increase in the leakage current leading to larger reverse blocking power loss. It has been demonstrated that the power loss can be minimized by reducing the Schottky barrier height at the expense of a reduced maximum operating temperature[1].

This optimization process is exacerbated by the rapid increase in the leakage current with increasing reverse bias voltage due to the Schottky barrier lowering phenomenon. The first method proposed to ameliorate the barrier lowering effect in vertical silicon Schottky rectifiers utilized shielding by incorporation of a P-N junction[2,3]. Since the basic concept was to create a potential barrier to shield the Schottky contact against high electric fields generated in the semiconductor by using closely spaced P$^+$ regions around the contact, this structure was named the *'Junction-Barrier controlled Schottky (JBS) rectifier'*. In the JBS rectifier, the on-state current was designed to flow in the un-depleted gaps between the P$^+$ regions when the diode is forward biased to preserve unipolar operation. In addition to reducing the leakage current, the presence of the P$^+$ regions was also shown to enhance the ruggedness of the diodes. Detailed optimization of the silicon JBS rectifier characteristics was achieved with sub-micron dimensions between the P$^+$ regions[4].

Subsequently, a novel silicon structure that utilizes the charge-coupling concept was proposed to reduce the resistance in the drift region[5]. Since these structures utilized an electrode embedded within an oxide coated trench region that surrounds the metal-semiconductor contact, this structure was named the *'Trench MOS Barrier controlled*

Schottky (TMBS) rectifier'. The performance of this structure was further enhanced by using a graded doping profile to create a device named the *'Graded-Doped Trench MOS Barrier controlled Schottky (GD-TMBS) rectifier'*[6]. This has allowed extending the breakdown voltage of silicon Schottky rectifiers to the 200 volt range[7,8].

Yet another method proposed to create a potential barrier under the Schottky contact was the utilization of a second metal-semiconductor contact with large barrier height surrounding the main Schottky contact with a low barrier height[9]. Since a stronger potential barrier could be created by locating the high barrier metal with a trench, this structure was named the *'Trench Schottky Barrier controlled Schottky (TSBS) rectifier'*.

In the previous chapter it was demonstrated that the increase in leakage current with reverse bias voltage is much stronger for the silicon carbide Schottky rectifiers. This is due to a larger Schottky barrier lowering effect associated with the larger electric field in silicon carbide drift regions and the onset of field emission (or tunneling) current. It is therefore obvious that the methods proposed to suppress the electric field at the Schottky contact in silicon diodes will have even greater utility in silicon carbide rectifiers.

This chapter discusses the application of shielding techniques to reduce the electric field at the metal-semiconductor contact in silicon carbide Schottky rectifiers. The physics of operation of these structures is analyzed by using two-dimensional numerical analysis. It is demonstrated that the JBS and TSBS concepts can be extended to silicon carbide to achieve significant improvement in performance. However, the TMBS and GD-TMBS approaches are not appropriate for silicon carbide because of the high electric field generated in the oxide leading to its rupture. Experimental results on relevant structures are provided to define the state of the development effort on shielded silicon carbide Schottky rectifiers.

6.1 Junction Barrier Schottky (JBS) Rectifier Structure

The Junction Barrier controlled Schottky (JBS) rectifier structure is illustrated in Fig. 6.1. It consists of a P^+ region placed around the Schottky contact to generate a potential barrier under the metal-semiconductor contact in the reverse blocking mode. The space between the P^+ regions is chosen so that there is an un-depleted region below the

Schottky contact to enable unipolar conduction through the structure. The voltage drop across the diode is usually not sufficient to forward bias the P-N junction. Low voltage silicon devices operate with on-state voltage drops around 0.4 volts which is well below the 0.7 volts needed for inducing strong injection across the P-N junction. The margin is even larger for silicon carbide due to its much larger band gap. Typical silicon carbide Schottky rectifiers have on-state voltage drops of less than 1.5 volts due to low specific resistance of the drift region. This is well below the 3 volts required to induce injection from the P-N junction. Consequently, the JBS concept is well suited for development of silicon carbide structures with very high breakdown voltages.

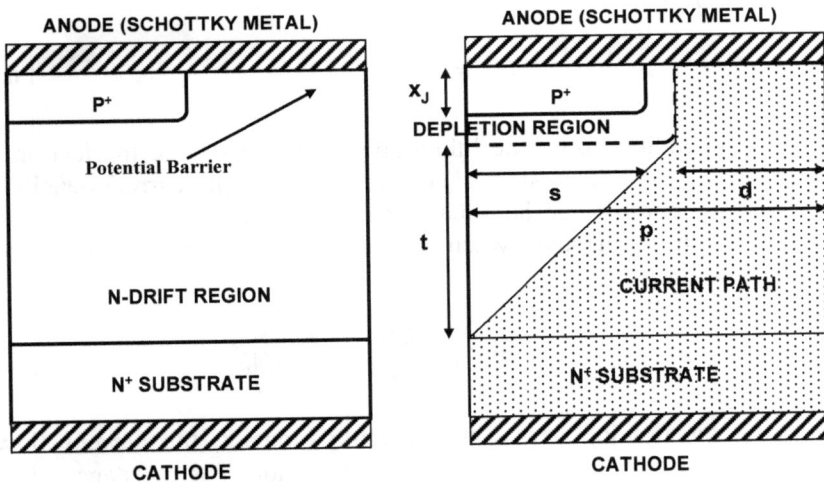

Fig. 6.1 Junction Barrier controlled Schottky Rectifier Structure.

The P^+ region is usually formed by ion-implantation of P-type dopants using a mask with appropriate spacing to leave room for the Schottky contact. For processing convenience, the same metal layer is used to make an ohmic contact to the highly doped P^+ region as well as to make the Schottky barrier contact to the N- drift region. The space between the P^+ regions must be optimized to obtain the best compromise between the on-state voltage drop, which increases as the space is reduced, and the leakage current, which decreases as the space is reduced.

6.1.1 JBS Rectifier Structure: Forward Conduction Model

Analysis of the on-state voltage drop of the JBS rectifier requires taking
into consideration the current constriction at the Schottky contact and the
spreading resistance of the drift region[10]. The current flow pattern in the
forward conduction mode is illustrated in Fig. 6.1 with the dotted area.
The on-state voltage drop for the JBS rectifier is sufficiently low so that
any injection from the P-N junction can be neglected resulting in no
current flow through the junction. The current through the Schottky
contact flows only within the un-depleted portion (with dimension 'd') of
the drift region at the top surface. Consequently, the current density at
the Schottky contact (J_{FS}) is related to the cell (or cathode) current
density (J_{FC}) by:

$$J_{FS} = \left(\frac{p}{d}\right) J_{FC}$$
[6.1]

Depending up on the lithography used for device fabrication to
minimize the size (dimension 's') of the P^+ region, the current density at
the Schottky contact can be enhanced by a factor of ten or more. This
must be taken into account when computing the voltage drop across the
Schottky contact:

$$V_{FS} = \phi_B + \frac{kT}{q}\ln\left(\frac{J_{FS}}{AT^2}\right)$$
[6.2]

After flowing across the Schottky contact, the current flows
through the un-depleted portion of the drift region. The resistance of the
drift region is larger than the specific on-resistance discussed in the
previous chapters due to current spreading from the Schottky contact to
the N^+ substrate:

$$R_{drift} = \rho_D(x_J + W_D)\left(\frac{p}{d}\right) + \rho_D\left(\frac{pt}{p-2d}\right)\ln\left(\frac{p}{2d}\right)$$
[6.3]

The first term accounts for the resistance of the drift region between the
P^+ regions. The dimension 'd' is determined by the cell pitch (p):

$$d = p - s - W_D$$
[6.4]

where W_D is the depletion width. For silicon carbide Schottky rectifiers it
is even more important to take the depletion width into account due to

the large built-in potential (V_{bi}) for P-N junctions. As an approximation, the zero bias depletion width (W_0) can be used:

$$W_0 = \sqrt{\frac{2\varepsilon_S V_{bi}}{qN_D}}$$ [6.5]

For a doping concentration of 1×10^{16} cm^{-3}, the zero bias depletion width is about 0.6 microns, which is comparable to the cell dimensions.

In addition, the resistance associated with the thick, highly doped N^{+} substrate (R_{subs}) can be included by taking the product of its resistivity and thickness. For the available substrates with a typical resistivity of 0.02 Ω-cm and thickness of 200 microns, the substrate contribution is 4 x 10^{-4} Ω-cm^2. The on-state voltage drop for the JBS rectifier at a forward cell current density J_{FC}, including the substrate contribution, is then given by:

$$V_F = \phi_B + \frac{kT}{q}\ln\left(\frac{J_{FS}}{AT^2}\right) + \left(R_{drift} + R_{subs}\right)J_{FC}$$ [6.6]

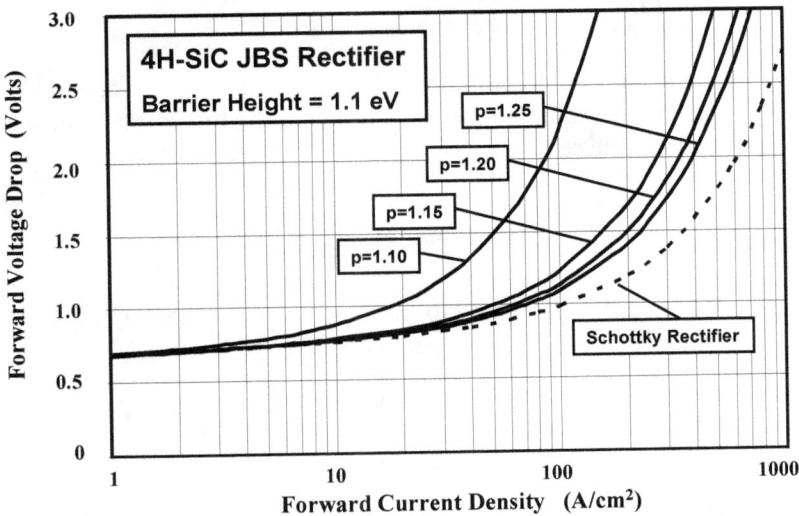

Fig. 6.2 Forward Characteristics of 3kV JBS Rectifiers.

The forward characteristics of 3kV JBS rectifiers calculated using the above analytical approach are shown in Fig. 6.2 with the cell pitch (p) as a parameter. The width of the P$^+$ region (s) was kept at 0.5 microns for this analysis. In comparison with the Schottky rectifier characteristics (shown by the dashed line in the figure), the increase in on-state voltage drop at a nominal forward current density of 100 A/cm^2 is small (less than 0.2 volts) as long as the cell pitch is more than 1.15 microns. This cell pitch is sufficient to obtain substantial reduction of the electric field at the metal-semiconductor contact as shown later in the chapter.

6.1.2 JBS Rectifier Structure: Reverse Leakage Model

The reverse leakage current in the JBS rectifier is reduced when compared with the Schottky rectifier due to the smaller electric field at the metal-semiconductor interface. In addition, the area of the Schottky contact is a fraction of the total cell area resulting in a smaller reverse current contribution. The leakage current in the JBS rectifier can be calculated using the same expression as for the Schottky rectifier with these adjustments:

$$J_L = \left(\frac{d}{p}\right) AT^2 \exp\left(-\frac{q\phi_b}{kT}\right) . \exp\left(\frac{q\Delta\phi_{bJBS}}{kT}\right) . \exp\left(C_T E_{JBS}^2\right) \qquad [6.7]$$

where C$_T$ is a tunneling coefficient (8 x 10^{-13} cm^2/V^2). In contrast with the Schottky rectifier, the barrier lowering and the tunneling contribution is determined by the reduced electric field E$_{JBS}$ at the contact:

$$\Delta\phi_{bJBS} = \sqrt{\frac{qE_{JBS}}{4\pi\varepsilon_S}} \qquad [6.8]$$

The electric field E$_{JBS}$ is related to the reverse bias voltage (V$_P$) at which the space between the P$^+$ regions gets depleted by:

$$E_{JBS} = \sqrt{\frac{2qN_D}{\varepsilon_S}(\beta V_P + V_{bi})} \qquad [6.9]$$

where β is a coefficient used to account for the build up in the electric field after pinch-off. The pinch-off voltage V$_P$ can be obtained from the device cell parameters:

$$V_P = \frac{qN_D}{2\varepsilon_S}(p-s)^2 - V_{bi} \qquad \text{[6.10]}$$

In the case of a cell pitch (p) of 1.25 microns with a P^+ region with dimension 's' of 0.5 microns, the pinch-off voltage is 5.3 volts for a drift region with doping concentration of 1×10^{16} cm^{-3}. Using a β of 10, the electric field at the Schottky contact is found to be reduced to about 4.5×10^5 V/cm at a reverse bias of 50 volts. The calculated leakage current for this electric field after accounting for the reduced contact area is about 8×10^{-11} A/cm^2. Even if the electric field at the Schottky contact increases to 1×10^6 V/cm corresponding to a reverse bias of 300 volts, the calculated leakage current is only 1×10^{-9} A/cm^2, which is far smaller than the values of about 1×10^{-4} A/cm^2 near breakdown for the Schottky rectifier. Thus, limiting the electric field by creating a potential barrier below the Schottky contact can be very beneficial to reducing the leakage current in silicon carbide Schottky rectifiers. In order to determine the magnitude of the electric field reduction at the Schottky contact it is necessary to resort to two dimensional numerical simulations.

6.1.3 JBS Rectifier Structure: Simulation Results

The case of a JBS rectifier with breakdown voltage of 3kV using a drift region with doping concentration of 1×10^{16} cm^{-3} will be considered here for comparison with the Schottky rectifier described in chapter 5. The two-dimensional numerical simulations were performed with various spacing between the P^+ regions. The depth of the P^+ region was also varied to examine its impact on the electric field reduction at the Schottky contact. In each case, the forward and reverse characteristics were analyzed to look at the trade-off between on-state voltage drop and electric field reduction at the Schottky contact. The P^+ dimension 's' was kept at 0.5 microns for all the simulation structures.

The forward characteristics of a 3kV JBS 4H-SiC rectifier is shown in Fig. 6.3 for the case of a cell pitch of 1.25 microns. The on-state voltage drop at a current density of 100 A/cm^2 is about 0.1 volts larger than that for the Schottky rectifier. This is similar to the results obtained using the analytical model in the previous section. The diode exhibits the desirable positive temperature coefficient with no kinks in the characteristics.

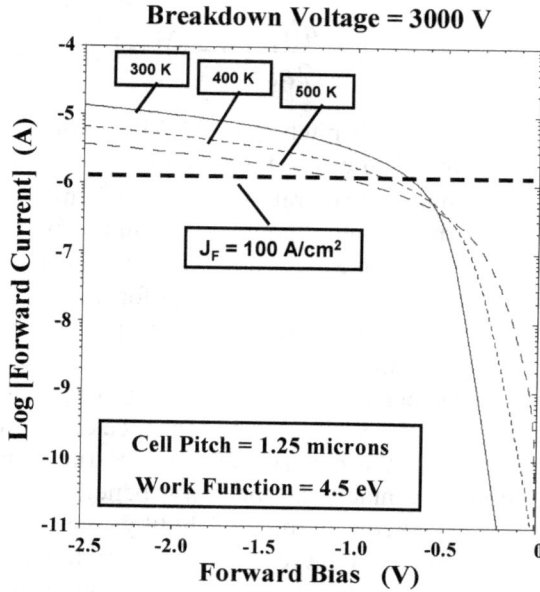

Fig. 6.3 Forward Characteristics of 3 kV 4H-SiC JBS Rectifier.

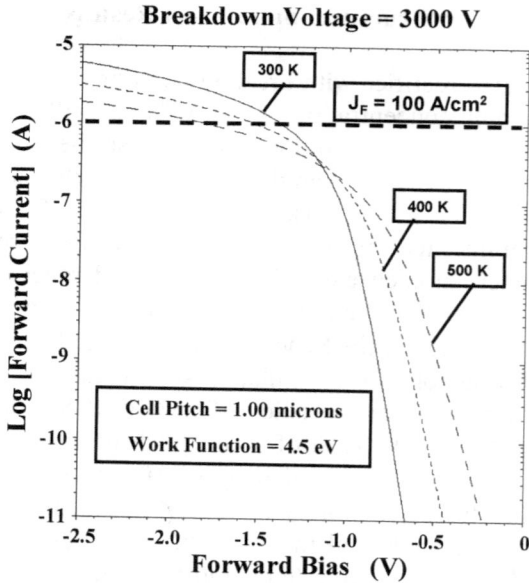

Fig. 6.4 Forward Characteristics of 3 kV 4H-SiC JBS Rectifier.

However, as expected, the forward characteristics were found to degrade substantially when the cell pitch was reduced to 1.0 microns as shown in Fig. 6.4. The on-state voltage drop increases by 0.7 volts when the cell pitch is reduced, which is consistent with the analytical model. It will be shown later in this section that a cell pitch of 1.25 microns is adequate to suppress the electric field at the Schottky contact.

The reason for the increase in the on-state voltage drop can be understood by examining the current flow in the cell under forward bias conditions. The current flow pattern for the structure with 1.00 micron cell pitch is compared with that in the structure with 1.25 micron cell pitch in Fig. 6.5. The much greater constriction of the current at the Schottky contact is obvious for the smaller cell pitch

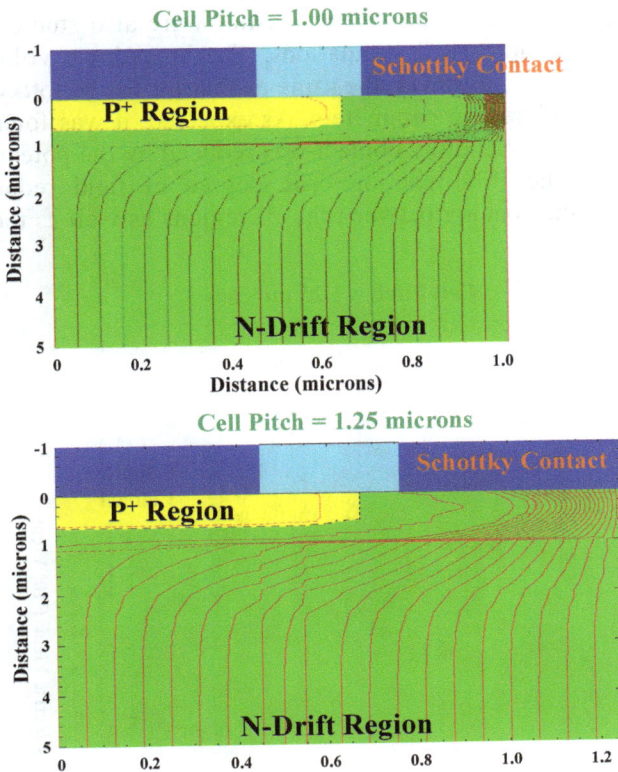

Fig. 6.5 On-state Current Flow in 3 kV 4H-SiC JBS Rectifiers.

It is also worth noting that the current spreading occurs quite rapidly and becomes uniform at a depth of about 2 microns. This indicates an even lower series resistance for the drift region for the JBS rectifier than obtained by the model in the previous section. By using a 45 degree spreading angle from the Schottky contact region into the drift region, the lower drift region resistance can be modeled using:

$$R_{drift} = \rho_D\left(x_J + W_D\right)\left(\frac{p}{d}\right) + \rho_D\left(\frac{p(s+W_D)}{p-2d}\right)\ln\left(\frac{p}{2d}\right)$$
$$+ \rho_D\left(t - x_J - s - 2W_D\right)$$

[6.11]

This lower drift region resistance favors silicon carbide JBS rectifier on-state performance approaching that of Schottky rectifiers.

The reverse blocking characteristics were also studied by using two-dimensional numerical simulations. The breakdown voltage of the JBS structure was found to be the same as that for the Schottky rectifiers for the same drift region parameters. As expected, it was found that the electric field at the Schottky contact was reduced by the potential barrier created using the P-N junction. The degree of field reduction was dependent on the spacing between the P⁺ regions as well as the depth of the P⁺ region.

Cell Pitch = 1.25 microns

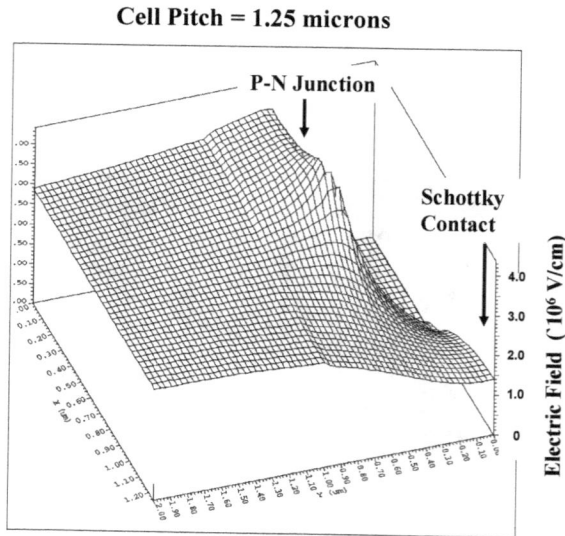

Fig. 6.6 Electric Field distribution in a 4H-SiC JBS Rectifier.

A three-dimensional view of the electric field in a JBS structure with cell pitch of 1.25 microns is shown in Fig. 6.6 at a reverse bias of 3000 volts (just prior to breakdown). It can be seen that the highest electric field occurs at the P-N junction and that the electric field is suppressed at the Schottky contact. It is worth pointing out that the maximum electric field at the Schottky contact occurs at the location furthest away from the P-N junction (at x = 1.25 microns for the structure in Fig. 6.6). Consequently, the largest leakage due to barrier lowering and tunneling components will occur at this location. For this reason, the highest electric field at the Schottky contact will be used to analyze the reverse leakage characteristics for the JBS rectifiers.

Cell Pitch = 1.25 microns

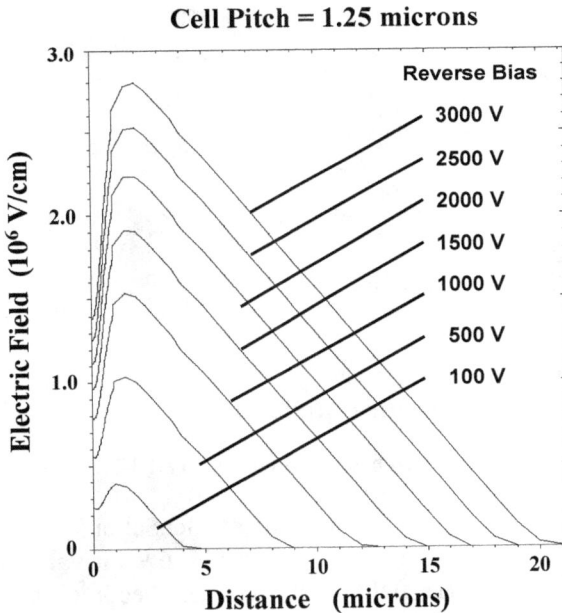

Fig. 6.7 Electric Field variation with Reverse Voltage in a 4H-SiC JBS Rectifier.

The electric field profile at the center of the Schottky contact is shown in Fig. 6.7 for the case of a cell pitch of 1.25 microns. It can be seen that the electric field at the surface under the contact is significantly reduced when compared the peak electric field in the bulk. The peak of the electric field occurs at a depth of about 2 microns. At a reverse bias of 3000 volts, the electric field at the Schottky contact is only 1.4 x 10[6]

V/cm compared with 2.8 x 10^6 V/cm at the maxima. An even greater reduction of the electric field at the Schottky contact can be achieved by reducing the cell pitch. This is illustrated in Fig. 6.8 for the case of a cell pitch of 1.00 microns. Here, the electric field at the Schottky contact is only 7 x 10^5 V/cm compared with 2.8 x 10^6 V/cm at the maxima.

Cell Pitch = 1.00 microns

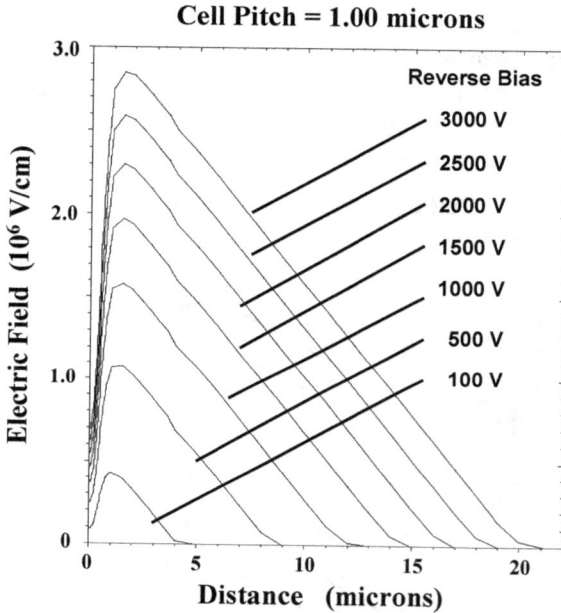

Fig. 6.8 Electric Field variation with Reverse Voltage in a 4H-SiC JBS Rectifier.

The increase in the electric field at the Schottky contact with increasing reverse bias voltage is charted in Fig. 6.9 for various cases of the cell pitch. The electric field at the contact becomes close to the maximum value in the bulk when the pitch is over 2 microns. Thus, the benefits of using the JBS concept to suppress the electric field at the Schottky contact can be obtained only with carefully optimized spacing. With proper spacing, the reduction of the electric field provides the benefit of reducing the Schottky barrier lowering and the leakage current. This is quantified in the graphs shown in Fig. 6.10 and Fig. 6.11, respectively, where the values were determined using the analytical formulae with the electric field extracted from the simulations. With a pitch of 1.25 microns, the Schottky barrier lowering is reduced from 0.22 eV to 0.14 eV.

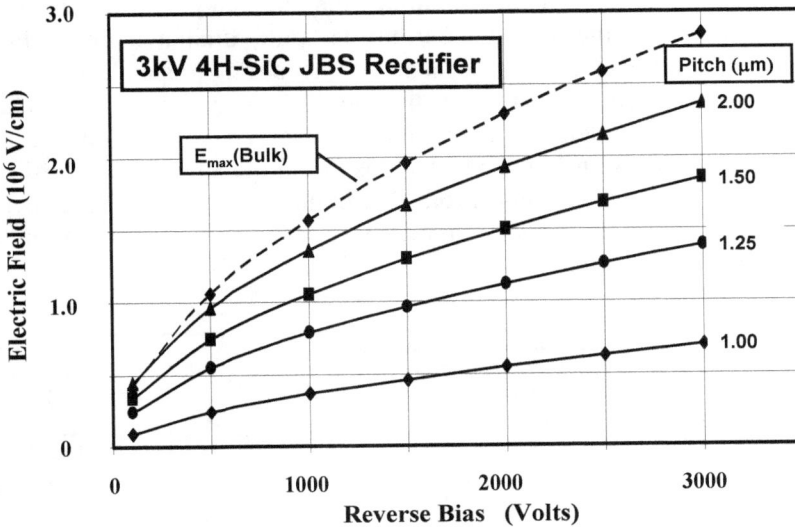

Fig. 6.9 Electric Field variation with Reverse Voltage in 4H-SiC JBS Rectifiers.

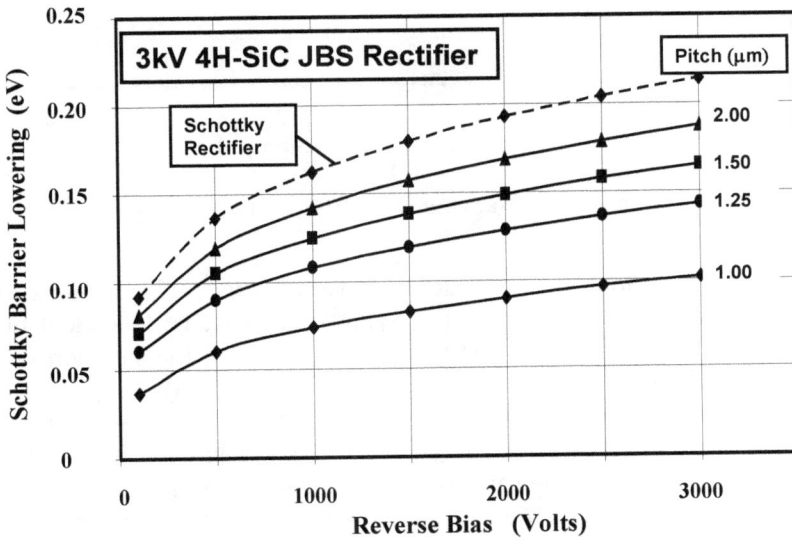

Fig. 6.10 Schottky Barrier Lowering in 4H-SiC JBS Rectifiers.

The impact of ameliorating the Schottky barrier lowering phenomenon on the leakage current is quite dramatic from silicon carbide Schottky rectifiers because of the strong dependence of the tunneling component on the electric field. A reduction in the leakage current by five orders of magnitude is observed at high reverse bias voltages, as shown in Fig. 6.11, demonstrating the advantage of utilizing the junction barrier concept. From this point of view, the JBS structure originally proposed for silicon rectifiers is even more utilitarian for silicon carbide rectifiers.

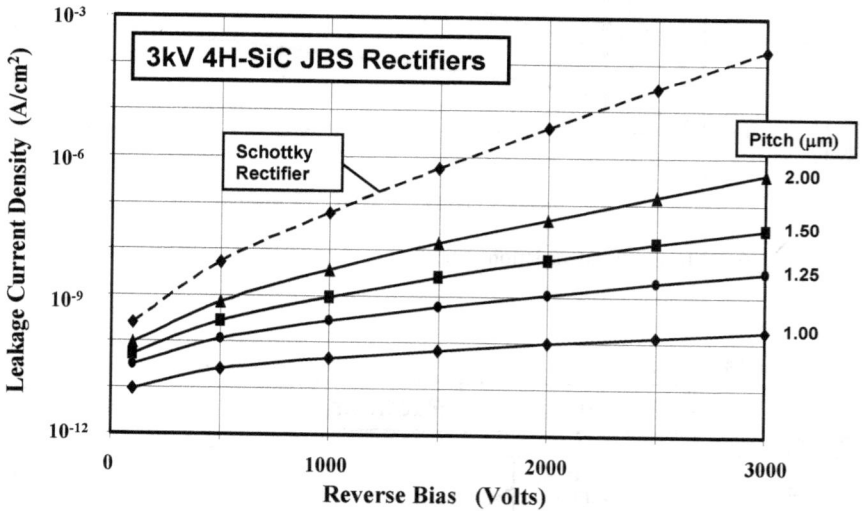

Fig. 6.11 Leakage Current in 4H-SiC JBS Rectifiers.

In addition to the spacing between the P-N junctions, the potential barrier formed under the Schottky contact is dependent upon the depth of the P$^+$ region. The effect of changing the junction depth from 0.5 microns in the previous structures to a larger (0.9 microns) and smaller (0.1 microns) value was analyzed using two-dimensional numerical simulations. It was found that the breakdown voltage reduced to 2550 volts for the larger junction depth. The reason for this is evident in the three dimensional view of the electric field in this structure shown in Fig. 6.12. An enhancement in the electric field occurs at the corner of the P$^+$ region leading to reduction of the breakdown voltage

Junction Depth = 0.9 microns

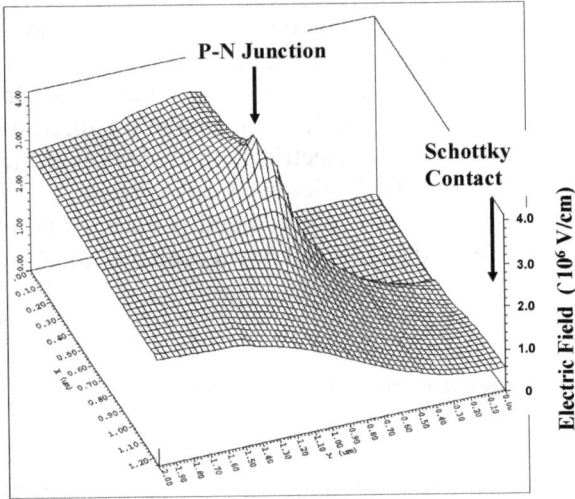

Fig. 6.12 Electric Field distribution in a 4H-SiC JBS Rectifier.

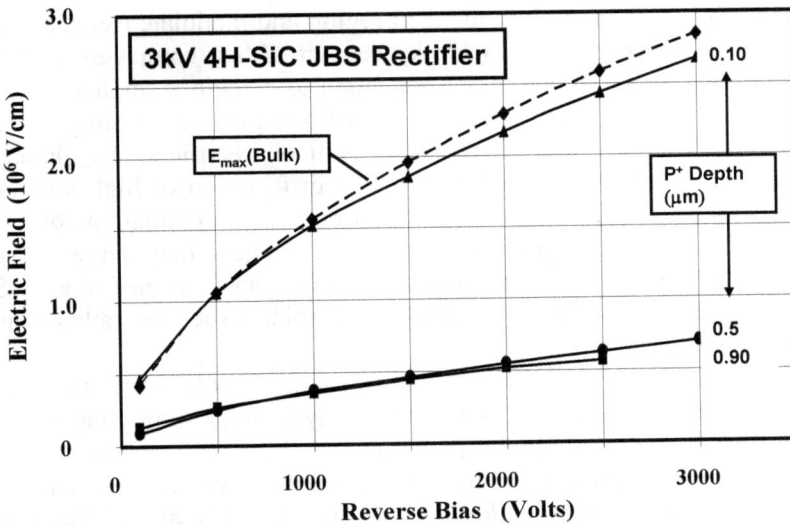

Fig. 6.13 Electric Field variation with Reverse Voltage in 4H-SiC JBS Rectifiers.

The electric field at the Schottky contact for the various junction depths is compared in Fig. 6.12 for a cell pitch of 1.25 microns. In the case of the 0.9 micron junction depth, the data extends to 2500 volts because of the premature breakdown discussed above. On the one hand, it is apparent that there is very little suppression of the Schottky barrier lowering when the junction depth is reduced to 0.1 microns. On the other hand, it can be seen that the electric field reduction is the same for the junction depths of 0.5 and 0.9 microns. Consequently, a junction depth of more than 0.5 microns is not required and is detrimental to both the on-state voltage drop and the breakdown voltage of the JBS rectifier structure.

6.1.4 JBS Rectifiers: Experimental Results

The first high voltage silicon carbide JBS rectifiers were reported in 1997 after the evolution of adequate technology for ion-implantation and activation of P-type dopants[11,12]. These structures were sometimes erroneously labeled as Merged PN Schottky (MPS) rectifiers[11]. The MPS concept was originally proposed for improving the performance of high voltage silicon rectifiers[13]. In the MPS concept, the P-N junction is used to inject minority carriers into the drift region and modulate (reduce) the resistance in series with the Schottky contact. When compared to PiN rectifiers, the MPS structure has been found to contain a smaller stored charge leading to an improved trade-off between on-state voltage drop and reverse recovery characteristics for silicon rectifiers. As already pointed out, the very low resistance in the drift region of high voltage silicon carbide rectifiers does not warrant the modulation of its conductivity with the injection of minority carriers that degrade the switching performance. It is therefore appropriate to use the JBS nomenclature for the high voltage silicon carbide structures rather than the MPS moniker.

In the first JBS rectifiers[12] with breakdown voltage of 700 volts achieved by using 9 micron thick epitaxial layers with doping level of 3 x 10^{15} cm^{-3}, the addition of the P$^+$ regions to the Schottky contact was found to result in an improvement in the breakdown voltage and leakage current. In spite of a reduction of the electric field at the titanium Schottky contact consistent with that shown in Fig. 6.9, the reduction of the leakage current was only by two orders of magnitude. However, the increase in the on-state voltage drop was small, similar to that shown in

Fig. 6.3. Subsequently, the breakdown voltage of the JBS rectifier was extended to 2.8 kV[14] by using 27 micron thick epitaxial layers with doping concentration of 3×10^{15} cm^{-3}. The on-state voltage drop for these JBS diodes was 2.0 volts versus 1.8 volts observed for the Schottky diodes. In this study, the leakage current for the Schottky diode was reported to increase with reverse voltage by a factor of 10^6 while that for the JBS diode increased by only a factor of 10^3. This behavior is consistent with the simulation results shown in Fig. 6.11.

Subsequent work[15,16] on JBS rectifiers confirmed the benefits of using the junction barrier concept for reducing the leakage current. In addition, it was found that the surge current capability of the diodes was enhanced when compared with Schottky diodes. In Schottky rectifiers, the on-state voltage drop becomes extremely large (~ 30 volts) under surge current levels (current density above 1000 A/cm^2) leading to the observation of destructive failures. In contrast, the injection of minority carriers at applied voltages above 3.5 volts in the JBS rectifier enables reduction of the on-state voltage drop under surge current levels preventing destructive failures. Further, unlike Schottky rectifiers, the JBS rectifiers were found to exhibit a positive temperature coefficient for the on-state voltage drop at a current density of 100 A/cm^2 as shown in Fig. 6.3. Thus, the increased resistance created by the P$^+$ regions in the JBS rectifiers had the benefit of compensating for the reduction in voltage drop across the Schottky contact with increasing temperature.

Most recently[17], the breakdown voltage of the JBS rectifier has been extended to 4.3 kV using a 30 micron thick drift layer with doping concentration of 2×10^{15} cm^{-3}. The leakage current for these diodes was reported to increase by 3 orders of magnitude with reverse bias when compared with 6 orders of magnitude for the Schottky diode. This behavior is consistent with the results of the modeling discussed above based upon the reduced electric field at the metal-semiconductor interface. For this reduced doping concentration, an optimum spacing of 9 microns between the P$^+$ regions was observed to provide the best characteristics.

In conclusion, the JBS concept has been found to be very valuable for reducing the leakage current in silicon carbide Schottky rectifiers with a modest increase in the on-state voltage drop. The presence of the P$^+$ regions used to form the potential barrier under the Schottky contact, creates the added benefit of improving the surge current handling capability.

6.2 Trench MOS Barrier Schottky (TMBS) Rectifier Structure

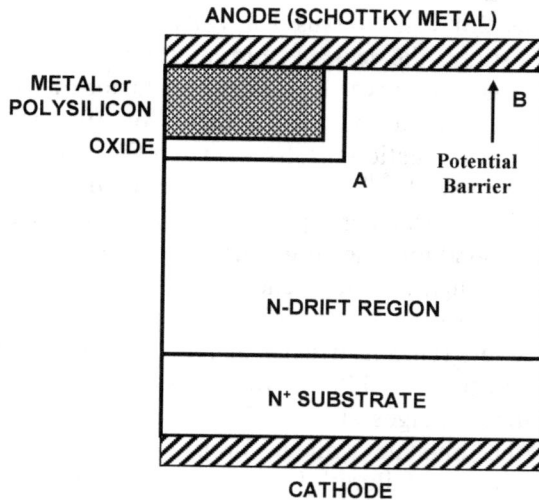

Fig. 6.14 Trench MOS Barrier controlled Schottky (TMBS) Rectifier Structure.

The Trench MOS barrier Schottky Rectifier (TMBS) structure, shown in Fig. 6.14, was originally proposed for silicon Schottky rectifiers[5,18] to achieve charge coupling between the electrode in the trench and the dopant charge in the drift region. The charge coupling allows for very high doping concentration in the drift region while maintaining a breakdown voltage above the theoretically predicted values for one-dimensional parallel plane junctions. The high drift region doping reduces the series resistance of the drift region resulting in low on-state voltage drop for Schottky rectifiers with higher breakdown voltages. This concept, particularly when enhanced with a linearly graded doping profile[6], has been shown to produce excellent silicon Schottky rectifiers with breakdown voltages of 50 to 100 volts[19]. In addition, the electric field at the Schottky contact can be reduced by the formation of a potential barrier due to extension of the depletion region from the sidewalls of the trench.

　　　Recently, there have been attempts to apply the TMBS concept to silicon carbide Schottky rectifiers[20,21]. This approach is misdirected because of two problems. Firstly, the doping concentration in the drift region for high voltage silicon carbide structures is very high when compared with silicon as shown earlier. It is therefore unnecessary to utilize charge coupling to enhance the doping unless the operating voltages are above 5000 volts. Secondly, the electric field within the silicon carbide drift regions is an order of magnitude larger than within silicon devices. This creates an equally higher electric field inside the oxide used for the TMBS structure. The oxide can consequently be subjected to field strengths that induce failure by rupture. These issues are discussed with the aid of numerical simulations in this section of the chapter.

6.2.1 TMBS Rectifier Structure: Simulation Results

Two dimensional numerical simulations of the TMBS structure were performed using a drift region with doping concentration of 1×10^{16} cm^{-3} and thickness of 20 microns for comparison with the Schottky and JBS structures already discussed in this chapter. The device cell, shown in Fig. 6.14, had a pitch of 1 micron. The trench width and mesa width were also 1 micron. Note that only half the trench and mesa widths are illustrated in Fig. 6.14. A trench depth of 1 micron was used with an oxide thickness of 0.1 microns (1000 Angstroms).

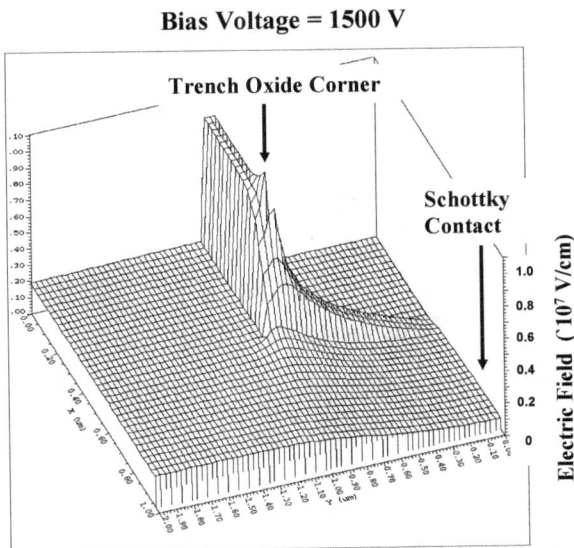

Fig. 6.15 Electric Field distribution in a 4H-SiC TMBS Rectifier.

A three dimensional view of the electric field distribution in the upper part of the TMBS structure is shown in Fig. 6.15 at a reverse bias of 1500 volts. As expected, there is a reduction of the electric field at the Schottky contact due to the potential barrier created by the trench MOS structure. This reduction of the electric field at the Schottky contact can be clearly observed in Fig. 6.16 where the electric field profile is shown at the center of the mesa region (location B in Fig. 6.14). The peak of the electric field in the silicon carbide occurs at a depth of about 1 micron with much lower values at the surface. Thus, the TMBS concept is effective in terms of reducing the Schottky barrier lowering and hence the leakage current. Unfortunately, this improvement cannot be utilized because of the very high electric field in the oxide.

Cell Pitch = 1.00 microns

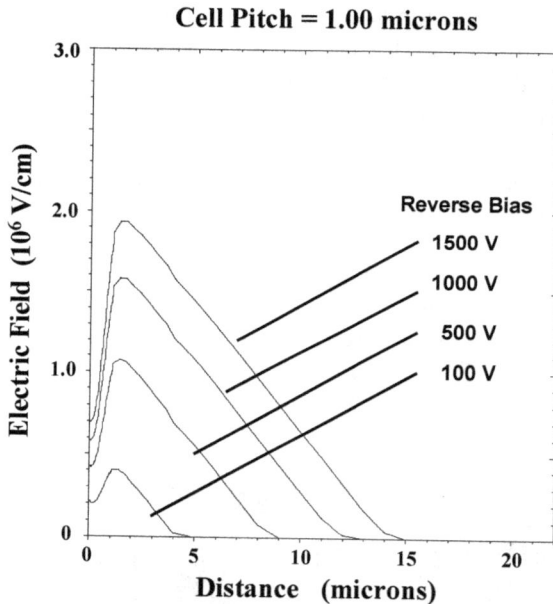

Fig. 6.16 Electric Field variation with Reverse Voltage in a 4H-SiC TMBS Rectifier.

The high electric field in the oxide is evident in Fig. 6.15 at a reverse bias of only 1500 volts for a drift region capable of supporting 3000 volts. The oxide field has reached a value close to 1×10^7 V/cm, which makes its rupture likely. This problem is exacerbated by the field enhancement at the trench corner (location A in Fig. 6.14). The magnitude of the electric field in the oxide can be observed more clearly in Fig. 6.17. There is a jump in the electric field at the oxide-semiconductor interface due to the difference in dielectric constants. At a reverse bias of 1500 volts, the electric field in the oxide has reached a value of almost 1×10^7 V/cm, the breakdown strength of silicon dioxide. The oxide fields at other reverse bias voltages are indicated by the arrows in the figure. For reliable operation, the electric field in the oxide will have to be kept below 3×10^6 V/cm. This will limit the maximum operating reverse voltage to less than 500 volts. Such a severe reduction of the reverse blocking capability removes most the advantages of using silicon carbide as the semiconductor material. Consequently, the ad hoc application of a structure developed for silicon Schottky rectifiers is not warranted for silicon carbide. In fact, the reported breakdown voltages of the experimentally fabricated devices[20] were only 100 volts even for an epitaxial layer doping concentration of 6×10^{15} cm^{-3}.

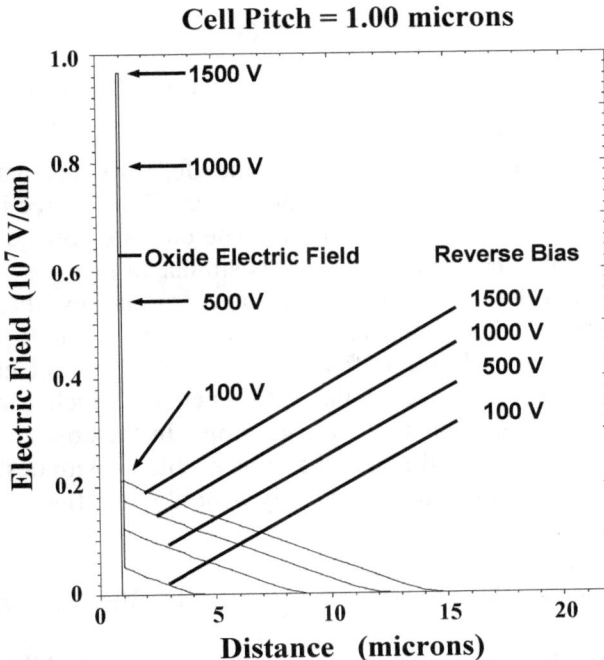

Fig. 6.17 Electric Field variation with Reverse Voltage in a 4H-SiC TMBS Rectifier.

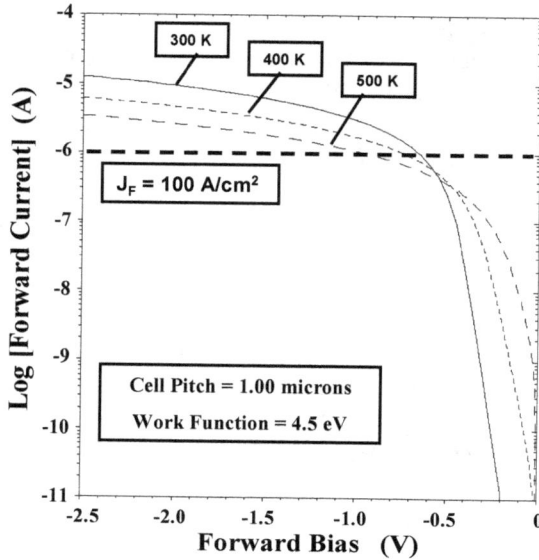

Fig. 6.18 Forward Characteristics of 3 kV 4H-SiC TMBS Rectifier.

For the sake of completeness, the forward characteristics obtained by numerical simulations for the TMBS 4H-SiC rectifier are shown in Fig. 6.18 at various temperatures. The on-state voltage drop at the nominal current density of 100 A/cm² is similar to that observed for the JBS rectifier. Although the on-state voltage drop is low, there is no particular advantage when compared with the JBS rectifier. In fact, as already pointed out, the JBS rectifier has good surge current handling capability due to the presence of the P-N junctions which can inject minority carriers at higher on-state voltage drops. In the absence of this feature, the TMBS rectifier will have the same problems with destructive failure at surge current levels as the ordinary Schottky rectifier.

6.3 Trench Schottky Barrier Schottky (TSBS) Rectifier Structure

In the previous section, it has been demonstrated that the shielding of the metal-semiconductor contact to reduce the electric field suppresses leakage current at high voltages. The potential barrier that must be created for the shielding of the contact can also be produced by using another Schottky contact placed adjacent to the main Schottky contact. The second Schottky contact must have a larger barrier height than the main Schottky contact so that the leakage current from this contact is small despite the high electric field at its interface with the semiconductor. In addition, it is preferable to locate the high barrier height metal contact within a trench surrounding the main low barrier metal Schottky contact as illustrated in Fig. 6.19. The trench architecture enhances the potential barrier under the main Schottky contact in comparison with a planar structure.

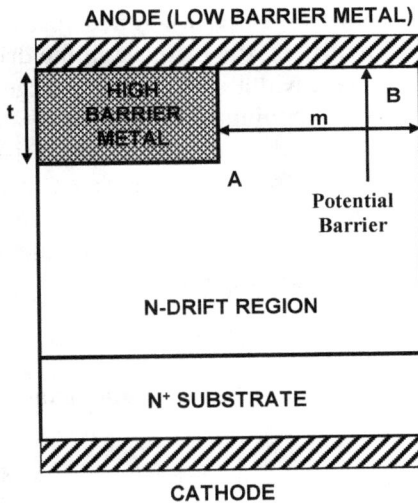

Fig. 6.19 Trench Schottky Barrier controlled Schottky (TSBS) Rectifier Structure.

All of the above features were first proposed for improvement of Schottky barrier rectifiers in a patent[9] issued in 1993. This patent describes devices with a high barrier metal located in trenches etched around the main low barrier metal Schottky contact with various geometries, such a stripes or cellular configurations. A simple method for fabrication of the device is also described. It consists of depositing the low barrier metal and patterning it to expose areas for etching the trenches. The trenches are then formed using the metal as the masking layer. The high barrier metal is then deposited into the trenches and over the low barrier metal to complete the cell structure. The TSBS concept was not pursued for silicon devices due to the success of the JBS and TMBS concepts. However, it is particularly suitable for silicon carbide devices because it eliminates the need for formation of P-N junctions which require very high temperature anneals that can degrade the semiconductor surface. Due to the large band gap of silicon carbide, it is possible to produce high barrier Schottky contacts by using metals such as nickel in conjunction with low barrier main Schottky contacts such as titanium.

In the TSBS structure, the on-state current flow occurs via the low barrier main metal contact. The same analytical model developed earlier for the JBS rectifier can also be applied to the TSBS structure. The increase in resistance of the drift region due to current constriction at the main Schottky contact must be accounted for. However, as already discussed for the JBS rectifiers, the increase in the drift region resistance is small as long as the mesa width is sufficiently large to prevent the zero bias depletion width from pinching it off completely.

In the same vein, the reverse characteristics of the TSBS structure can be analyzed by using the same methodology already described for the JBS rectifier. However, the leakage current from the high barrier metal ($J_{L(HB)}$) must to be added to the suppressed leakage current from the main low barrier metal ($J_{L(LB)}$):

$$J_{L(TSBS)} = J_{L(LB)} + J_{L(HB)} \qquad \text{[6.12]}$$

The leakage current from the low barrier main metal contact is given by:

$$J_{L(LB)} = \left(\frac{d}{p}\right) A T^2 \exp\left(-\frac{q\phi_b}{kT}\right).\exp\left(\frac{q\Delta\phi_{bTSBS}}{kT}\right).\exp\left(C_T E_{TSBS}^2\right) \text{[6.13]}$$

where C_T is a tunneling coefficient (8×10^{-13} cm^2/V^2). The barrier lowering and the tunneling contributions are determined by the reduced electric field E_{TSBS} at the main contact:

$$\Delta\phi_{bTSBS} = \sqrt{\frac{qE_{TSBS}}{4\pi\varepsilon_S}} \qquad \text{[6.14]}$$

The electric field E_{TSBS} is related to the reverse bias voltage (V_P) at which the space between the P$^+$ regions gets depleted by:

$$E_{TSBS} = \sqrt{\frac{2qN_D}{\varepsilon_S}(\beta V_P + V_{bi})} \qquad \text{[6.15]}$$

where β is a coefficient used to account for the build up in the electric field after pinch-off. In contrast, the leakage current from the high barrier metal contact is given by:

$$J_{L(HB)} = \left(\frac{d}{p}\right) A T^2 \exp\left(-\frac{q\phi_b}{kT}\right) . \exp\left(\frac{q\Delta\phi_{bHB}}{kT}\right) . \exp\left(C_T E_{HB}^2\right) \quad [\,6.13\,]$$

where C_T is a tunneling coefficient (8 x 10^{-13} cm^2/V^2). The barrier lowering and the tunneling contributions are determined by the electric field E_{HB} at the unprotected high barrier contact:

$$\Delta\phi_{bHB} = \sqrt{\frac{qE_{HB}}{4\pi\varepsilon_S}} \quad [\,6.14\,]$$

The electric field E_{HB} is determined by the applied reverse bias voltage (V_A):

$$E_{HB} = \sqrt{\frac{2qN_D}{\varepsilon_S}(V_A + V_{bi})} \quad [\,6.15\,]$$

It has been found that the leakage current contribution from the high barrier metal can become comparable to that from the main low barrier metal as the reverse bias voltage approaches the breakdown voltage.

6.3.1 TSBS Rectifier Structure: Simulation Results

For comparison with previously described structures, TSBS rectifiers with breakdown voltage of 3kV using a drift region with doping concentration of 1 x 10^{16} cm^{-3} will be considered here. The two-dimensional numerical simulations of the structures were performed with various mesa widths (spacing 'm' in Fig. 6.19). The depth of the trench (dimension 't' in Fig. 6.19) was also varied to examine its impact on the electric field reduction at the Schottky contact. In each case, the forward and reverse characteristics were analyzed to look at the trade-off between on-state voltage drop and electric field reduction at the Schottky contact. The trench width was kept at 1 micron for all the simulation structures. Note that only half the trench is shown in Fig. 6.19. A work function of 4.5 eV, corresponding to a barrier height of 0.8 eV, was used as before for the main Schottky contact. A work function of 5.0 eV, corresponding to a barrier height of 1.3 eV, was used for the high barrier Schottky contact within the trenches.

Trench Depth = 0.5 microns

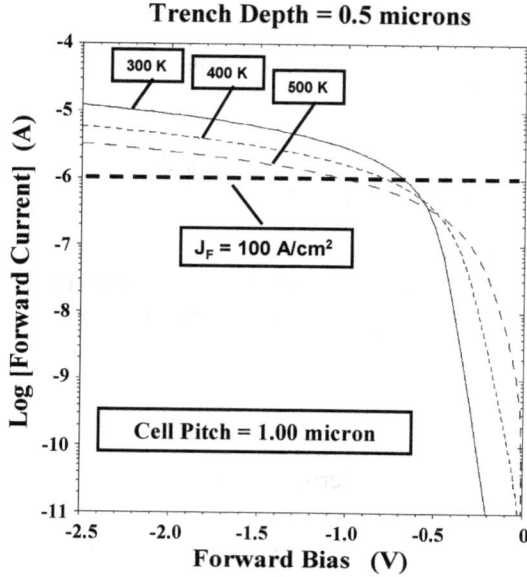

Fig. 6.20 Forward Characteristics of 3 kV 4H-SiC TSBS Rectifier.

Trench Depth = 0.5 microns

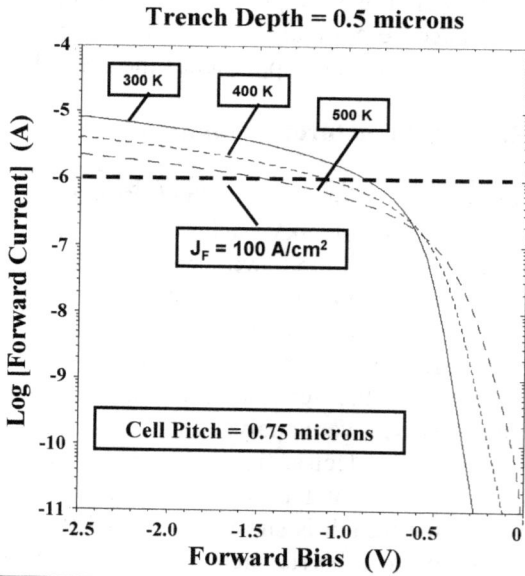

Fig. 6.21 Forward Characteristics of 3 kV 4H-SiC TSBS Rectifier.

The forward characteristics of a 3kV JBS 4H-SiC rectifier are shown in Fig. 6.20 for the case of a cell pitch of 1.00 micron at various temperatures. The on-state voltage drop at a current density of 100 A/cm^2 is about 0.1 volts larger than that for the Schottky rectifier. The on-state voltage drop increases by 0.3 volts, as shown in Fig. 6.21, when the cell pitch is reduced to 0.75 microns (by decreasing the mesa width to 0.25 microns). This is due to the stronger constriction of current flow under the main Schottky contact. Note that the increase in on-state voltage drop is not as severe for the TSBS structure as for the JBS structure for the same cell pitch. This is because of the reduced zero bias depletion width in the TSBS structure due to the smaller potential across the high barrier to semiconductor interface when compared with that across the P-N junction.

The reason for the increase in the on-state voltage drop in the TSBS structure can be understood by examining the current flow in the cell under forward bias conditions. The current flow pattern for the structure with 1.00 micron cell pitch is shown in Fig. 6.22. The constriction of the current at the main Schottky contact is not as strong as that observed in Fig. 6.5 for the JBS rectifier structure with the same 1.00 micron cell pitch. It is worth pointing out that the current spreading occurs quite rapidly and becomes uniform at a depth of 1 micron. Consequently, the model for the drift resistance described by Eq [6.11] is suitable for the TSBS rectifier structure as well. This lower drift region resistance favors silicon carbide TSBS rectifier on-state performance approaching that of Schottky rectifiers.

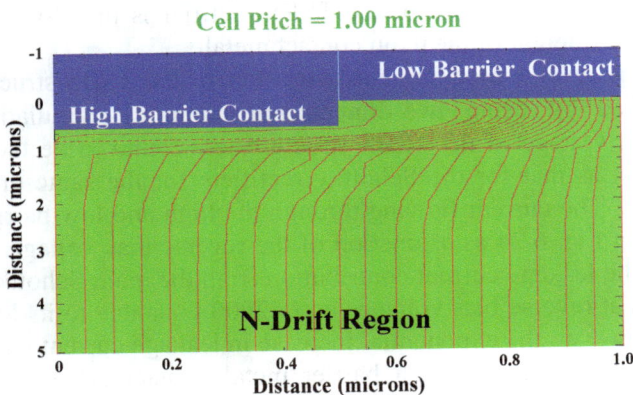

Fig. 6.22 On-state Current Flow in a 3 kV 4H-SiC TSBS Rectifier.

Trench Depth = 0.5 microns

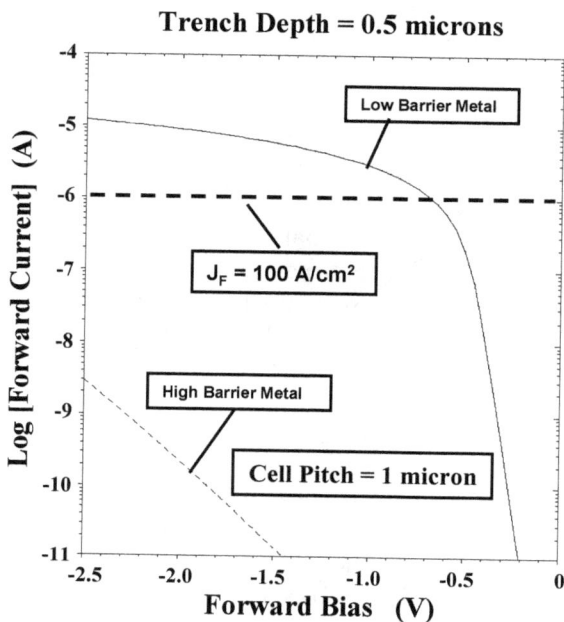

Fig. 6.23 Current Distribution in a 3 kV 4H-SiC TSBS Rectifier.

The relative current flow between the low and high barrier metal contacts in the TSBS structures can be observed in Fig. 6.23 for the case of a cell pitch of 1 micron. The current through the high barrier metal is about 5 orders of magnitude less than that through the low barrier metal. The on-state voltage drop for the TSBS rectifier is therefore controlled by the barrier height of the main contact metal.

The reverse blocking characteristics of the TSBS structure were also studied by using two-dimensional numerical simulations. The breakdown voltage of the TSBS structure was found to be the same (~ 3000 volts) as that for the Schottky rectifiers for the same drift region parameters. The current flowing through the high and low barrier metals is shown in Fig. 6.24 as a function of the reverse bias voltage. It can be seen that the leakage current contribution from the main Schottky contact dominates at reverse bias voltages up to 2000 volts due to its low barrier height. However, the exponential increase in leakage current with reverse bias on the unprotected high barrier metal contact allows its leakage current to catch-up and then exceed the leakage current of the main contact at a reverse bias of 2900 volts.

Trench Depth = 0.5 microns

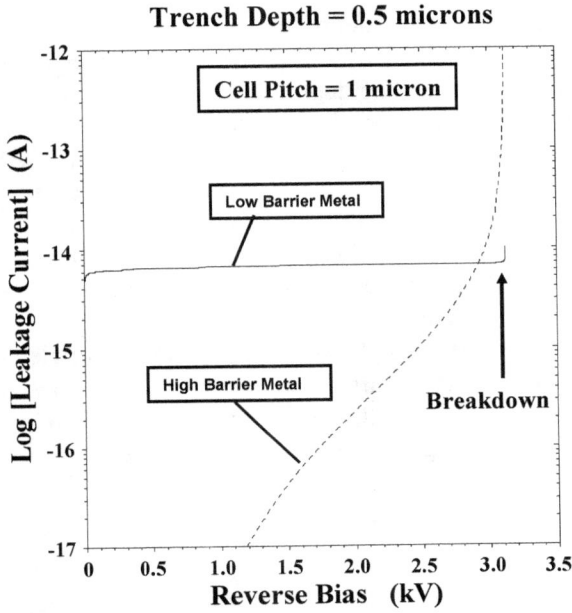

Fig. 6.24 Reverse Blocking Characteristics of a 3 kV 4H-SiC TSBS Rectifier.

Bias Voltage = 3000 V

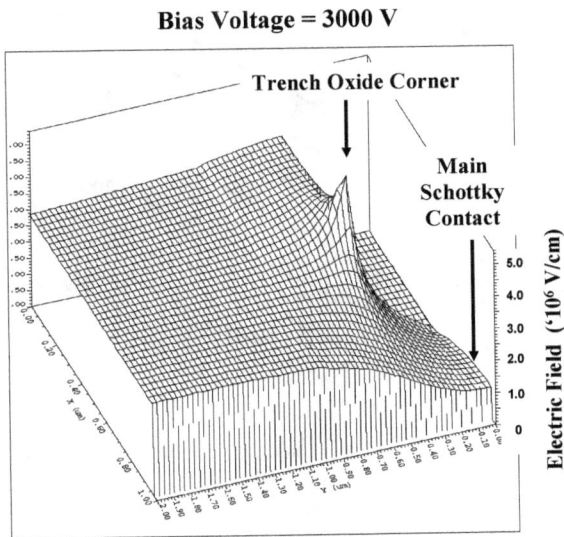

Fig. 6.25 Electric Field distribution in a 4H-SiC TSBS Rectifier.

As expected, it was found that the electric field at the main Schottky contact was reduced by the potential barrier created using the high Schottky barrier metal. The degree of field reduction was dependent on the spacing between the trench regions as well as the depth of the trench. A three-dimensional view of the electric field in a TSBS structure with cell pitch of 1.00 micron is shown in Fig. 6.25 at a reverse bias of 3000 volts (just prior to breakdown). It can be seen that the highest electric field occurs at the high barrier metal contact with some enhancement at the trench corner, and that the electric field is suppressed at the main Schottky contact.

As in the case of the JBS rectifier, the maximum electric field at the main Schottky contact occurs at the location furthest away from the trench (at x = 1.00 micron for the structure in Fig. 6.25). Consequently, the largest barrier lowering and tunneling components will occur at this location. For this reason, the highest electric field at the Schottky contact will be used to analyze the reverse leakage characteristics for the TSBS rectifiers.

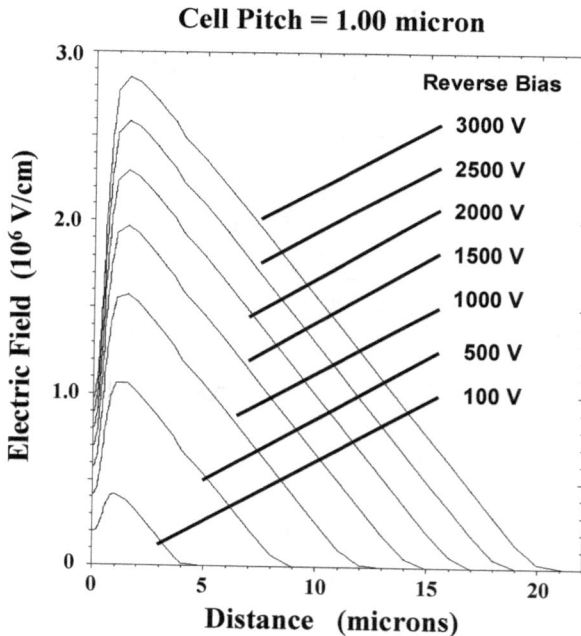

Fig. 6.26 Electric Field variation with Reverse Voltage in a 4H-SiC TSBS Rectifier.

The electric field profile at the center of the Schottky contact is shown in Fig. 6.26 for the case of a cell pitch of 1 micron. It can be seen that the electric field at the surface under the contact is significantly reduced when compared the peak electric field in the bulk. The peak of the electric field occurs at a depth of about 1 micron. At a reverse bias of 3000 volts, the electric field at the Schottky contact is only 1×10^6 V/cm compared with 2.85×10^6 V/cm at the maxima in the bulk. An even greater reduction of the electric field at the Schottky contact can be achieved by reducing the cell pitch. This is illustrated in Fig. 6.27 for the case of a cell pitch of 0.75 microns. Here, the electric field at the Schottky contact is only 2×10^5 V/cm compared with 2.85×10^6 V/cm at the maxima in the bulk.

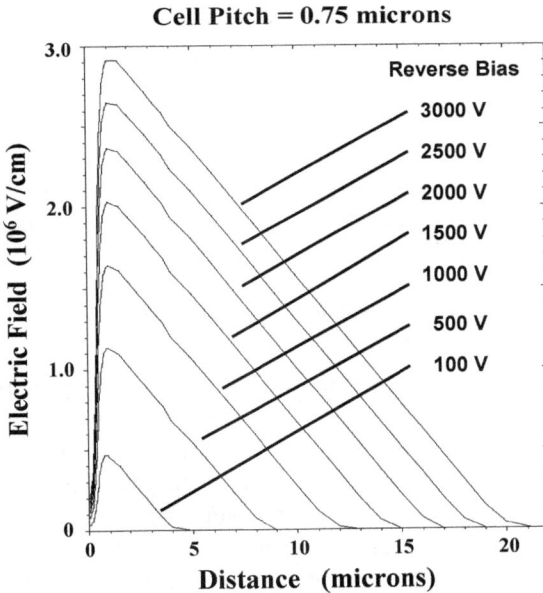

Fig. 6.27 Electric Field variation with Reverse Voltage in a 4H-SiC TSBS Rectifier.

The increase in the electric field at the Schottky contact with increasing reverse bias voltage is charted in Fig. 6.28 for various cases of the cell pitch with the trench depth held constant at 0.5 microns. The electric field at the contact becomes close to the maximum value in the bulk when the pitch is over 2 microns. Thus, the benefits of using the TSBS concept to suppress the electric field at the Schottky contact can be obtained only with carefully optimized spacing which is slightly smaller than for the JBS rectifier[22]. With proper spacing, the reduction of the electric field provides the benefit of reducing the Schottky barrier lowering and the leakage current.

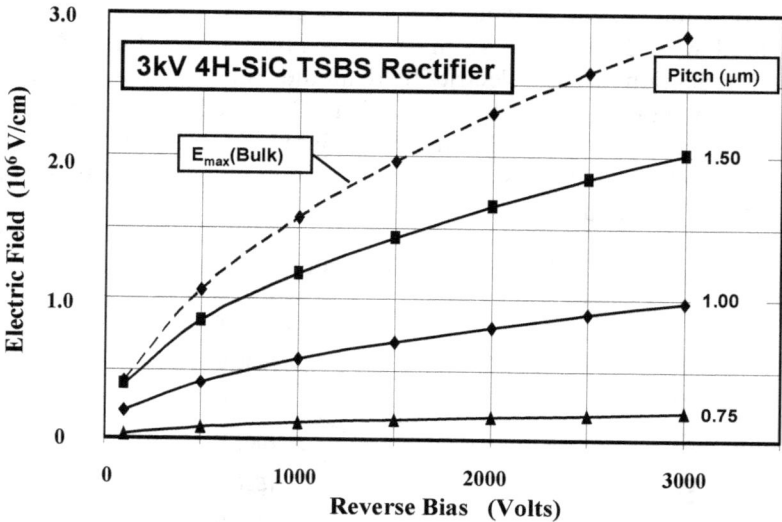

Fig. 6.28 Electric Field variation with Reverse Voltage in 4H-SiC TSBS Rectifiers.

The protection of the main Schottky contact against barrier lowering is quantified in the graphs shown in Fig. 6.29 and Fig. 6.30, respectively, where the values were determined using the analytical formulae with the electric field extracted from the simulations. With a pitch of 1 micron, the Schottky barrier lowering is reduced from 0.22 eV to 0.12 eV. The impact of ameliorating the Schottky barrier lowering phenomenon on the leakage current is quite dramatic for silicon carbide Schottky rectifiers because of the strong dependence of the tunneling component on the electric field. A reduction in the leakage current by five orders of magnitude is observed at high reverse bias voltages, as shown in Fig. 6.30, demonstrating the advantage of utilizing the potential barrier to suppress the leakage current from the main contact. (Note that the leakage current from the high barrier metal was neglected in these plots.) The TSBS structure is therefore quite effective for improving the reverse blocking characteristics of high voltage silicon carbide rectifiers.

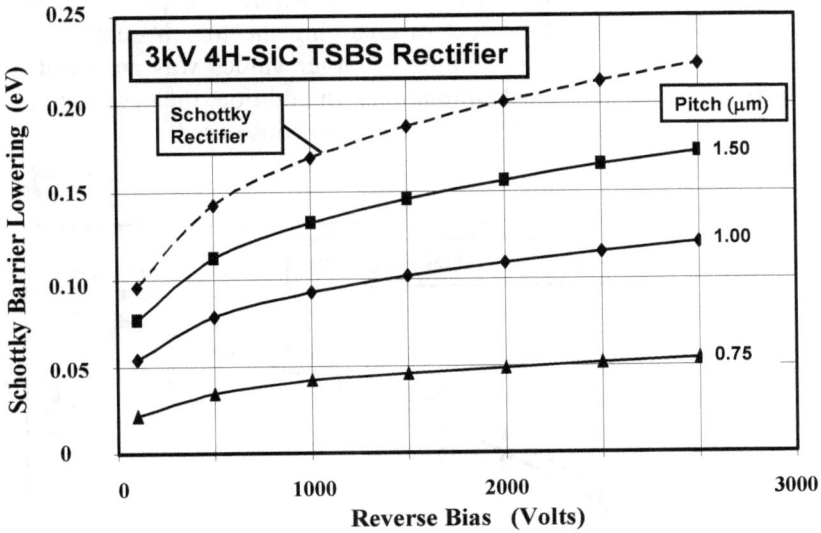

Fig. 6.29 Schottky Barrier Lowering in 4H-SiC TSBS Rectifiers.

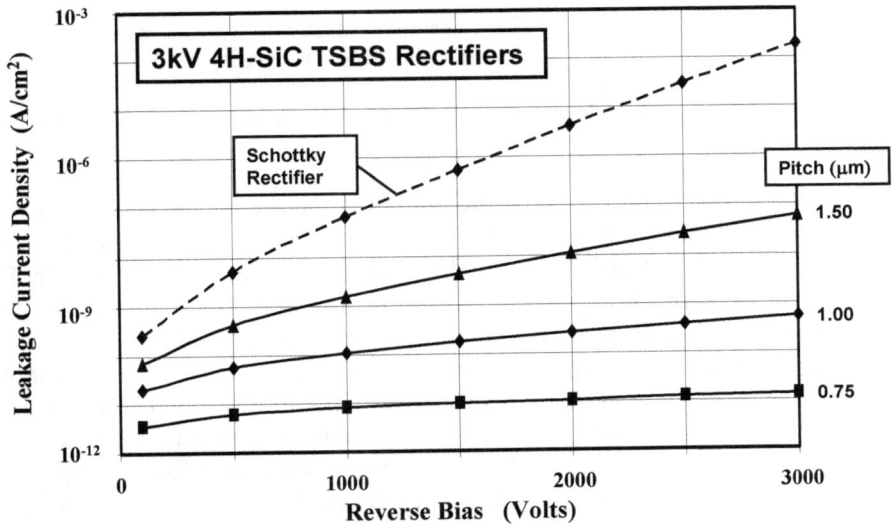

Fig. 6.30 Leakage Current in 4H-SiC TSBS Rectifiers.

In addition to the mesa width, the potential barrier formed under the main Schottky contact in the TSBS structure is dependent upon the depth of the trench. The effect of changing the trench depth from 0.5 microns in the previous structures to larger values was analyzed using two-dimensional numerical simulations. In addition, the case with a zero trench depth (or a planar structure) was also considered.

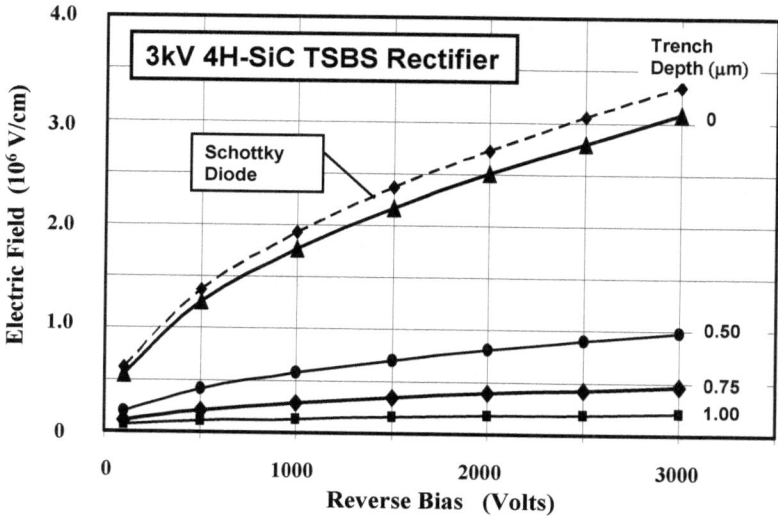

Fig. 6.31 Electric Field variation with Reverse Voltage in 4H-SiC TSBS Rectifiers.

The electric field at the Schottky contact for the different trench depths is compared in Fig. 6.31 for a cell pitch of 1 micron. It can be seen that the electric field at the Schottky contact (location B in Fig. 6.19) for a trench depth of 1 micron is significantly (~ 5x) smaller than for the trench depth of 0.5 microns. This is beneficial for reducing the Schottky barrier lowering and consequently the leakage current as shown in Fig. 6.32 and Fig. 6.33, respectively. The improvement in the leakage current by many orders of magnitude, when compared with the Schottky rectifier, is quite evident. However, in the case of reducing the trench depth to zero (i.e. a planar structure), it was found that there was only a slight reduction of the electric field at the main Schottky contact for the case of a cell pitch of 1 micron, as shown in Fig. 6.31. It is therefore necessary to create a trench to house the high barrier metal adjacent to the main low barrier Schottky contact to get the full benefits of the TSBS concept.

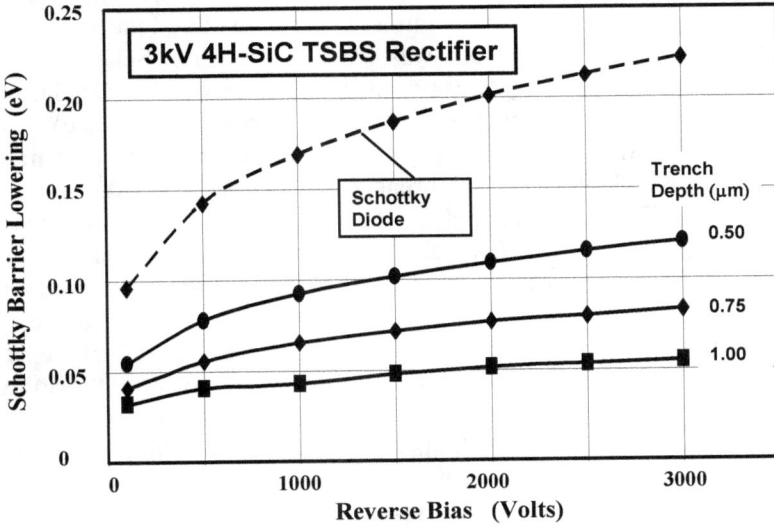

Fig. 6.32 Schottky Barrier Lowering in 4H-SiC TSBS Rectifiers.

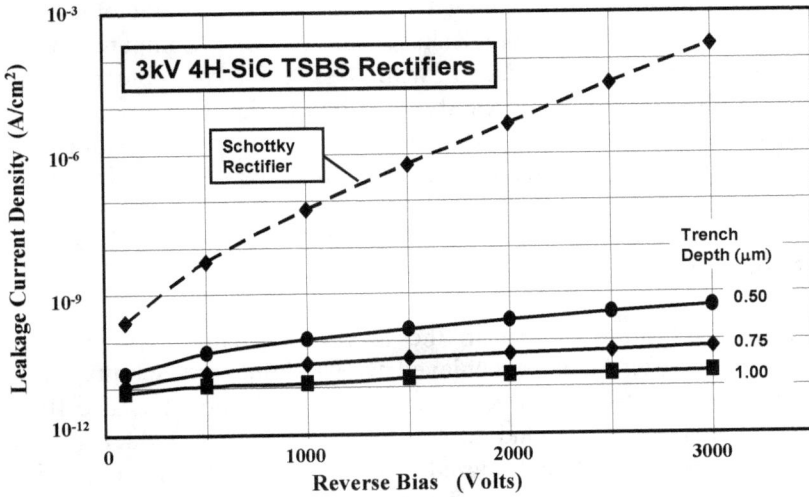

Fig. 6.33 Leakage Current in 4H-SiC TSBS Rectifiers.

6.3.2 TSBS Rectifiers: Experimental Results

The TSBS concept for improving the performance of silicon carbide Schottky rectifiers was first simultaneously explored at PSRC[23] and Purdue University[24]. Titanium was used as the main Schottky contact metal due to its relatively low barrier height (1.1 eV) on 4H-SiC while nickel was used in the trenches as the high barrier height (1.7 eV) metal. For an epitaxial layer with doping concentration of 3 x 10^{15} cm^{-3} and thickness of 13 microns, a breakdown voltage of 400 volts was observed[25]. This was far below the capability of the material and compares poorly with 1720 volts observed for the planar nickel Schottky rectifier. The analysis discussed in the previous section indicates that this breakdown problem is not endemic to the TSBS structure but may be related to the chip design and layout. However, a reduction of the leakage current by a factor of 75x was observed confirming the ability of the TSBS concept in mitigating the barrier lowering and tunneling currents from the main Schottky contact. In addition, the on-state characteristics were found to be similar to those for a planar titanium Schottky contact as expected from the description of the simulated structures in the previous section.

More recently[25], a reduction of leakage current by a factor of 30x has also been reported by using the planar TSBS structure (i.e. zero trench depth) with little impact on the forward characteristics. The smaller improvement in the leakage current when compared with the trench structure is consistent with less effective potential barrier observed in the simulation results described in the previous section. The breakdown voltages of these diodes ware far below the capability of the underlying drift region due to absence of adequate edge termination.

6.4 Summary

A significant improvement in the leakage current of silicon carbide Schottky rectifiers can be achieved by shielding the Schottky contact against high electric fields generated in the semiconductor. The JBS and TSBS structures are found to be very effective in reducing leakage current without significant degradation of the forward characteristics. In contrast, the TMBS concept, although very effective for improvement of silicon Schottky rectifiers, is inappropriate for silicon carbide due to the generation of high electric fields in the oxide.

References

[1] B. J. Baliga, "Power Semiconductor Devices", PWS Publishing Company, 1996.

[2] B. J. Baliga, "The Pinch Rectifier: A Low Forward Drop High Speed Power Diode", IEEE Electron Device Letters, Vol. 5, pp. 194-196, 1984.

[3] B. J. Baliga, "Pinch Rectifier", U. S. Patent 4,641,174, Issued February 3, 1987.

[4] M. Mehrotra and B. J. Baliga, "Very Low Forward Drop JBS Rectifiers Fabricated using Sub-micron Technology", IEEE Transactions on Electron Devices, Vol. 41, pp. 1655-1660, 1994.

[5] M. Mehrotra and B. J. Baliga, "Schottky Barrier Rectifier with MOS Trench", U. S. Patent 5,365,102, Issued November 15, 1994.

[6] B.J. Baliga, "Schottky Barrier Rectifiers and Methods of Forming the Same", U.S. Patent 5,612,567, March 18, 1997.

[7] S. Mahalingam and B.J. Baliga, "The Graded Doped Trench MOS Barrier Schottky Rectifiers", Solid State Electronics, Vol. 43, pp. 1-9, 1999.

[8] B.J. Baliga, "The Future of Power Semiconductor Technology", Proceedings of the IEEE, Vol. 89, pp. 822-832, 2001.

[9] L. Tu and B. J. Baliga, "Schottky Barrier Rectifier including Schottky Barrier regions of Differing Barrier Heights", U. S. Patent 5,262,668, November 16, 1993.

[10] B. J. Baliga, "Analysis of Junction Barrier controlled Schottky Rectifier Characteristics", Solid State Electronics, Vol. 28, pp. 1089-1093, 1985.

[11] R. Held, N. Kaminski and E. Niemann, "SiC Merged p-n/Schottky Rectifiers for High Voltage Applications", Silicon Carbide and Related Materials – 1997, Material Science Forum, Vol. 264-268, pp. 1057-1060, 1998.

[12] F. Dahlquist, et al, "Junction barrier Schottky Diodes in 4H-SiC and 6H-SiC", Silicon Carbide and Related Materials – 1997, Material Science Forum, Vol. 264-268, pp. 1061-1064, 1998.

[13] B. J. Baliga, "Analysis of a High Voltage Merged PiN/Schottky (MPS) Rectifier", IEEE Electron Device Letters, Vol. 8, pp. 407-409, 1987.

[14] F. Dahlquist, et al, "A 2.8kV JBS Diode with Low Leakage", Silicon Carbide and Related Materials – 1999, Material Science Forum, Vol. 338-342, pp. 1179-1182, 2000.

[15] D. Peters, et al, "Comparison of 4H-SiC pn, Pinch, and Schottky Diodes for the 3kV Range", Silicon Carbide and Related Materials – 2001, Material Science Forum, Vol. 389-393, pp. 1125-1128, 2002.

[16] F. Dahlquist, H. Lendenmann and M. Ostling, "A JBS Diode with controlled Forward Temperature Coefficient and Surge Current Capability", Silicon Carbide and Related Materials – 2001, Material Science Forum, Vol. 389-393, pp. 1129-1132, 2002.

[17] J. Wu, et al, "4,308V, 20.9 mO-cm2 4H-SiC MPS Diodes Based on a 30 micron Drift Layer", Silicon Carbide and Related Materials – 2003, Material Science Forum, Vol. 457-460, pp. 1109-1112, 2004.

[18] M. Mehrotra and B. J. Baliga, "The Trench MOS Barrier Schottky Rectifier", IEEE International Electron Devices Meeting, Abstract 28.2.1, pp. 675-678, 1993.

[19] S. Mahalingam and B. J. Baliga, "A Low Forward Drop High Voltage Trench MOS Barrier Schottky Rectifier with Linearly Graded Doping Profile", IEEE International Symposium on Power Semiconductor Devices and ICs, Paper 10.1, pp. 187-190, 1998.

[20] V. Khemka, V. Ananthan, and T. P. Chow, "A 4H-SiC Trench MOS Barrier Schottky (TMBS) Rectifier", IEEE International Symposium on Power Semiconductor Devices and ICs, pp. 165-168, 1999.

[21] Q. Zhang, M. Madangarli, and T. S. Sudarshan, "SiC Planar MOS-Schottky Diode", Solid State Electronics, Vol. 45, pp. 1085-1089, 2001.

[22] B. J. Baliga, "High Voltage Silicon Carbide Devices", Material Research Society Symposium, Vol. 512, pp. 77-88, 1998.

[23] M. Praveen, S. Mahalingam, and B. J. Baliga, "Silicon Carbide Dual Metal Schottky Rectifiers", PSRC Technical Working Group Meeting, Report TW-97-002-C, 1997

[24] K. J. Schoen, et al, "A Dual Metal Trench Schottky Pinch-Rectifier in 4H-SiC", IEEE Electron Device Letters, Vol. 19, pp. 97-99, 1998.

[25] F. Roccaforte, et al, "Silicon Carbide Pinch Rectifiers using a Dual-Metal Ti-NiSi Schottky Barrier", IEEE Transactions on Electron Devices, Vol. 50, pp. 1741-1747, 2003.

Chapter 7

Metal-Semiconductor Field Effect Transistors

As already pointed out in earlier chapters, the main advantage of wide band gap semiconductors for power device applications is the very low resistance of the drift region even when it is designed to support large voltages. This favors the development of high voltage unipolar devices which have much superior switching speed than bipolar structures. The *Junction Field Effect Transistor (JFET)* and the *Metal Semiconductor Field Effect Transistor (MESFET)* are potential candidates as unipolar switches for power applications. These structures have also been called *Static Induction Transistors*[1]. In the case of silicon, the maximum breakdown voltage of JFETs has been limited by the increase in the resistance of the drift region[2]. This limitation does not apply to silicon carbide due to the much larger doping concentration within the drift region for high voltage structures. However, as in the case of silicon structures, the normally-on behavior of the high voltage JFETs has been found to be a serious impediment to circuit applications. When powering up any electronic system, it is impossible to ensure that the gate voltage required to block current flow in the JFET is provided to the structure prior to the incidence of the drain voltage on the structure. This situation can result in shoot-through currents between the power rails resulting in destructive failure of the devices. This has discouraged the use of normally-on devices in power electronic applications.

The normally-on behavior of silicon JFETs motivated the development of MOS-gated bipolar structures leading to the evolution of the Insulated Gate Bipolar Transistor (IGBT)[3]. Consequently, after a vigorous period of development in the 1980s, interest in silicon power JFETs declined. Problems encountered with quality of the oxide-semiconductor interface in silicon carbide brought renewed interest in JFETs for power switching applications. In addition, the invention of the

141

Baliga Pair circuit configuration[4] enabled achieving a normally-off power switch function with a high speed integral diode that is ideally suitable for H-bridge applications. The Baliga-Pair utilizes a high voltage normally-on silicon carbide JFET or MESFET in a cascode configuration[5] with a low voltage silicon power MOSFET to achieve a normally-off function. Since this approach provides many advantages as discussed in the next chapter, it was an important conceptual break-through that has motivated the development of silicon carbide high voltage JFET/MESFET structures.

       This chapter reviews the basic principles of operation of the Junction (or Metal-Semiconductor) Field Effect Transistor. These structures must be designed to deplete the channel region by application of a reverse bias voltage to the gate-source junction. To prevent current flow through the device, a potential barrier for electron transport must be created in the channel to suppress electron transport even under large drain bias voltages. Concurrently, the channel must remain un-depleted (preferably at zero gate bias) to allow on-state current flow with low on-state resistance. Unique issues that relate to the wide band gap of silicon carbide must be given special consideration. The results of the analysis of these structures by using two-dimensional numerical simulations are described in this chapter. Experimental results on relevant structures are provided to define the state of the development effort on high voltage silicon carbide JFETs and MESFETs.

       The operating principles for the Baliga-Pair configuration that provide the desired normally-off function are discussed in the next chapter. The results of two-dimensional numerical simulations of a concatenated MOSFET and MESFET structure are described to provide insight into the operation of this concept. The results of experimental work on the Baliga-Pair are summarized validating the utility of this idea for high power system applications.

7.1 Trench Junction (Metal-Semiconductor) Field Effect Transistor Structure

The basic structure of the vertical trench Junction Field Effect Transistor (JFET) and the trench Metal-Semiconductor Field Effect Transistor (MESFET) are shown in Fig. 7.1. The structures contain a drift region (usually N-type) between the drain and the source regions designed to support the desired maximum operating voltage. As discussed in detail later in the chapter, the drift region must be capable of supporting the sum of the drain bias and the reverse gate bias potentials without undergoing breakdown. To prevent current flow under forward blocking conditions, a gate region must be incorporated in the drift region. A P-N junction gate region is used for the JFET structure and a Schottky (metal-semiconductor) contact is used for the MESFET structure. In order to produce a strong potential barrier for suppressing the transport of electrons between the drain and the source, it is preferable to create a gate with vertical sidewalls. For the JFET structure, this can be achieved by utilizing multiple P-type ion implants with increasing energy using a common mask edge. For the MESFET structure, a trench is etched with vertical sidewalls followed by the selective deposition of the Schottky barrier gate metal to fill the trench.

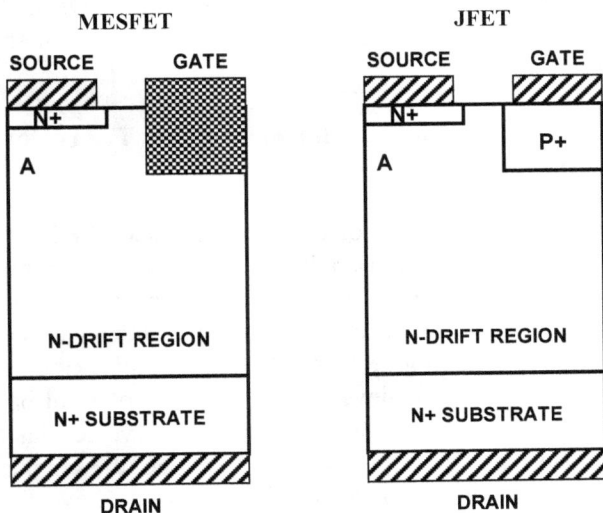

Fig. 7.1 The Vertical Trench JFET and MESFET Structures.

When a negative bias is applied to the gate electrode, a depletion layer extends from the gate into the drift region. With sufficient gate bias, the entire space between the gate regions becomes depleted. The gate bias required to deplete the space between the gate regions is referred to as the *pinch-off voltage*. At a gate bias above the pinch-off voltage, a potential barrier forms under the source region at location A. This barrier suppresses the transport of electrons between the drain and the source. However, the potential barrier is reduced with application of the drain bias. Consequently, as the drain bias increases, drain current flow commences at voltages well below the breakdown voltage capability of the drift region. Since a larger potential barrier is created for larger gate bias voltages, it is possible to support a larger drain bias voltage before the observation of drain current flow.

The spacing between the gate regions is usually designed to be more than twice the zero bias depletion width. Consequently, an un-depleted portion of the drift region remains under the source region at zero gate bias. Current flow between the drain and source occurs through this region with the amount limited by the resistance of the channel region (between the gates) and the drift region below the gate. A large un-depleted region is favored for reducing the on-resistance but this reduces the magnitude of the potential barrier when the device operates in the forward blocking mode. Thus, a trade-off between the on-state resistance and blocking characteristics must be made when designing these structures.

7.1.1 Trench Metal-Semiconductor Field Effect Transistor Structure: Forward Blocking

Since the operating principles for the MESFET and JFET structures are similar, these names will be used interchangeably in this chapter. In a normally-on device structure, an un-depleted portion of the channel exists at zero gate bias. Drain current flow can therefore occur via this un-depleted region at zero gate bias. The forward blocking capability in the MESFET structure is achieved by creating a potential barrier in the channel between the gate regions by applying a reverse bias to the gate region. At a gate bias above the pinch-off voltage, the channel becomes completely depleted. As the gate bias is increased above the pinch-off voltage, the potential barrier increases in magnitude. This will be shown later in the chapter with the aid of two-dimensional numerical analysis of typical structures. Current transport between the drain and source is suppressed because electrons must overcome the potential barrier. As the drain voltage is increased, the potential barrier reduces allowing injection of electrons over it. An exponential increase in the drain current is observed with increasing drain bias with 'triode-like' characteristics. This behavior can be described by[6]:

$$J_D = \frac{qD_nN_D}{L}\sqrt{\frac{q}{\pi kT}(\alpha V_G - \beta V_D)}\exp\left\{-\left[\frac{q}{kT}(\alpha V_G - \beta V_D)\right]\right\} \quad [7.1]$$

where α and β are constants that depend upon the gate geometry. In this equation, D_n is the diffusion coefficient for electrons, N_D is the doping concentration in the drift region, L is the gate length (in the direction of current flow), q is the charge of the electron, k is Boltzmann's constant, T is the absolute temperature, V_G is the gate bias voltage and V_D is the drain bias voltage. The exponential variation of drain current with variation of gate and drain bias has been observed in silicon vertical channel JFETs[7].

On the one hand, if the space between the gate regions in the JFET structure is reduced so that it becomes completely depleted by the zero bias depletion width, the structure can exhibit a normally-off behavior upto a limited drain voltage. This type of design will exhibit purely triode-like characteristics. On the other hand, if the space between the gate regions is larger than the maximum depletion width at breakdown for the drift region, the structure will exhibit purely pentode-like characteristics. If the space between the gate regions falls between these extremes, the device will exhibit a mixed triode-pentode like characteristics. High voltage normally-on JFETs are usually designed to operate in this mixed triode-pentode mode to obtain a good compromise between low on-state resistance and high blocking voltage capability.

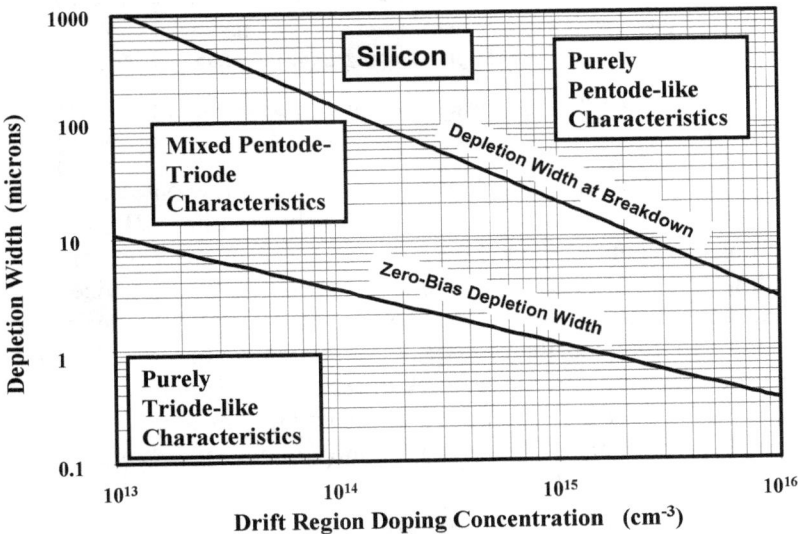

Fig. 7.2 Design Space for High Voltage Silicon JFETs.

The design space for vertical high voltage JFETs is bounded by the depletion widths at breakdown and the zero-bias depletion width. These boundaries are shown in Fig. 7.2 for the case of silicon devices, and in Fig. 7.3 for the case of 4H-SiC devices. For any given doping concentration, a much larger spacing is required for 4H-SiC devices due to the bigger depletion width at breakdown. In general, a greater latitude exists for the design of 4H-SiC devices to optimize the characteristics.

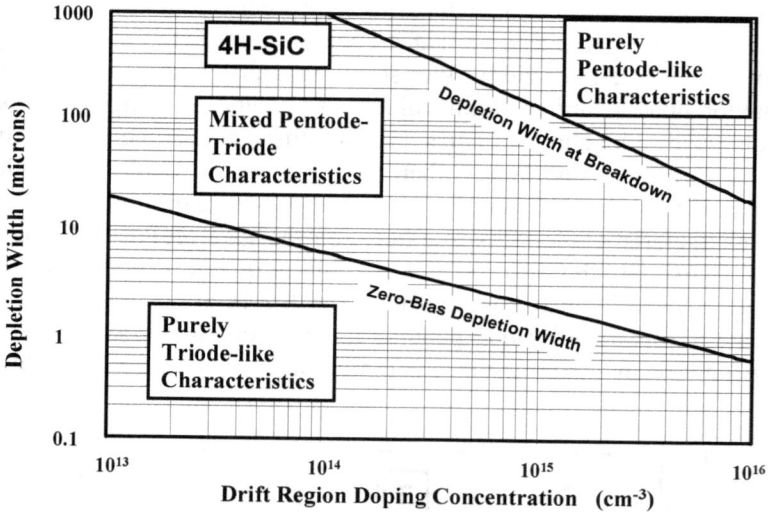

Fig. 7.3 Design Space for High Voltage 4H-SiC JFETs.

A good compromise between achieving a low on-state resistance and a good blocking voltage capability for high voltage JFETs requires designing the channel to operate in the mixed pentode-triode regime. At low drain current levels and high drain voltages, these devices exhibit triode-like characteristics. In this mode, it is useful to define a DC blocking gain (G_{DC}) as the ratio of the drain voltage to the gate voltage at a specified leakage current level. A differential blocking gain (G_{AC}) can also be defined as the increase in drain voltage, at a specified leakage current level, for an increase in the gate voltage by 1 volt. From Eq. [7.1], it can be shown that:

$$G_{AC} = \frac{dV_D}{dV_G} = \frac{\alpha}{\beta}$$

[7.2]

The parameters α and β are dependent upon the channel aspect ratio. The channel aspect ratio is defined as the ratio of the length of the gate (dimension 'L' in Fig. 7.4) in the direction of current flow and the space between the gate regions (dimension 'a' in Fig. 7.4). A large aspect ratio favors obtaining a high blocking gain, which is beneficial for reducing the gate voltage required to block high drain voltages. Theoretical analysis[8] and empirical observations[9] on high voltage silicon JFETs indicate that the blocking gain can be described by:

$$G_{DC} = \frac{V_D}{V_G} = A.\exp\left(B\frac{L}{a} \right)$$

[7.3]

where A and B are constants. In silicon JFETs, a blocking gain of about 10 is obtained for a vertically walled structure when the channel aspect ratio is approximately equal to unity. Since the gate junction must support the sum of the negative gate bias voltage and the applied positive drain voltage, the drift region parameters must be chosen to account for the finite blocking gain.

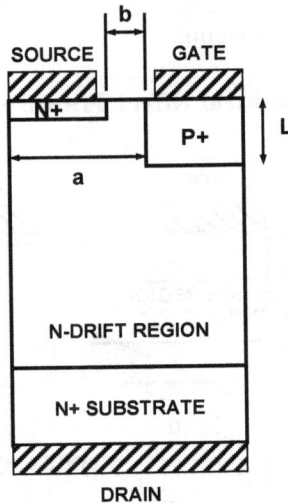

Fig. 7.4 Channel Aspect Ratio (L/a) for JFET structures.

The largest drain voltage that can be supported by the JFET structure before the on-set of significant current flow is determined by several factors. Firstly, it is limited by the intrinsic breakdown voltage capability of the drift region as determined by the doping concentration and thickness. The breakdown voltage can be obtained by using the graphs and equations provided in chapter 3. Secondly, the maximum drain voltage that can be supported without significant current flow can be limited by the applied gate bias and the blocking gain of the structure. In addition, the largest gate bias that can be applied is limited by the on-set of breakdown between the gate and the source regions. The breakdown voltage between the gate and the source regions is determined by the depletion layer punch-through from the gate junction to the highly doped N^+ source region. Since the space between these regions must be kept small in order to obtain a high channel aspect ratio leading to high blocking gain, the electric field between the gate and source regions can be assumed to be uniform. Under this approximation, the gate-source breakdown voltage can be calculated using:

$$BV_{GS} = b.E_C \qquad\qquad [7.4]$$

where 'b' is the space between the gate and the source as shown in Fig. 7.4 and E_C is the critical electric field for breakdown. Fortunately, the critical electric field for breakdown in silicon carbide is much larger than for silicon, allowing high gate-source breakdown voltages in spite of using small gate-source spacing.

7.1.2 Trench Junction Field Effect Transistor Structure: On-State

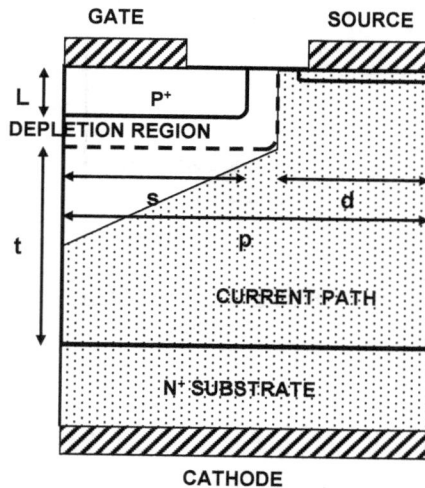

Fig. 7.5 Current Flow in On-State for JFET structures.

For the normally-on JFET design with mixed Pentode-Triode mode of operation, the on-state current flow pattern is as indicated in Fig. 7.5 by the dotted lines. The current flows from the source through a uniform cross section between the gate regions with a width (d) determined by the space between the gate regions and the zero-bias depletion width. The current then spreads at a 45 degree angle into the drift region to a depth of 's' under the gate region, and then becomes uniform throughout the cross-section. This current flow pattern can be used to model the on-state resistance:

$$R_{on,sp} = \rho_D (L + W_0) \left(\frac{p}{d} \right) + \rho_D \left(\frac{p(s + W_0)}{p - 2d} \right) \ln \left(\frac{p}{2d} \right)$$
$$+ \rho_D (t - s) \qquad\qquad [7.5]$$

where W_0 is the zero bias depletion width.

The specific on-resistance of the JFET can be much larger than the ideal specific on-resistance for any particular blocking voltage capability of the JFET structure. Firstly, this is because the drift region must support the sum of the gate and drain bias voltages. Secondly, it is increased by the contributions from the channel region and the spreading resistance. These contributions become more significant as the space between the gates is reduced. Thus, a compromise must be made between obtaining a low specific on-state resistance and a high blocking gain.

When a negative gate voltage is applied to reverse bias the gate junction, the depletion region extends further into the channel producing an increase in the on-resistance. Further, if the drain voltage becomes comparable to the gate bias voltage, the depletion width at the bottom of the channel on the drain side becomes larger (as determined by V_{GS} plus V_{DS}). This alters the current flow pattern as illustrated in Fig. 7.6. Under the assumption of a field independent mobility and by using the gradual channel approximation, the drain current is determined by[2]:

$$I_D = 2a\rho_D \frac{Z}{L} \left\{ V_{Dch} - \frac{2}{3a} \left(\frac{2\varepsilon_S}{qN_D} \right)^{1/2} \left[\left(V_{Dch} + V_G + V_{bi} \right)^{3/2} - \left(V_G + V_{bi} \right)^{3/2} \right] \right\}$$
$$[7.6]$$

where Z is the length of the device in the direction orthogonal to the cross-section, and V_{Dch} is the drain voltage at the bottom edge of the channel.

The basic I-V characteristics determined by this equation can be described using three segments. In the first segment, the drain voltage is much smaller than the gate bias voltage. In this case, the channel resistance increases with reverse gate voltage as given by:

$$R_{ch} = \rho_D \frac{L}{2Z} \left[a - \sqrt{\frac{2\varepsilon_S}{qN_D}\left(V_{GS} + V_{bi}\right)} \right] \qquad [7.7]$$

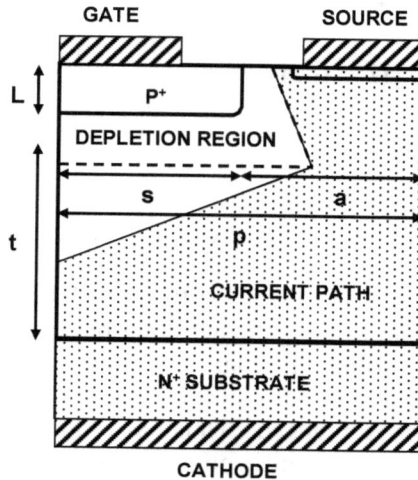

Fig. 7.6 Current Flow at larger V_{DS} for JFET structures.

In the second segment, the drain voltage is comparable to the gate voltage. This produces a non-linear characteristic with the resistance increasing with increasing drain bias. Eventually, the entire space between the gate regions becomes depleted at a certain drain bias. This condition is described by:

$$V_P = \left(V_{Dch} + V_{GS} + V_{bi}\right) = \frac{qN_D a^2}{2\varepsilon_S} \qquad [7.8]$$

with V_P defined as the pinch-off voltage. The drain voltage, at which channel pinch-off occurs, decreases linearly with increasing gate voltage.

In the third segment, the drain current becomes constant because the channel is completely pinched off. The saturated drain current is given by:

$$I_D = 2a\rho_D \frac{Z}{L}\left\{\frac{qN_D a^2}{6\varepsilon_S} - (V_{GS} + V_{bi}) + \frac{2}{3a}\left(\frac{2\varepsilon_S}{qN_D}\right)^{1/2}\left[(V_G + V_{bi})^{3/2}\right]\right\}$$

[7.9]

A family of drain current-voltage curves is formed with the saturated drain current decreasing with increasing gate voltage. Although these equations predict a constant drain current beyond the channel pinch-off point, in practice, the drain current can increase after pinch-off due to electron injection over the potential barrier formed at the bottom of the channel. This phenomenon leads to the mixed Pentode-Triode characteristics.

7.2 Trench Metal-Semiconductor Field Effect Transistor Structure: Simulation Results

The case of a vertical MESFET structure with drift region concentration of 1 x 10^{16} cm^{-3} and thickness of 20 microns will be used to illustrate the operating principles of these devices. The two dimensional numerical simulations were performed with various spacing between the gate regions as well as by varying the depth of the gate region (trench depth). In each case, the on-state resistance was extracted at zero gate bias, and the blocking characteristics were obtained by application of various negative gate bias voltages. A gate metal work-function of 5.2 eV was used corresponding to a barrier height of 1.5 eV.

7.2.1 Trench MESFET Structure: On-State Characteristics

First, consider the case of a MESFET with gate spacing (dimension 'a' in Fig. 7.4) of 0.6 microns with a space 'b' of 0.35 microns between the N$^+$ source region and the gate metal. The width of the trench (dimension 's' in Fig. 7.5) was kept constant for all structures at a value of 0.5 microns. The impact of changing the trench depth was studied by using values of 0.5, 1.0, and 1.5 microns. The on-state current flow pattern for the structures at zero gate bias is shown in Fig. 7.7. It can be seen that the current flow is uniform in the channel region within a width determined by the un-depleted portion. The current then spreads rapidly (at an angle of 45 degrees) and then becomes uniform for most of the drift region.

The specific on-resistance for the device with 1 micron trench depth was found to be 1.83 mΩ-cm^2. This is extremely small for a structure capable of supporting nearly 3000 volts. The ideal specific on-resistance for the drift region is 1.0 mΩ-cm^2 as previously discussed in chapter 3 (see Fig. 3.6). Thus, the MESFET structure provides a specific on-resistance approaching the ideal value. The specific on-resistance was found to have a weak dependence on the trench depth. It reduced to 1.62 mΩ-cm^2 for a trench depth of 0.5 microns and increased to 2.04 mΩ-cm^2 for a trench depth of 1.6 microns. Consequently, the blocking gain can be increased by increasing the trench depth without a strong adverse impact on the on-resistance.

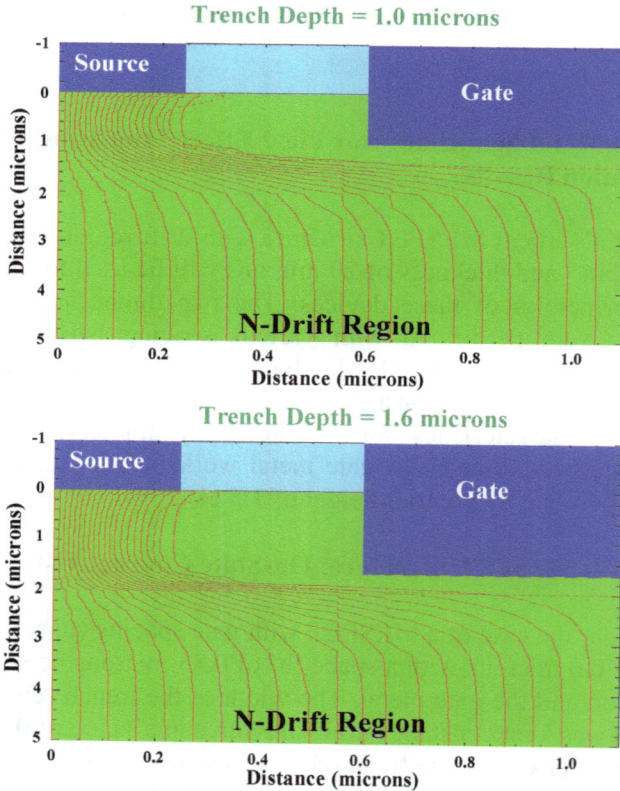

Fig. 7.7 On-State Current Flow in 4H-SiC MESFETs:

Impact of Trench Depth.

The blocking gain of MESFETs can be increased by reducing the spacing between the gate regions. The impact of this space reduction on the on-resistance was studied while keeping the trench depth at 1 micron. It was found that the specific on-resistance reduced from 1.83 mΩ-cm^2 at a spacing (dimension 'a') of 0.6 microns to 1.66 mΩ-cm^2 at a spacing of 0.8 microns. This is due to the larger conduction area in the channel region with increasing gate spacing as shown in Fig. 7.8. However, when the spacing was further increased to1 micron, the specific on-resistance increased slightly to 1.68 mΩ-cm^2. This occurs because the improvement due to reduction in absolute cell on-resistance is counteracted by the enlargement of the cell pitch. Consequently, it is advantageous to keep the gate spacing below 0.8 microns to obtain both a low specific on-resistance and a high blocking gain. The blocking gain for these structures is discussed in the next section.

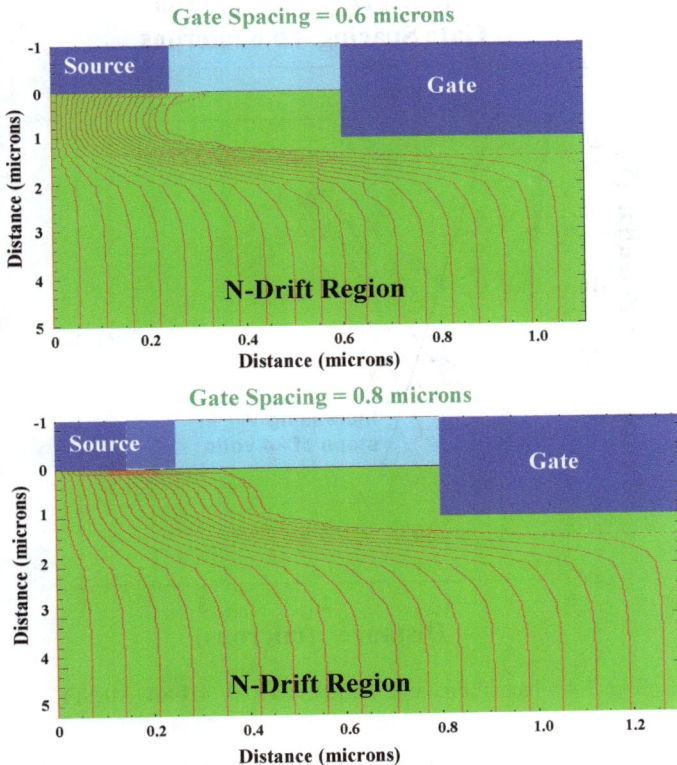

Fig. 7.8 On-State Current Flow in 4H-SiC MESFETs:

Impact of Gate Spacing.

7.2.2 Trench MESFET Structure: Blocking Characteristics

In a normally-on MESFET structure, the ability to block current flow with positive drain voltage can be achieved by application of a negative gate bias. The gate bias depletes the space between the gate regions and produces a potential barrier to the flow of electrons from the source to the drain. The minimum voltage required to deplete the space between the gate regions is the pinch-off voltage which reduces with smaller gate spacing. For a given gate bias voltage, a larger potential barrier can be formed with smaller gate spacing and a larger channel aspect ratio as discussed earlier. These features are discussed with the aid of results from two-dimensional numerical simulations in this section.

Trench Depth = 1 micron
Gate Spacing = 0.6 microns

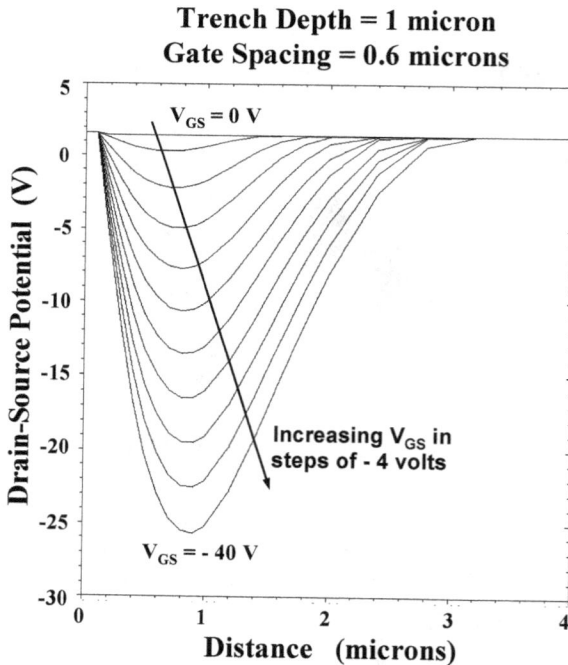

Fig. 7.9 Channel Potential Barrier in a 3 kV 4H-SiC MESFET.

In the MESFET structure, electron injection can occur over the portion of the channel with the smallest potential barrier leading to the on-set of current flow. The smallest potential barrier occurs at a location in the channel furthest removed from the edge of the gate region. For this reason, the potential variation between the drain and source at the center of the channel region (i.e. at x = 0 in Fig. 7.4 and Fig. 7.7) is shown in Fig. 7.9. The pinch-off voltage for a structure with a gate spacing of 0.6 microns is 3.35 volts for the drift region doping concentration of 1 x 10^{16} cm^{-3}. Consequently, it can be seen that a negative potential develops in the channel after the gate bias exceeds -4 volts. The magnitude of the potential barrier increases with gate bias and reaches a value of -26 volts at a gate bias of -40 volts. A linear increase in the potential barrier is observed with increasing gate bias as shown in Fig. 7.10.

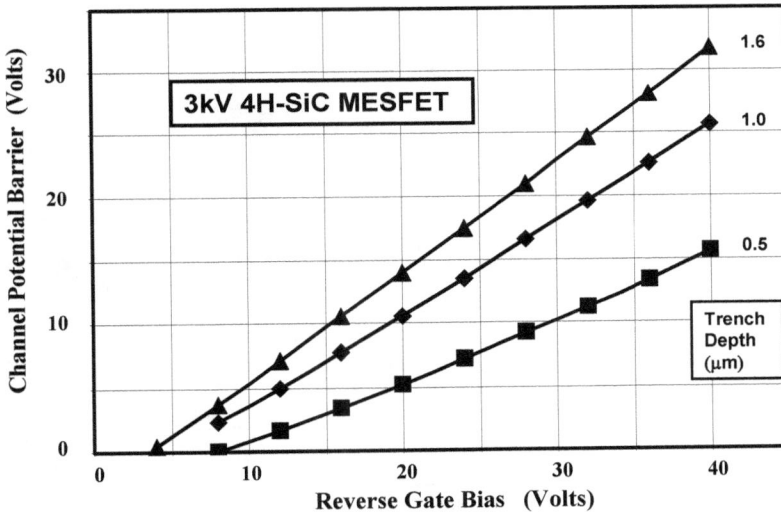

Fig. 7.10 Channel Potential Barriers in 3 kV 4H-SiC MESFETs.

The development of the channel potential barrier for the case of increasing the trench depth from 1 to 1.6 microns is shown in Fig. 7.11. The potential barrier becomes larger with a value of -32 volts at a gate bias of -40 volts. In addition, the barrier becomes broader between the drain and the source creating additional difficulty for injection of electrons and thus suppressing drain current flow. The increase in the barrier height as a function of gate bias for this case is also shown in Fig. 7.10. In addition, the barrier height for the case of a trench depth of 0.5 microns is shown in this figure for comparison. With the reduced trench depth a much smaller potential barrier develops in the channel. This

reduces the blocking gain and consequently the ability of the device to
support high drain voltages.

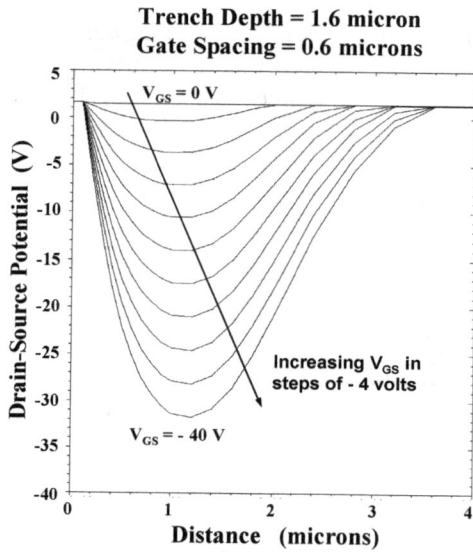

Fig. 7.11 Channel Potential Barrier in a 3 kV 4H-SiC MESFET.

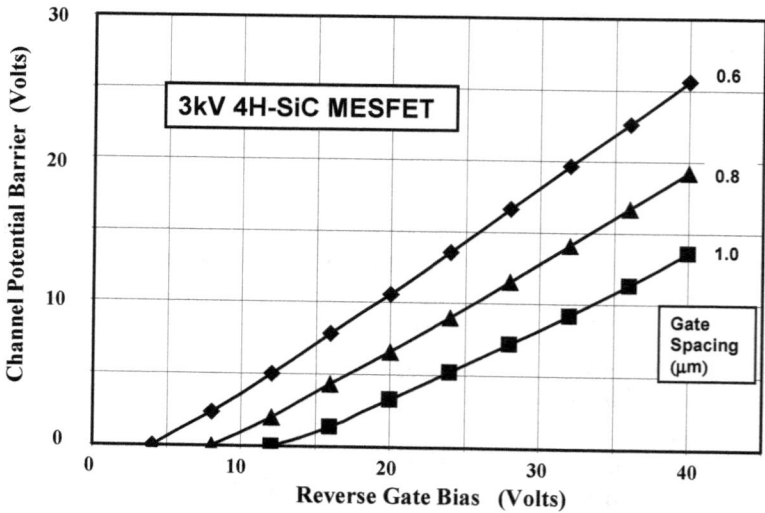

Fig. 7.12 Channel Potential Barriers in 3 kV 4H-SiC MESFETs.

The height of the potential barrier is also dependent upon the gate spacing. A smaller gate spacing not only reduces the pinch-off voltage but also increases the potential barrier height resulting is larger blocking gain. The change in the potential barrier height as a function of the reverse gate bias is shown in Fig. 7.12 when the trench depth is maintained at 1 micron. Note that the potential barrier begins to develop at larger gate bias voltages with increased gate spacing because of the larger pinch-off voltage, namely, 9.3 volts for the 1 micron gate spacing versus 3.3 volts for the 0.6 micron gate spacing. The potential barrier height for the 1 micron gate spacing is about half that observed with the gate spacing of 0.6 microns. It is therefore important to obtain precise control over the geometry of the MESFET structure during fabrication. The variation of the potential barrier for the structure with 1 micron gate spacing is provided in Fig. 7.13 for comparison with the simulation results shown in Fig. 7.9 for the structure with spacing of 0.6 microns.

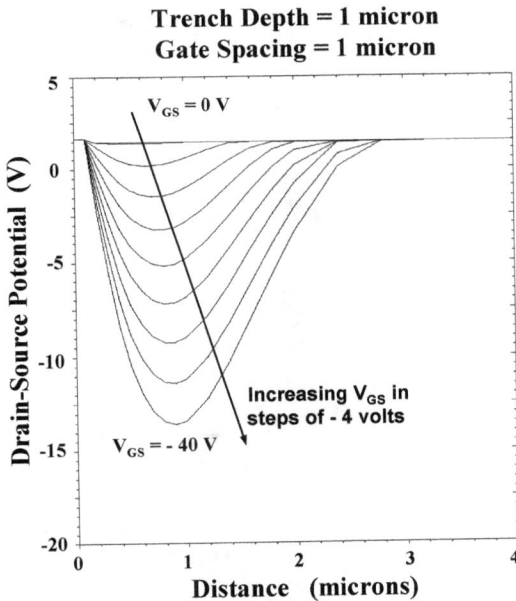

Fig. 7.13 Channel Potential Barrier in a 3 kV 4H-SiC MESFET.

The blocking characteristics of the MESFET structure are determined by the triode-like behavior at low drain bias voltages. In this domain of operation, current flow occurs when the electrons overcome the potential barrier created by the gate bias. As the drain voltage increases, the potential barrier is reduced promoting the injection of electrons. This produces triode-like characteristics with an exponential increase in drain current with increasing drain bias as described by Eq. [7.1]. The change in the potential barrier height with increasing drain bias is shown in Fig. 7.14 for the case of the structure with 1 micron trench depth and 0.6 micron gate spacing when a gate bias of -40 volts is applied. It can be seen that the potential barrier becomes smaller with increasing drain voltage. A -3 volt barrier is still preserved even at a drain bias of 3000 volts indicating that this structure can support 3000 volts with low drain leakage current flow. Note that the peak of the potential barrier moves towards the source with increasing drain bias indicating encroachment of the drain potential into the channel.

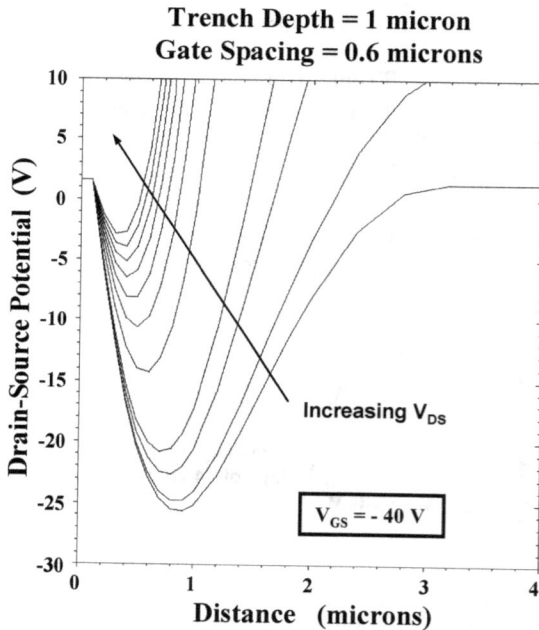

Fig. 7.14 Channel Potential barrier in a 3 kV 4H-SiC MESFET.

(V_{DS} = 0,10,50,100,500,1000,1500,2000,2500,3000 volts)

    The reduction of the potential barrier height with increasing drain voltage is shown in Fig. 7.15 for the case of devices with different trench depths. It can be seen that the barrier height reduces to a greater extent for the shallower trench depth. For the case of a trench depth of 0.5 microns, the potential barrier is completely removed with a drain bias of above 500 volts. Thus, this structure cannot support high drain voltages with a gate bias of -40 volts.

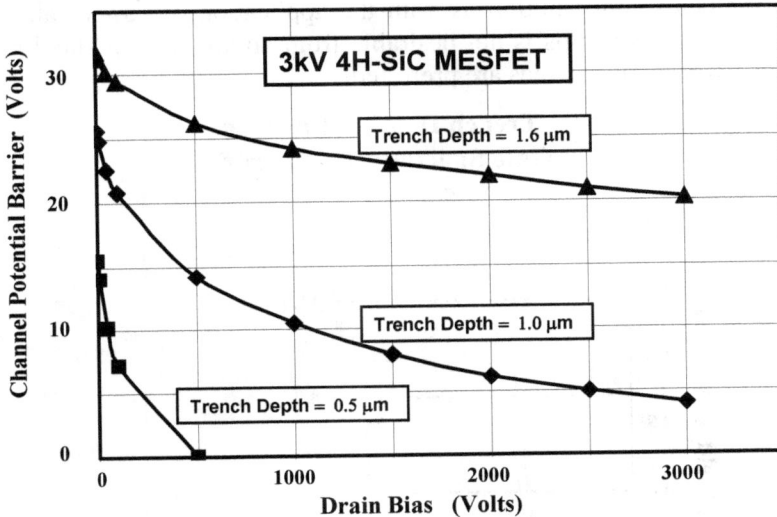

Fig. 7.15 Channel Potential Barriers in 3 kV 4H-SiC MESFETs.

In general, the normally-on MESFET structure can support a larger drain bias voltage without significant drain current flow if the gate voltage is increased. However, the reverse gate bias cannot be arbitrarily increased because of the on-set of gate-source breakdown. The breakdown voltage between the gate and source regions is decided by the depletion layer punch-through process. Under strong punch-through conditions, the electric field can be calculated using the approximation that it is uniform between the gate and the source regions:

$$E_{GS} = \frac{(V_{GS} + V_{bi})}{b} \qquad [7.10]$$

where 'b' is the space between the gate and the source as shown in Fig. 7.4 and V_{bi} is the built-in potential for the gate region. The maximum gate voltage that can be applied (BV_{GS}) is then given by the condition where the electric field becomes equal to the critical electric field for breakdown:

$$BV_{GS} = b.E_C - V_{bi} \qquad [7.11]$$

For a spacing 'b' of 0.35 microns, the gate-source breakdown voltage for 4H-SiC is about 100 volts if a critical electric field of 3×10^6 V/cm is used. Thus, the MESFET structures with shallower trenches could block drain voltages above 500 volts with the application of larger gate bias voltages. However, this is not desirable from an application stand point where lower input voltages are preferred.

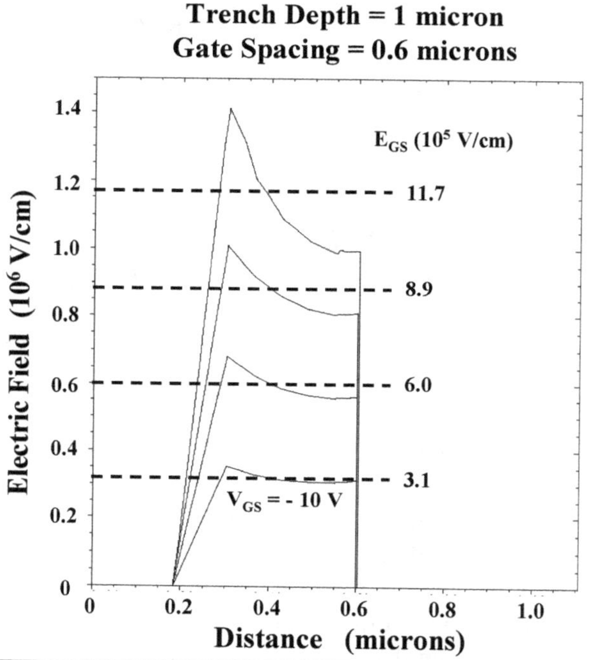

Trench Depth = 1 micron
Gate Spacing = 0.6 microns

Fig. 7.16 Electric Field between Gate and Source in a 3 kV 4H-SiC MESFET.

The build-up of the electric field between the gate and source is shown in Fig. 7.16 for a typical simulated structure with gate-source spacing 'b' of 0.35 microns. As anticipated, the electric field is approximately uniform between the regions. The electric field calculated using Eq. [7.10] with a built-in potential of 1 volt and a spacing of 0.35 microns is also shown by the dashed lines in the figure. A good agreement is observed between the simple analytical method of calculation of the electric field and the simulated results.

When the gate bias is changed, the on-set of drain current up to any given leakage current level occurs at different drain bias voltages. This produces the triode-like characteristics between the drain current and voltage. A typical set of I-V characteristics is shown in Fig. 7.17 for the case of the structure with 1 micron trench depth and 0.6 micron gate spacing. Note that the drain current increases exponentially with increasing drain bias consistent with the electron potential barrier model discussed in the previous sections until the drain current exceeds 10^{-6} A corresponding to a current density of 100 A/cm^2.

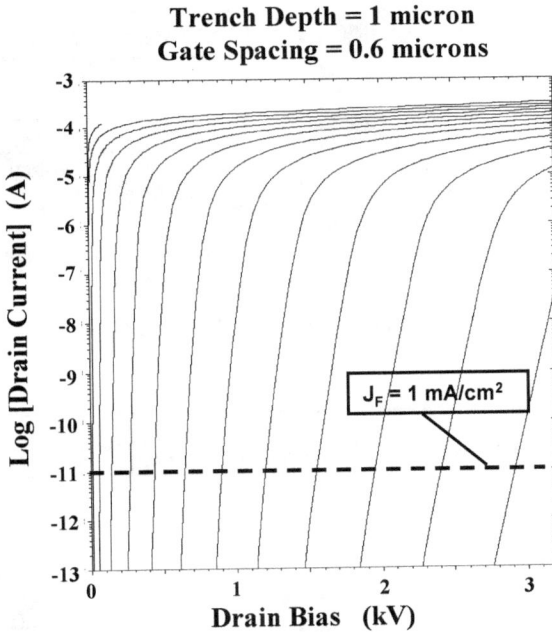

Fig. 7.17 Blocking Characteristics for a 3 kV 4H-SiC MESFET.

(V_{GS} = 0,-2,-4,-6,...,-26 volts)

By choosing a suitable magnitude for the drain leakage current under forward blocking conditions, it is possible to track the drain voltage supported at each gate bias. The resulting *drain blocking voltage* is plotted in Fig. 7.18 as a function of the magnitude of the reverse gate bias. From this graph, the benefits of increasing the trench depth to enhance the channel aspect ratio become quite obvious. Similarly, the adverse impact of increasing the gate spacing to reduce the channel aspect ratio is shown in Fig. 7.19 for the case of a trench depth of 1 micron.

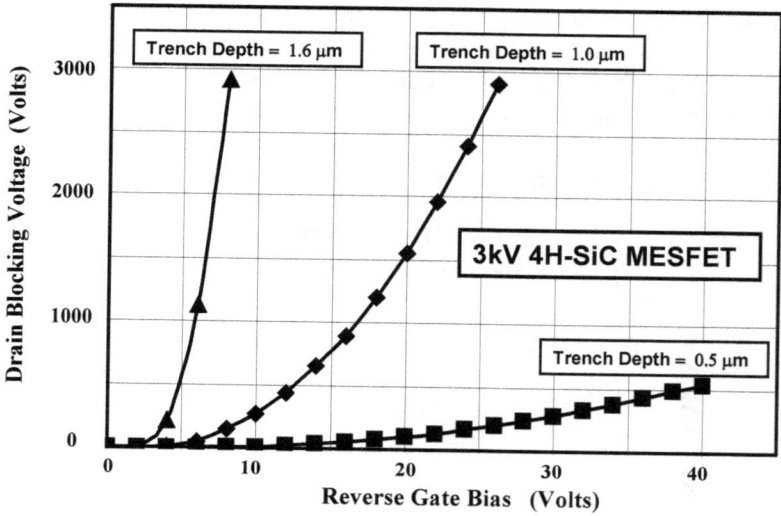

Fig. 7.18 Drain Blocking Voltage for 3 kV 4H-SiC MESFETs.

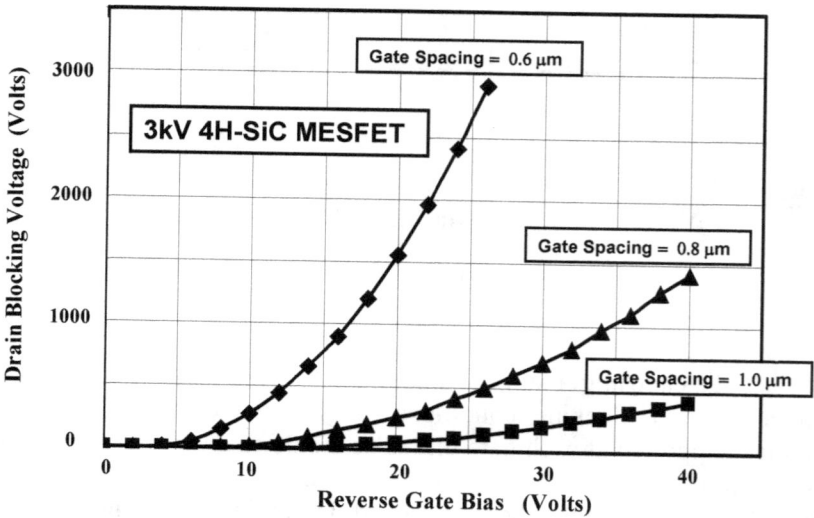

Fig. 7.19 Drain Blocking Voltage for 3 kV 4H-SiC MESFETs.

In general, the voltage blocking behavior of the vertical MESFET structure can be examined by looking at the blocking gain as a function of the channel aspect ratio. The DC blocking gain is defined as the ratio of the drain blocking voltage to the corresponding gate bias. This gain varies with drain and gate bias as observed in the above figures. If the gain is taken at a drain bias of 500 volts, then the various structures can be compared in terms of the channel aspect ratio. This is done in Fig. 7.20 for the structures simulated with various trench depths and gate spacing. The exponential increase in blocking gain with increasing channel aspect ratio is observed when either the trench depth or the gate spacing is changed. However, the trend lines are not identical due to the difference in the width of the potential barrier.

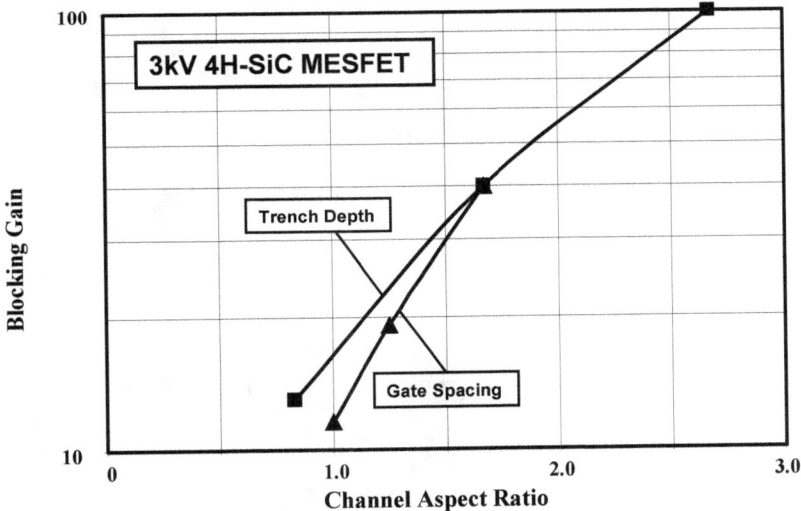

Fig. 7.20 Blocking Gain for 3 kV 4H-SiC MESFETs.

7.2.3 Trench MESFET Structure: Output Characteristics

The output characteristics of the vertical MESFET structure determine the safe operating area of the device. In a typical pulse width modulated control scheme, the power transistor undergoes a switching transient from either the on-state to off-state or vice versa. The loci for these transients must be kept within the bounds of the safe operating area for the transistor.

Trench Depth = 1 micron
Gate Spacing = 0.6 microns

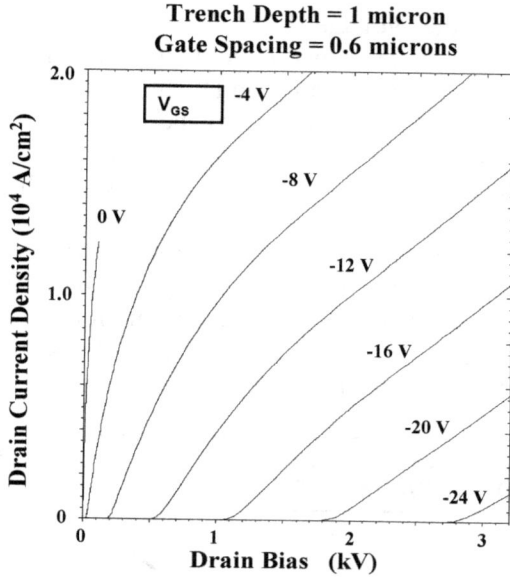

Fig. 7.21 Output Characteristics for a 3 kV 4H-SiC MESFET.

Trench Depth = 0.5 micron
Gate Spacing = 0.6 microns

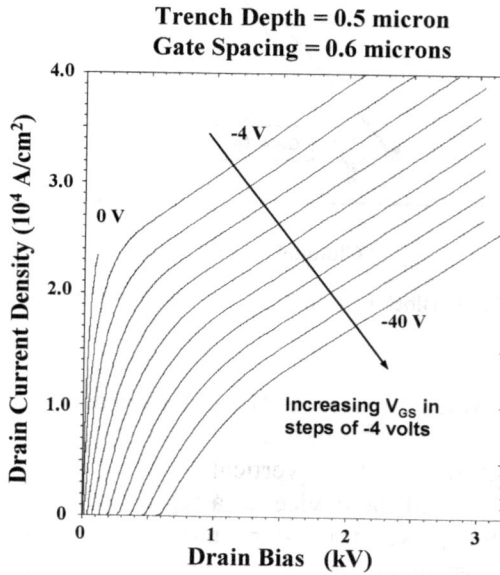

Fig. 7.22 Output Characteristics for a 3 kV 4H-SiC MESFET.

The channel design has a strong impact on the output characteristics of the vertical MESFET structure. For the structure with trench depth of 1 micron and gate spacing of 0.6 microns, the output characteristics exhibit a mixed triode-pentode behavior as shown in Fig. 7.21. This structure is capable of supporting a drain bias of 3000 volts with a gate bias of -26 volts. On the other hand, the structure with reduced trench depth of 0.5 microns is unable to support a drain bias of more than 600 volts with a gate bias of -40 volts as shown by the output characteristics in Fig. 7.22. A careful optimization of the structure is required to ensure that it can support the desired drain blocking voltage with moderate (less than 40 volts) gate voltages. When this criterion is satisfied, the vertical MESFET structure can be utilized in the *Baliga-Pair* circuit configuration to create a very efficient normally-off power switch with high input impedance MOS interface. This is discussed in detail in the next chapter.

7.3 Junction Field Effect Transistor Structure: Simulation Results

Although the Junction Field Effect Transistor operates on the same basic principles as the Metal-Semiconductor Field Effect Transistor, the design of the two devices is slightly altered by the larger built-in potential for the P-N junction when compared with the metal-semiconductor contact. The case of a vertical JFET structure with drift region concentration of 1 x 10^{16} cm^{-3} and thickness of 20 microns will be used to illustrate these differences. The two dimensional numerical simulations were performed with gate spacing (dimension 'a' in Fig. 7.4) of 0.6 microns with a space 'b' of 0.35 microns between the N$^+$ source region and the gate junction. The width of the P$^+$ region (dimension 's' in Fig. 7.5) was kept at a value of 0.5 microns.

The on-state current flow pattern for the JFET structure at zero gate bias is compared in Fig. 7.23 with the MESFET structure having the same gate spacing. Due to the larger built-in potential of the P-N junction, a much stronger constriction of the current is observed for the JFET structure. This leads to an increase in the specific on-resistance. The specific on-resistance for the JFET structure was found to be 3.93 mΩ-cm^2 when compared with 1.83 mΩ-cm^2 for the MESFET structure and 1 mΩ-cm^2 for the ideal drift region alone. However, this value is still respectable for a device capable of supporting 3000 volts making the JFET structure an alternative option to the MESFET structure.

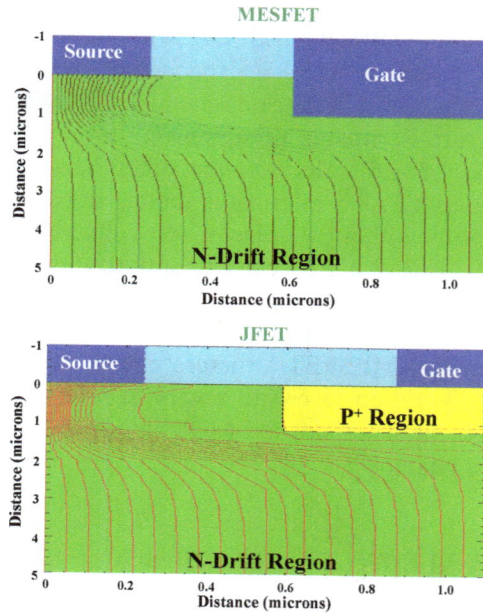

Fig. 7.23 On-State Current Flow in a 4H-SiC JFET versus a MESFET.

The blocking capability of the JFET structure is predicated upon formation of a potential barrier in the channel to suppress the injection of electrons from the source to the drain. The barrier height created by the application of various reverse gate bias voltages for the JFET structure is shown in Fig. 7.24. A slightly larger potential barrier is created in the JFET structure when compared with the MESFET structure for the same channel design because of the larger built-in potential. This is more obvious in the plot of the potential barrier as a function of gate bias provided in Fig. 7.25. The barrier is about 1 volt larger for the JFET structure at all gate bias voltages.

The larger potential barrier in the channel for the JFET structures promotes higher drain blocking voltage capability. This can be observed in the I-V characteristics for the device shown in Fig. 7.26. The structure is able to block drain current flow upto a drain bias of 3000 volts with a gate bias of -26 volts. The improvement in the blocking gain is less than 10 percent. Consequently, the MESFET structure is preferable to the JFET structure in order to obtain a lower specific on-resistance while keeping approximately the same blocking gain.

Junction Depth = 1 micron
Gate Spacing = 0.6 microns

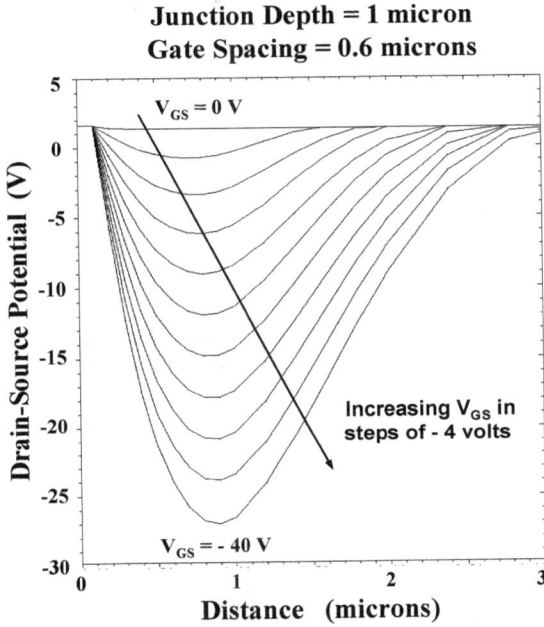

Fig. 7.24 Channel Potential Barrier in a 3 kV 4H-SiC JFET.

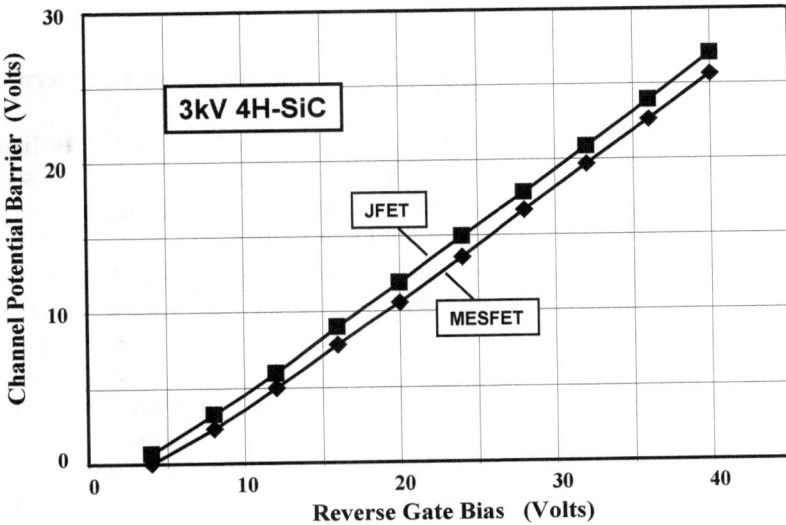

Fig. 7.25 Channel Potential Barrier in 3 kV 4H-SiC JFET and MESFET.

Junction Depth = 1 micron
Gate Spacing = 0.6 microns

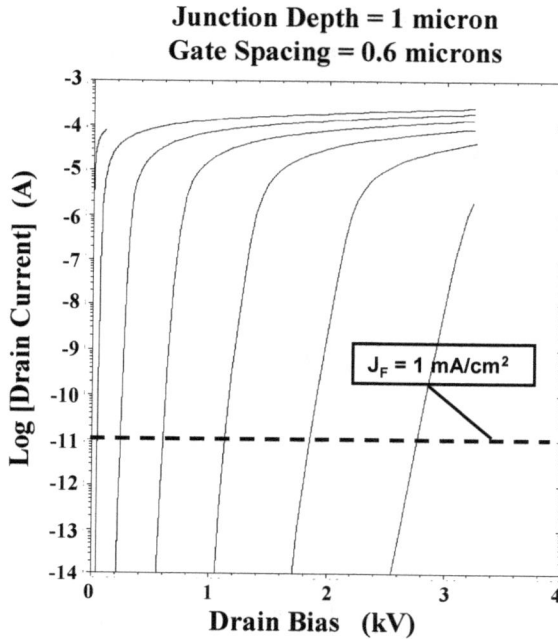

Fig. 7.26 Blocking Characteristics for a 3 kV 4H-SiC JFET.

(V_{GS} = 0,-4,-8,..,-24 volts)

7.4 Planar Metal-Semiconductor Field Effect Transistor Structure

An elegant high voltage silicon carbide MESFET structure that utilizes a planar gate architecture (in place of the trench gate structure discussed in the previous section) has also been proposed and demonstrated[10]. In this device structure, a sub-surface heavily doped P-type region is placed below the N^+ source region as illustrated in Fig. 7.27. The P^+ region acts as a barrier to current flow between the source and drain regions restricting the current to gaps between the P^+ regions. A gate region is placed over these gaps and overlapping the P^+ region to enable control over the transport of current between the drain and source regions. In general, the gate can be constructed as a Metal-Semiconductor contact (to form a MESFET structure), a P-N Junction (to form a JFET structure) or as a Metal-Oxide-Semiconductor sandwich (to form a MOSFET structure). The MESFET structure will be discussed in detail in this section. The operation of the JFET structure is similar but requires taking into account the larger built-in potential of the gate P-N junction. The MOSFET structure is discussed in a subsequent chapter.

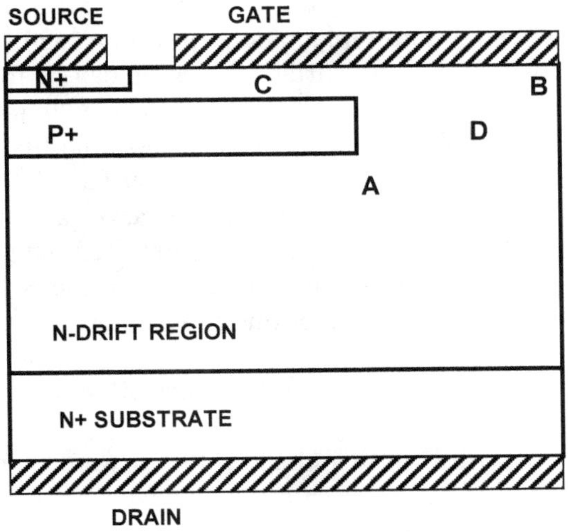

Fig. 7.27 The Planar Gate MESFET Structure.

The sub-surface P^+ region in the planar gate MESFET structure can be either connected to the gate or to the source region. If the P^+ region is electrically connected to the gate electrode, it collaborates with the metal-semiconductor contact to constrict current flow from the source to the drain region. However, when the gate is reverse biased, a low gate-source breakdown voltage can occur due to the close vertical proximity of the N^+ source region and the sub-surface P^+ region. It is therefore preferable to connect the P^+ region to the source electrode. During the fabrication of the MESFET structure, this can be achieved by interruption of the N^+ region in the orthogonal direction to the cross-section shown in Fig. 7.27 and placing a P^+ contact region from the surface down to the sub-surface P^+ region in these gaps. This approach avoids a potential breakdown problem between the N^+ source and P^+ sub-surface regions. However, the breakdown voltage between the gate and source electrodes can now occur due to depletion region reach-through from the gate contact to the underlying P^+ region. Fortunately, as shown below with numerical simulations, relatively small reverse gate bias voltages are required in the planar MESFET to achieve high drain blocking voltages. The low gate-source breakdown voltage as a result of reach-through is therefore not a serious limitation with proper design of the structure.

The sub-surface P$^+$ region in the planar MESFET structure can be created by using ion-implantation of boron with the appropriate energy[10]. Alternately, the sub-surface P$^+$ region can be formed by growth of an N-type epitaxial layer over a P$^+$ region formed in the drift region by lower energy ion-implantation[11]. In either case, the thickness of the N-type region between the gate and the sub-surface P$^+$ region must be sufficient to prevent complete depletion at zero gate bias. The doping concentration of the N-type region located between the gate and the sub-surface P$^+$ region can be increased above that for the N-type drift region if necessary by ion-implantation or during its epitaxial growth.

7.4.1 Planar MESFET Structure: Forward Blocking

As previously described in this chapter, the forward blocking regime of operation for the MESFET structure is achieved by creating a potential barrier for transport of electrons between the source and drain region by the application of a reverse gate bias. In the planar MESFET structure, this potential barrier is formed in the channel (at location 'C' shown in Fig. 7.27). If the thickness of the channel (shown as t$_{CH}$ in Fig. 7.28) is narrow, a potential barrier can be formed with relatively low reverse gate bias voltages. In addition, the planar MESFET structure contains a second JFET region formed between the adjacent P$^+$ regions (at location 'D' shown in Fig. 7.27). When the drain bias exceeds the pinch-off voltage for this JFET region, the potential at the surface under the gate becomes isolated from the potential applied at the drain electrode. Consequently, the channel potential barrier is shielded from the drain voltage enabling the support of high drain voltages without the on-set of drain current flow. These features favor producing a very high blocking gain with low reverse gate bias voltages as shown with two-dimensional numerical simulations later in this chapter.

7.4.2 Planar MESFET Structure: On-State Resistance

The planar MESFET can be designed to contain an un-depleted channel region at zero gate bias. Current can then flow between the source and drain regions through the channel and the gap between the P$^+$ regions down to the N-type drift region. The resistance of these regions must be included in the analysis of the total on-state resistance of the structure.

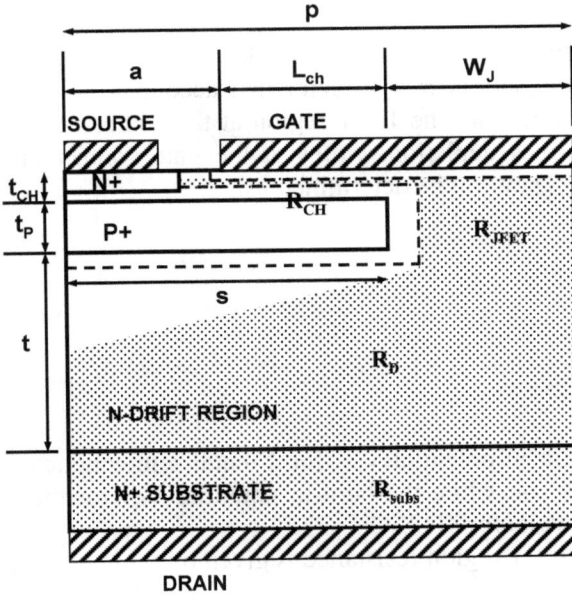

Fig. 7.28 Resistances in a Planar Gate MESFET Structures.

The total specific on-resistance is given by:

$$R_{on,sp} = R_{CH} + R_{JFET} + R_D + R_{subs} \qquad [7.12]$$

where R_{CH} is the channel resistance, R_{JFET} is the resistance of the JFET region, R_D is the resistance of the drift region after taking into account current spreading from the JFET region, and R_{subs} is the resistance of the N^+ substrate. These resistances can be analytically modeled by using the current flow pattern indicated by the shaded regions in Fig. 7.28. In this figure, the depletion region boundaries have also been shown using dashed lines. The drain current flows through a channel region with a small cross-section before entering the JFET region. The current spreads into the drift region from the JFET region at a 45 degree angle and then becomes uniform. The dimension 'a' in Fig. 7.28 is decided by alignment tolerances used during device fabrication. A typical value of 0.5 microns has been assumed for the analysis in this chapter.

The channel resistance is given by:

$$R_{CH} = \frac{\rho_D(L_{CH} + \alpha W_P)p}{(t_{CH} - W_G - W_P)}$$

[7.13]

where t_{CH} and L_{CH} are the channel thickness and length as shown in Fig. 7.28. In this equation, α is a factor that accounts for current spreading from the channel into the JFET region at their intersection. W_G and W_P are the zero-bias depletion widths at the gate contact and P^+ regions, respectively. They can be determined using:

$$W_G = \sqrt{\frac{2\varepsilon_S V_{biG}}{qN_D}}$$

[7.14]

$$W_P = \sqrt{\frac{2\varepsilon_S V_{biP}}{qN_D}}$$

[7.15]

where the built-in potential V_{biG} for a metal-semiconductor gate contact is typically 1 volt while the built-in potential V_{biP} for the P^+ junction is typically 3.3 volts for 4H-SiC.

The JFET region resistance is given by:

$$R_{JFET} = \rho_D(t_{CH} + t_P - W_G)\left(\frac{p}{W_J - W_P}\right)$$

[7.16]

where p is the cell pitch.

The drift region spreading resistance can be obtained by using:

$$R_D = \rho_D\left(\frac{2p}{W_J - W_P}\right)\ln\left(\frac{2p}{W_J - W_P}\right) + \rho_D(t - s - W_P)$$

[7.17]

where t is the thickness of the drift region below the P^+ region and s is the width of the P^+ region.

The contribution to the resistance from the N^+ substrate is given by:

$$R_{subs} = \rho_{subs}.t_{subs}$$

[7.18]

where ρ_{subs} and t_{subs} are the resistivity and thickness of the substrate, respectively. A typical value for this contribution is 4 x 10^{-4} Ω-cm^2.

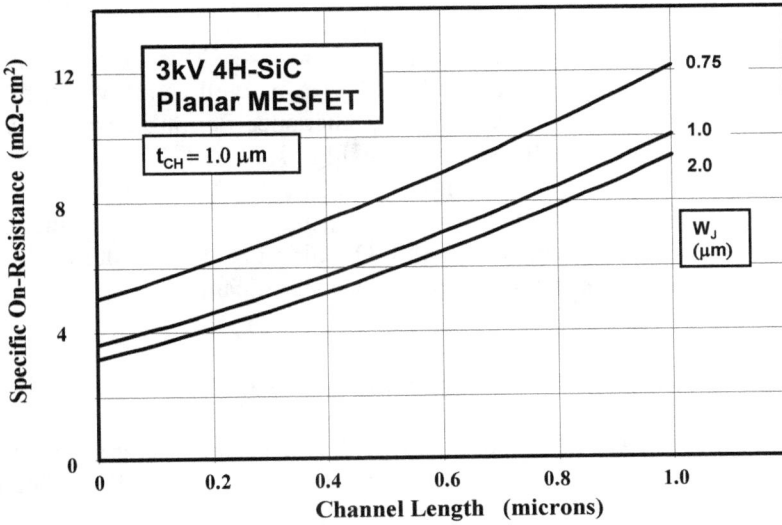

Fig. 7.29 Analytically Calculated On-Resistance for a 4H-SiC Planar MESFET.

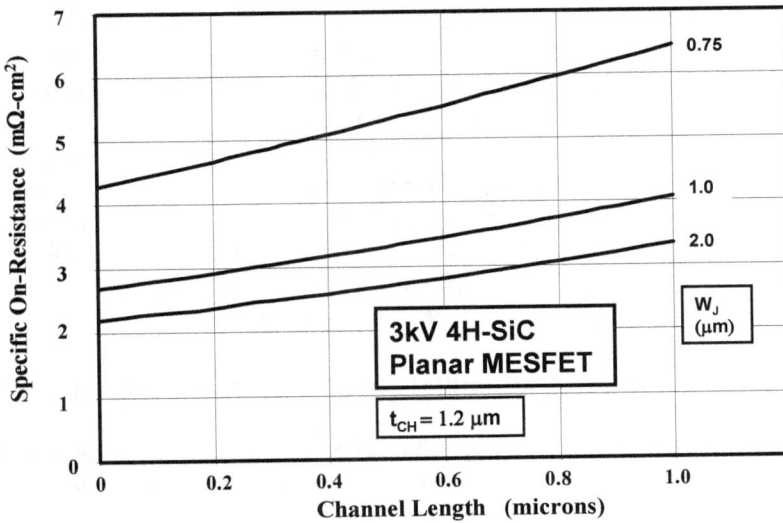

Fig. 7.30 Analytically Calculated On-Resistance for a 4H-SiC Planar MESFET.

Using the analytical expressions, it is possible to model the change in specific on-resistance with alterations of the cell design parameters. The specific on-resistance is most sensitive to variations of the channel length (L_{CH}) and thickness (t_{CH}), as well as the width of the JFET region (W_J). The variation of the specific on-resistance with increasing channel length is shown in Fig. 7.29 and 7.30 for cases of channel thickness of 1 and 1.2 microns, respectively. It can be seen that opening up the channel thickness reduces the specific on-resistance and its dependence on the channel length. Opening the JFET width beyond 1 micron does not improve the on-resistance significantly

7.5 Planar MESFET Structure: Numerical Simulations

Numerical simulations of the planar MESFET structure were performed to examine the on-state and blocking characteristics in detail. The impact of changing cell parameters was determined for the case of drift region with doping concentration of 1×10^{16} cm^{-3} and thickness of 20 microns. As in the case of the trench MESFET structure, the development of the potential barrier is critical to the ability for these devices to support high drain voltages without current flow. In the planar MESFET, the potential barrier is formed at point C in the channel. Its magnitude is determined not only by the channel dimensions but also the JFET dimensions because this impacts the potential developed under the gate with increasing drain bias. Since the analytical model indicated that the on-resistance is sensitive to the channel length and thickness, these parameters were varied to examine their impact on overall device performance. In each case, the on-state resistance was extracted at zero gate bias, and the blocking characteristics were obtained by application of various negative gate bias voltages. A gate metal work-function of 5.2 eV was used corresponding to a barrier height of 1.5 eV.

7.5.1 Planar MESFET Structure: On-State Characteristics

The on-state resistance of the planar MESFET structure was determined with the sub-surface P$^+$ region connected to the source electrode and the gate held at zero potential. First, consider the case of a planar MESFET with a JFET region width (W_J) of 1 micron. The width of the N$^+$ source region and its spacing from the gate edge (total dimension 'a' in Fig. 7.28) was kept constant for all structures at a value of 0.5 microns. The impact of changing the channel length (L_{CH}) was studied by using values of 0, 0.25, 0.75, and 1 micron, while keeping channel thickness (t_{CH}) of 1 micron between the P^+ region and the gate metal.

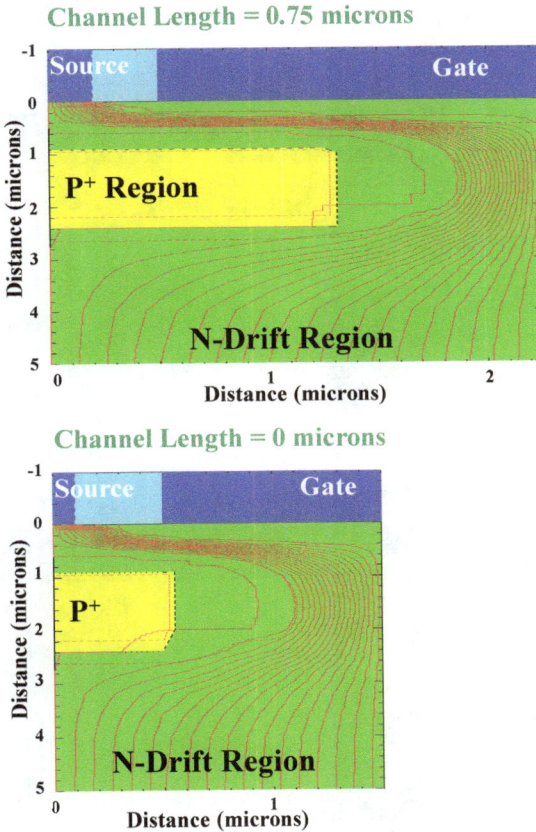

Fig. 7.31 On-State Current Flow in 4H-SiC Planar MESFETs:

The on-state current flow patterns for two of the structures at zero gate bias are shown in Fig. 7.31. It can be seen that the current flow is severely constricted in the channel of the device with channel length of 0.75 microns (see upper part of Fig. 7.31) because of the extension of depletion regions from both the P^+ region and the gate contact. Consequently, the channel contribution becomes significant unless the channel length is small. Even for the extreme case of a device structure with zero channel length, shown in the lower part of Fig. 7.31, there is still some constriction of the current flow in the channel region because of the lateral extension of the depletion region from the P$^+$ region. The current then flows into the JFET region and spreads rapidly (at an angle of 45 degrees) into the drift region eventually becoming uniform at a depth of 4 microns from the top surface.

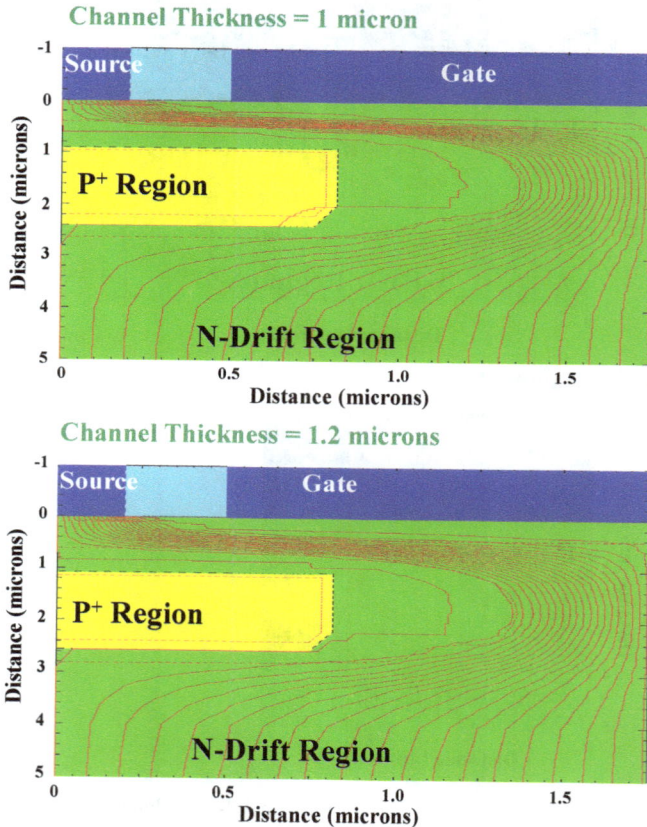

Fig. 7.32 On-State Current Flow in 4H-SiC Planar MESFETs.

Another method to reducing the channel resistance is by increasing the thickness of the channel. This provides a larger cross-section in the channel for the transport of the current as shown in Fig. 7.32. The improved current pattern produces a reduction of the specific on-resistance from 4.25 mΩ-cm^2 to 2.87 mΩ-cm^2. This is in good agreement with the predictions of the analytical model previously discussed in this chapter. The impact of the change in channel aspect ratio on the blocking gain for these structures is discussed in the next section.

In the simulations, the specific on-resistance for the device with 0.25 micron channel length was found to be 4.25 $m\Omega$-cm^2. This is in reasonable agreement with the value of 4.9 $m\Omega$-cm^2 obtained using the analytical model. However, it is about 4 times larger than the ideal specific on-resistance for the drift region for a structure capable of supporting 3000 volts. When the channel length was reduced to zero, the specific on-resistance was found to reduce to 2.93 $m\Omega$-cm^2. The impact of the channel length reduction on the blocking characteristics is discussed in the next section

7.5.2 Planar MESFET Structure: Blocking Characteristics

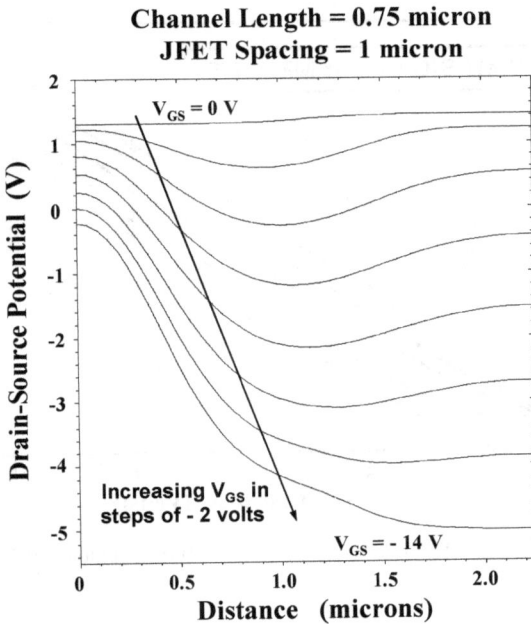

Fig. 7.33 Channel Potential Barrier in a 3 kV 4H-SiC Planar MESFET.

In the planar MESFET structure, the ability to block current flow with positive drain voltage can be achieved by application of a negative gate bias. The gate bias depletes the space between the gate region and the sub-surface P$^+$ region, which produces a potential barrier to the flow of electrons from the source to the drain. A larger potential barrier can be formed with larger gate bias voltage as discussed earlier. The behavior of the planar MESFET structure is discussed here with the aid of results from two-dimensional numerical simulations.

In the MESFET structure, electron injection can occur over the portion of the channel with the smallest potential barrier leading the on-set of current flow. The smallest potential barrier occurs at the center of the channel. For this reason, the potential variation along the center of the channel region (i.e. at y = 0.5 microns in Fig. 7.31 and Fig. 7.32) is shown in Fig. 7.33. Since the channel is nearly depleted at zero gate bias, a potential barrier develops in the channel even for a gate bias of -2 volts. The magnitude of the potential barrier increases with gate bias and reaches a value of -3 volts a gate bias of -10 volts.

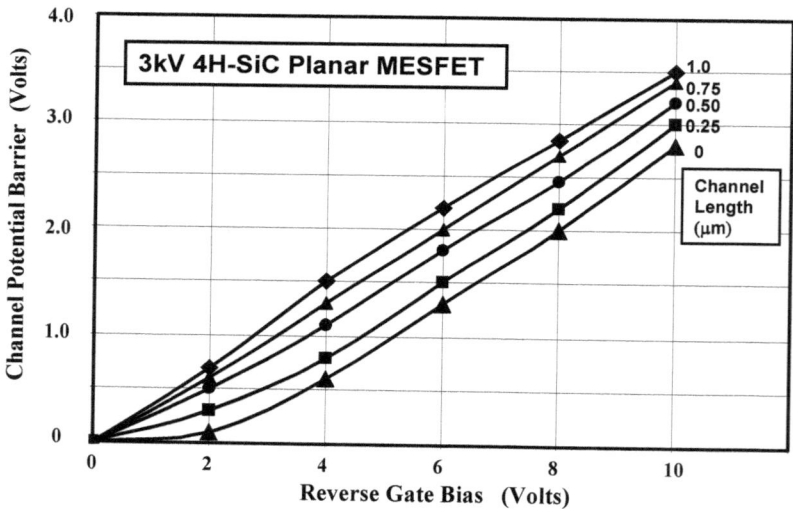

Fig. 7.34 Channel Potential Barriers in 3 kV 4H-SiC Planar MESFETs.

The magnitudes of the channel potential barrier for the cases of changes in the channel length ranging from 0 to 1 micron are shown in Fig. 7.34. Once the potential barrier is formed, a linear increase in the barrier height is observed for all cases with increasing gate bias. The potential barrier height is relatively small (only -3 volts at a reverse gate bias of -10 volts) when compared with the barrier heights observed in the trench MESFET structure. However, this is found to be adequate for providing a high blocking voltage capability because of shielding of the channel region by the JFET region in the planar MESFET structure. The magnitude of the potential barrier height created by the gate bias was found to be insensitive to the width of the JFET region.

The blocking voltage capability of the MESFET structure is dependent upon its ability to retain a potential barrier to the transport of electrons from the source to the drain as the drain bias is increased. The relatively small change in the barrier height with increasing drain voltage observed for the device structure with channel length of 0.75 microns and JFET spacing of 1 micron is shown in Fig. 7.35 when a channel thickness of 1 micron is used. A potential barrier is retained even when the drain bias is increased to 3000 volts indicating the ability to support high blocking voltages.

A larger reduction of the potential barrier is observed when the channel length is reduced from 0.75 microns to 0.25 microns as shown in Fig. 7.36. In this case, the potential barrier becomes small when the drain bias exceeds 1500 volts indicating problems with supporting high drain bias voltages without electron injection induced current flow.

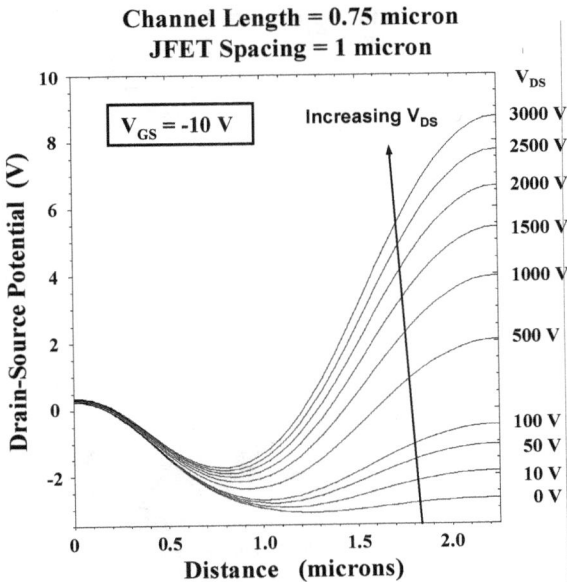

Fig. 7.35 Channel Potential distribution in a 3 kV 4H-SiC Planar MESFET.

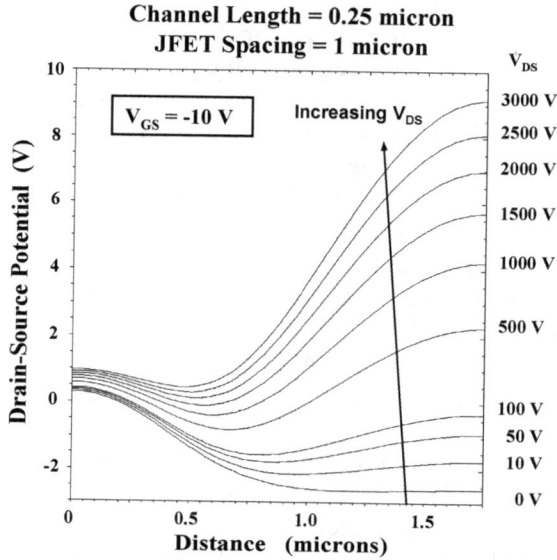

Fig. 7.36 Channel Potential distribution in a 3 kV 4H-SiC Planar MESFET.

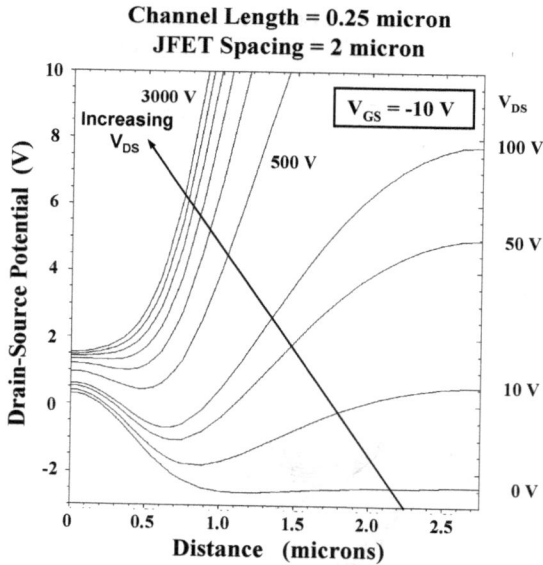

Fig. 7.37 Channel Potential distribution in a 3 kV 4H-SiC Planar MESFET.

The reduction of the potential barrier with increasing drain bias is relatively small for the above structures in which the channel is well shielded by the JFET regions whose width is 1 micron. When the width of the JFET region is enlarged to 2 microns, the shielding becomes much less effective as shown in Fig. 7.37. This structure has a much lower blocking gain than the structure with a JFET width of 1 micron. As discussed later in this section, the gate bias required to support 3000 volts for the structure with JFET width of 2 microns approaches the gate-source breakdown voltage. This compromises its ability to provide the full blocking voltage capability of the drift region

The reduction of the potential barrier height with increasing drain voltage is shown in Fig. 7.38 for the case of planar MESFETs with channel lengths of 0.25 and 0.75 microns with the same JFET width of 1 micron. It can be seen that the barrier height reduces to a greater extent when the channel length is smaller. When the JFET width is increased to 2 microns with a channel length of 0.25 microns, the potential barrier is completely removed at a drain bias of above 1500 volts. Thus, this structure cannot support high drain voltages with a gate bias of -10 volts.

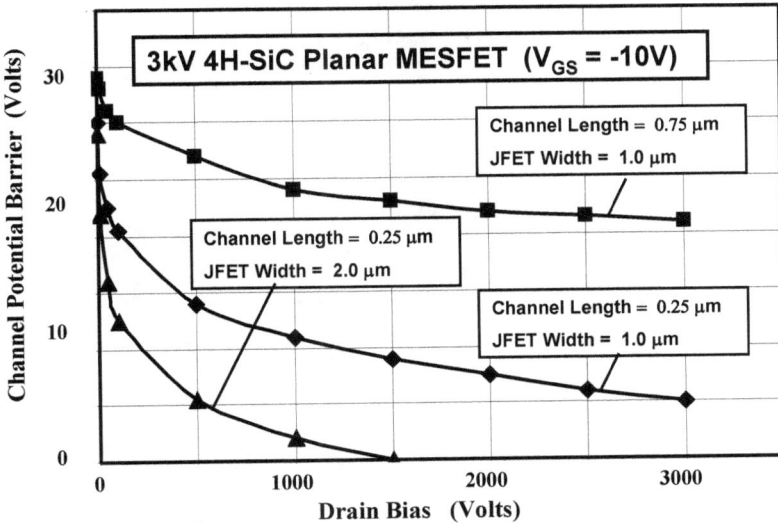

Fig. 7.38 Channel Potential Barriers in 3 kV 4H-SiC Planar MESFETs.

In general, the normally-on MESFET structure can support a larger drain bias voltage without significant drain current flow if the gate voltage is increased. However, the reverse gate bias cannot be arbitrarily increased because of the on-set of gate-source breakdown. The breakdown voltage between the gate and source regions is decided the depletion layer reach-through between the gate and the P^+ region. For a channel thickness of 1 micron, the gate breakdown voltage was found to be -12 volts. When the channel thickness was increased to 1.2 microns, the breakdown voltage improved to -16 volts.

When the gate bias is changed, the on-set of drain current up to any given leakage current level occurs at different drain bias voltages. This produces triode-like characteristics between the drain current and voltage. A typical set of I-V characteristics is shown in Fig. 7.39 for the case of the structure with channel length of 0.25 microns, channel thickness of 1 micron and JFET width of 1 micron. The structure is able to support a drain bias of over 3000 volts with a gate bias of only -10 volts. Thus, the planar MESFET structure exhibits an extremely high blocking gain even for a short channel length. In fact, the structure with channel length of 0.75 microns could support a drain bias of over 3000 volts with a gate bias of only -4 volts.

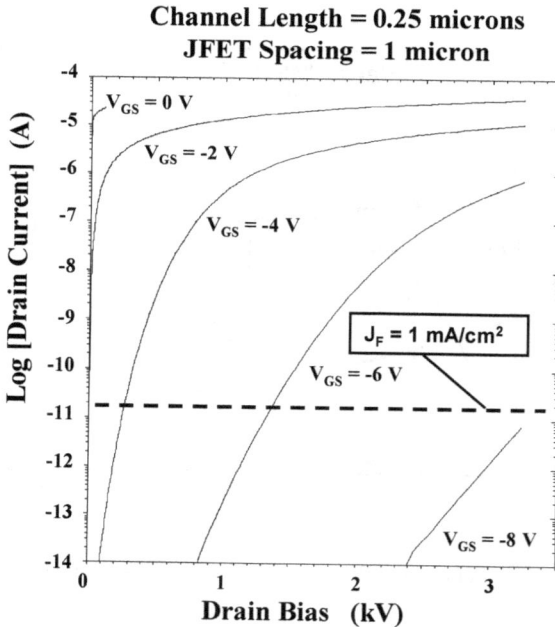

Fig. 7.39 Blocking Characteristics for a 3 kV 4H-SiC Planar MESFET.

Channel Length = 0 microns
JFET Spacing = 1 micron

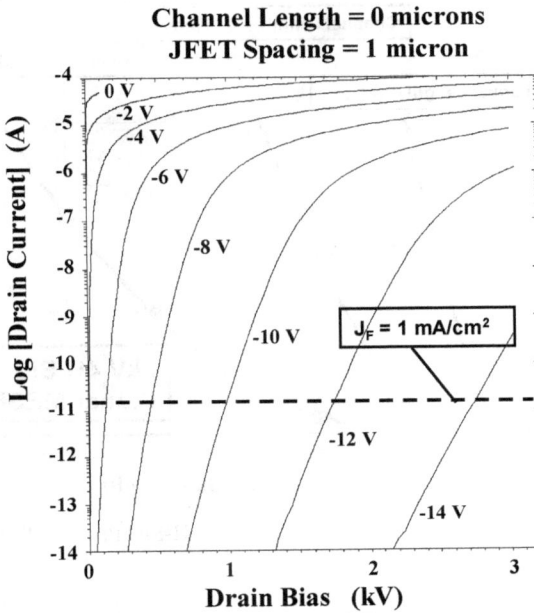

Fig. 7.40 Blocking Characteristics for a 3 kV 4H-SiC Planar MESFET.

When the channel length is reduced to zero, the blocking gain deteriorates as shown in Fig. 7.40 and the structure is unable to support a drain bias of 3000 volts unless the gate voltage is increased beyond -14 volts. Under these gate bias conditions, there is substantial gate current flow due to the on-set of reach-through induced breakdown. This demonstrates the need to retain some overlap between the edge of the sub-surface P$^+$ region and the edge of the gate electrode during the design of the planar MESFET structure. The amount of the overlap, which determines the channel length, must be decided by performing a trade-off between minimizing the on-state resistance and maximizing the blocking gain.

The *drain blocking voltage* at a leakage current density of 1 mA/cm^2 is plotted in Fig. 7.41 for the planar MESFET structure with channel thickness of 1 micron and JFET spacing of 1 micron as a function of the magnitude of the reverse gate bias. From this graph, the need to provide a small (0.25 micron) overlap between the P$^+$ region and the gate contact is quite clear.

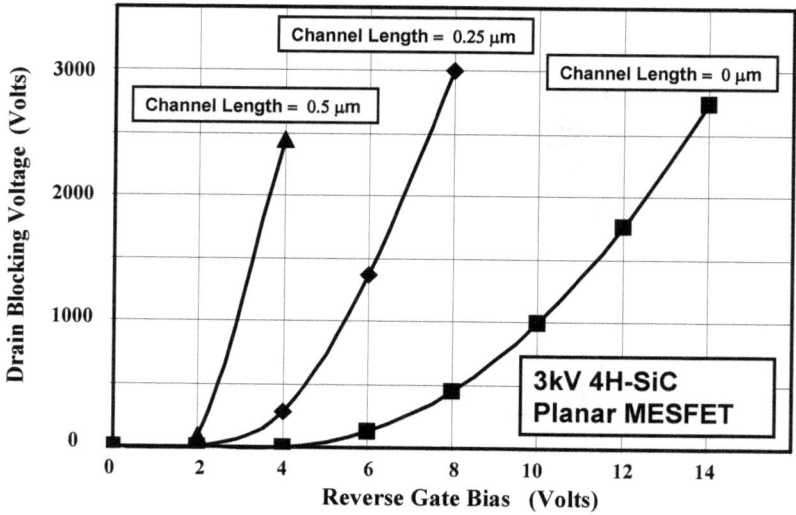

Fig. 7.41 Drain Blocking Voltage for 3 kV 4H-SiC Planar MESFETs.

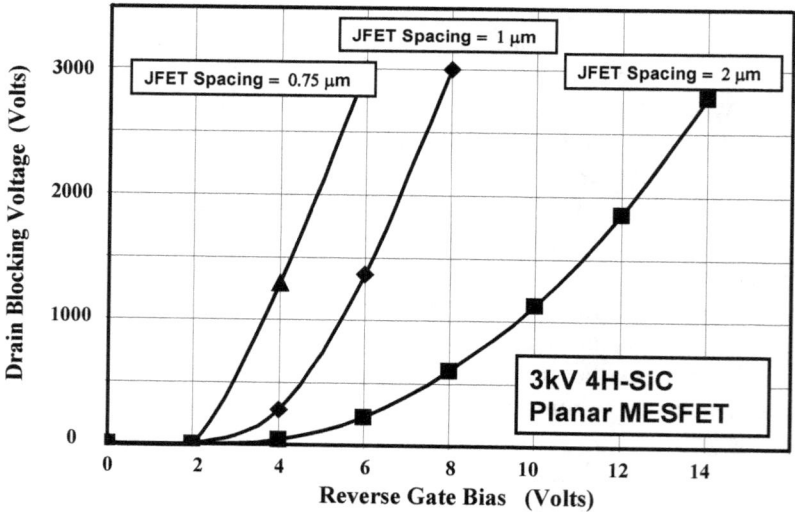

Fig. 7.42 Drain Blocking Voltage for 3 kV 4H-SiC Planar MESFETs.

Similarly, the impact of changing the JFET spacing is shown in Fig. 7.42 for the case of a channel length of 0.25 microns. When the JFET spacing is increased to 2 microns, the drain potential encroaches into the channel and the drain voltage blocking capability deteriorates. The blocking voltage capability of the planar MESFET structures becomes limited by the on-set of gate breakdown to about 2500 volts. When the JFET spacing is reduced to 0.75 microns, the channel becomes extremely well shielded and the structure is able to support a drain bias of over 3000 volts with a gate bias of only -6 volts.

The voltage blocking behavior of the planar MESFET structure can be examined by looking at the blocking gain as a function of the channel length. The DC blocking gain is defined as the ratio of the drain blocking voltage to the corresponding gate bias. This gain varies with drain and gate bias as observed in the above figures. If the gain is taken by using the gate voltage required to support a drain bias of 3000 volts, then the various structures can be compared in terms of the channel length. This is done in Fig. 7.43 for the planar MESFET structures simulated with various channel lengths using a channel thickness of 1 micron and a JFET spacing of 1 micron. An exponential increase in blocking gain is observed with increasing channel length. For completeness, the changes in the blocking gain with variations in the JFET spacing are included in this figure for the case of a channel length of 0.25 microns.

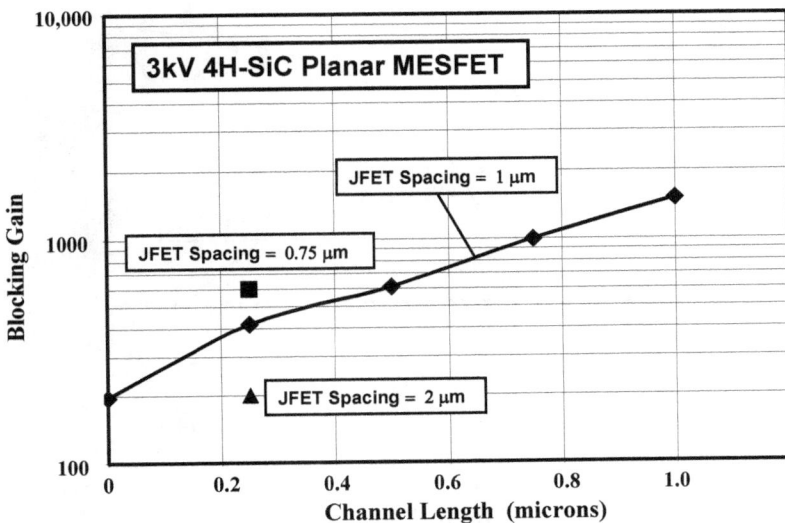

Fig. 7.43 Blocking Gain for 3 kV 4H-SiC Planar MESFETs.

In the previous section, it was demonstrated that the on-resistance for the planar MESFET structure could be significantly reduced by increasing the channel thickness. The impact of this change on the blocking characteristics is shown in Fig. 7.44 for the case of a channel thickness of 1.2 microns when the channel length is 0.25 microns and the JFET spacing is 1 micron. It can seen that this structure requires a gate bias of -14 volts to block a drain bias of 3000 volts in contrast with only -8 volts required for the structure with channel thickness of 1 micron (whose characteristics are shown in Fig. 7.39). The blocking gain is therefore reduced from about 400 to about 200. However, this is a good compromise because the structure has 50% lower specific on-resistance and still needs only 14 volts to block a drain bias of 3000 volts. This gate bias is also compatible with utilization of the planar MESFET in the *Baliga-Pair* circuit configuration with a silicon power MOSFET rated with blocking voltage of 20 to 30 volts.

Channel Length = 0.25 microns
Channel Thickness = 1.2 microns

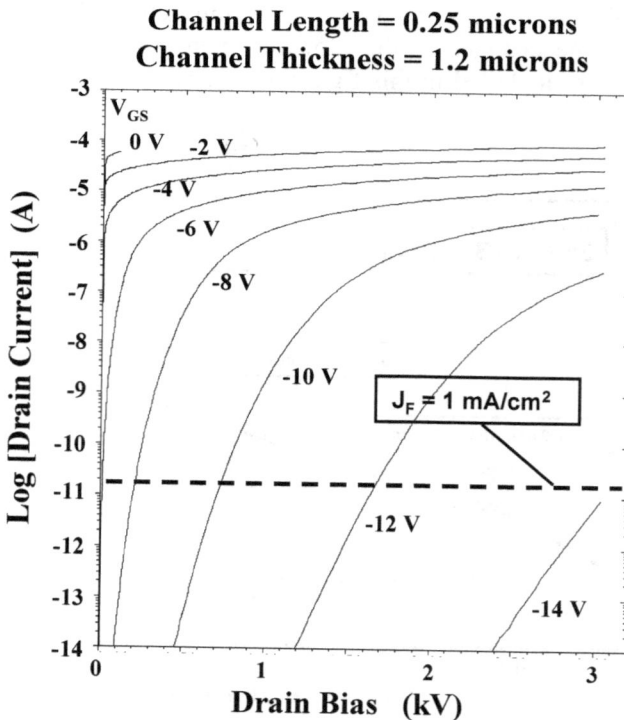

Fig. 7.44 Blocking Characteristics for a 3 kV 4H-SiC Planar MESFET.

7.5.3 Planar MESFET Structure: Output Characteristics

The output characteristics of the planar MESFET structure determine the safe operating area of the device. In a typical pulse width modulated control scheme, the power transistor undergoes a switching transient from either the on-state to off-state or vice versa. The loci for these transients must be kept within the bounds of the safe operating area for the transistor.

Channel Length = 0.25 microns
JFET Spacing = 1 micron

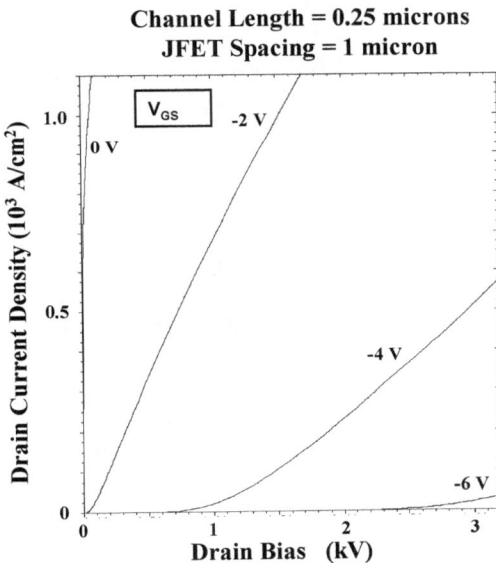

Fig. 7.45 Output Characteristics for a 3 kV 4H-SiC Planar MESFET.

As expected from the previous discussion, the channel design has a strong impact on the output characteristics of the planar MESFET structure. For the structure with channel length of 0.25 microns, channel thickness of 1 micron, and JFET spacing of 1 micron, the output characteristics exhibit a mixed triode-pentode behavior as shown in Fig. 7.45. When the channel length is reduced to zero, the characteristics are broadened to higher current levels as shown in Fig. 7.46. However, a larger gate bias is needed to traverse the safe-operating-area for this device. A similar outcome is observed when the channel thickness is increased from 1 micron to 1.2 microns with a channel length of 0.25 microns as shown in Fig. 7.47.

Channel Length = 0 microns
JFET Spacing = 1 micron

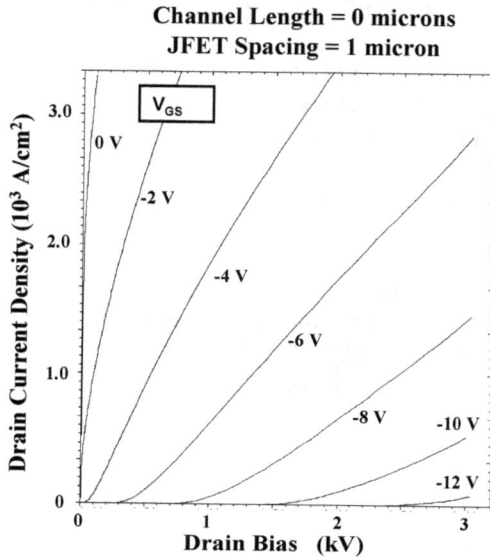

Fig. 7.46 Output Characteristics for a 3 kV 4H-SiC Planar MESFET.

Channel Length = 0.25 microns
Channel Thickness = 1.2 microns

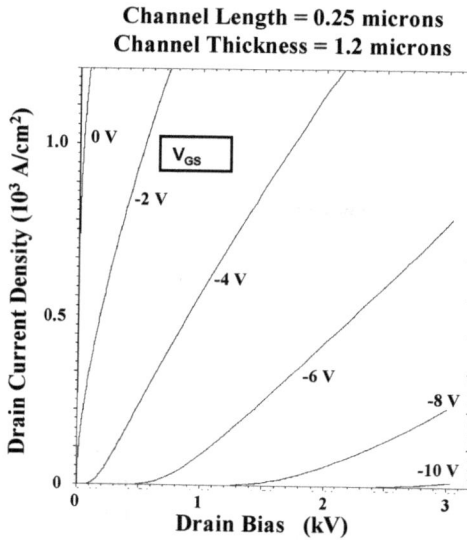

Fig. 7.47 Output Characteristics for a 3 kV 4H-SiC Planar MESFET.

7.5.4 Planar MESFET Structure: Gate Contact Shielding

One of the key advantages of the planar MESFET structure is the ability to shield the gate contact from the high electric fields developed within the semiconductor when the device is biased close to the breakdown voltage capability of the drift region. In the planar MESFET structure, the depletion of the JFET region (location D in Fig. 7.27) at a relatively low drain bias voltage produces a potential barrier. This barrier prevents increase in the electric field at the semiconductor surface under the gate contact (location B in Fig. 7.27) with further increase in the drain bias voltage. Consequently, the maximum electric field in the semiconductor occurs just below the bottom of the sub-surface P^+ region.

JFET Spacing = 1 micron

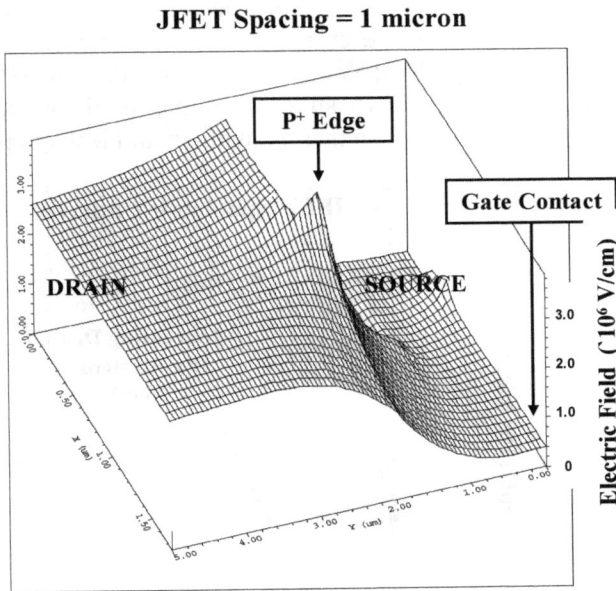

Fig. 7.48 Electric Field within a 3 kV 4H-SiC Planar MESFET.

A three-dimensional view of the electric field is shown in Fig. 7.48 for a structure with JFET spacing of 1 micron at a drain bias of 3000 volts and a gate bias of -10 volts. The highest electric field is observed at the bottom of the sub-surface P^+ region with some enhancement at its corner as indicated in the figure. The electric field at the semiconductor surface is low under the entire gate contact interface due to the shielding provided by the junction. The development of the electric field at the gate contact interface can be best observed at the right hand edge of the cell structure shown in Fig. 7.27 where it has the highest value. The change in the electric field at various drain bias voltages is shown in Fig. 7.49 for the case of a JFET spacing of 1 micron. As expected, this field distribution was found to be independent of the channel length and thickness as long as the JFET spacing is kept at 1 micron. It can be clearly seen that that the largest electric field at the gate contact interface (E_M(Surface)) is much smaller than the maximum electric field (E_M(Bulk)) in the bulk of the semiconductor. The electric field at the gate contact interface remains below 4×10^5 V/cm even for a drain bias of 3000 volts while the electric field in the bulk approaches the breakdown strength of 3×10^6 V/cm. The low electric field at the gate contact prevents high leakage currents associated with Schottky barrier lowering and tunneling phenomena. These results demonstrate the ability to fully utilize the breakdown strength of the semiconductor material without encountering a gate contact breakdown and leakage problem.

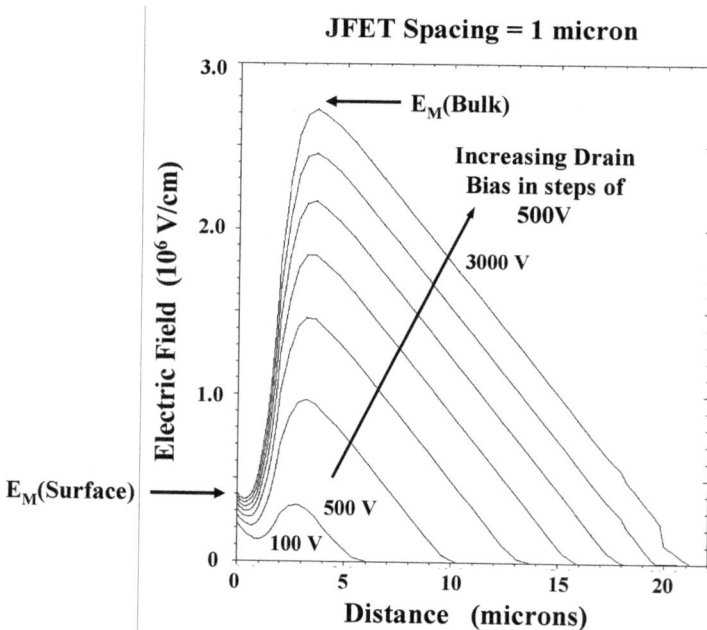

Fig. 7.49 Electric Field within a 3 kV 4H-SiC Planar MESFET.

The suppression of the electric field at the gate contact interface is strongly dependent upon the JFET spacing because this determines the drain bias at which a potential barrier is created in region D of the planar MESFET structure. The impact of changes to the JFET spacing on the electric field within the planar MESFET structure is shown in Fig. 7.50. For all the structures, the electric field at the gate contact interface is smaller than the maximum electric field developed in the bulk just below the bottom of the P^+ region. Considerable reduction of the gate contact field is observed when the JFET spacing is reduced from 2 to 1 micron. The improvement obtained with reducing the JFET spacing to 0.75 microns is not substantial. This indicates that a JFET spacing of 1 micron is optimal for this structure to suppress the electric field in the gate contact while simultaneously obtaining a low specific on-resistance in the on-state.

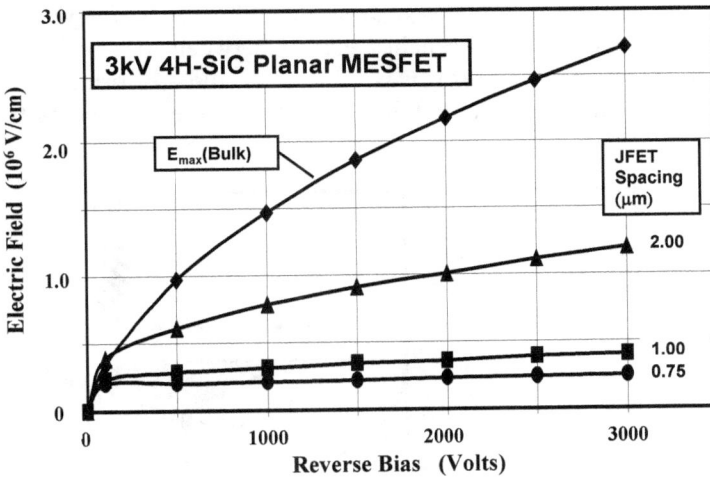

Fig. 7.50 Electric Field within 3 kV 4H-SiC Planar MESFETs.

7.6 Experimental Results: Trench Gate Structures

Several approaches can be taken to construct a vertical FET structure that utilizes a potential barrier along the vertical path induced by a gate bias. The gate region in the device can be formed by using ion implantation of P-type dopants to form a JFET structure or by placing a metal contact within a trench etched between the source regions to form a MESFET structure or by forming a hetero-junction gate region within the trench to form a HJFET structure. Prior to the development of a process for formation of heavily doped P-type regions in silicon carbide, it was more practical to construct the hetero-junction gate FET structure.

The formation of a hetero-junction between P-type polysilicon and N-type silicon carbide was first demonstrated at PSRC in 1996[12]. The band structure for the interface between P^+ Polysilicon and N-type 6H-SiC is shown in Fig. 7.51. Good rectification was experimentally confirmed at this hetero-junction[13] allowing consideration of this junction for the gate region in a HJFET structure, shown in Fig. 7.52. The fabrication process for this structure is simple because the polysilicon gate material can be deposited into the trenches and planarized due to the good selectivity between it and SiC during reactive-ion-etching. The HJFET structure was analyzed in detail in 1996[14] followed by experimental demonstration[15] of a structure with the P^+ polysilicon located within trenches etched between the source regions. Although some gate control was observed, the performance of the structure was poor due to the bad quality of the surface within the trenches after etching.

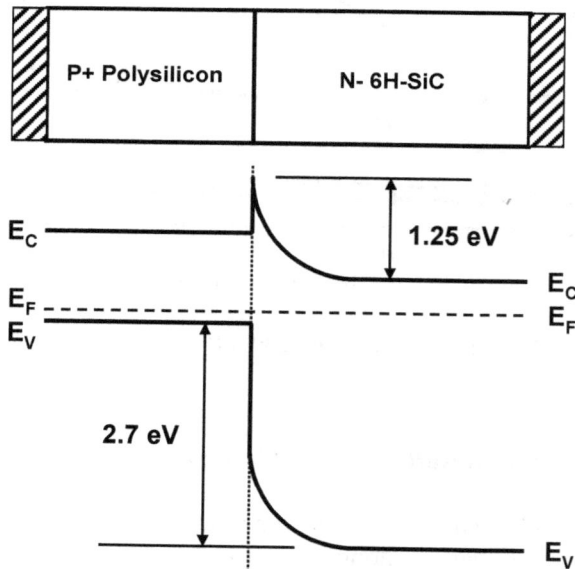

Fig. 7.51 P^+ Polysilicon/N- 6H-SiC Hetero-Junction Band Structure.

A trench JFET structure that contains a MOS sidewall (also illustrated in Fig. 7.52) has been proposed and demonstrated[16]. This structure requires ion implantation of the P^+ region at the bottom of the trenches after they have been etched. Since the authors masked the P^+ implant, the process is difficult to implement due to the poor topology for patterning photoresist in the presence of 1 micron deep trenches. In addition, it is not clear that the P-type implant will not occur on the trench sidewalls, an effect disregarded by the authors. The authors deposited an oxide after the P^+ implant and then refilled the trench with polysilicon to create an MOS-structure on the trench sidewalls. The process described by the authors precludes making contact to the P^+ gate regions along the trenches. Since the contact to the P^+ gate regions must be located at the periphery of the device, the structure would have poor switching characteristics due to the very high resistance of the P^+ fingers orthogonal to the cross-section. The authors report obtaining a specific on-resistance of 5 mΩ-cm^2 for a device that is able to block 600 volts using a reverse gate bias of -30 volts. A similar structure with contact made to the P^+ region by the P^+ polysilicon has been proposed[17] but the authors failed to describe how such a contact can be fabricated with low contact resistance.

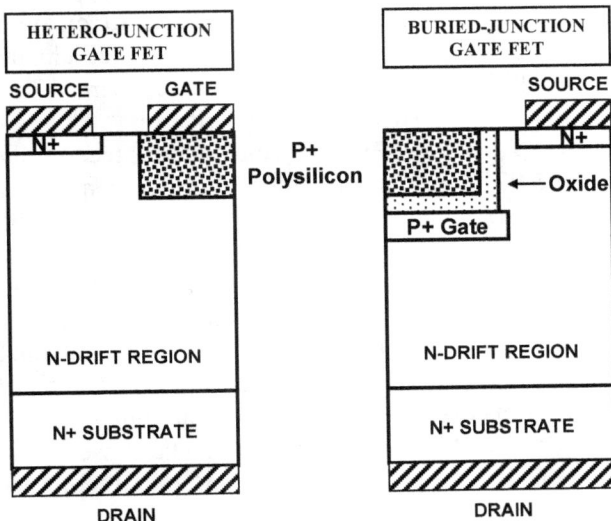

Fig. 7.52 Novel Vertical Trench Gate SiC FET Structures.

The vertical JFET structure can also be fabricated using ion implantation to form the P^+ gate region. In order to obtain a good channel aspect ratio, it is necessary to resort to very high energy (1.3 MeV) aluminum implants to form the P^+ gate regions[18]. The authors obtained a blocking voltage of 2000 volts with a reverse gate bias of -50 volts (blocking gain of 40). However, the on-resistance for the devices was very high unless a positive gate bias of 2.5 volts was applied. With the positive gate bias, the gate depletion region was reduced allowing drain current flow through the channel with a specific on-resistance of 70 mΩ-cm^2. These values indicate poor optimization of the structure which had a very large cell pitch of 32 microns.

7.7 Experimental Results: Planar Gate Structures

The planar gate FET structure was originally proposed and patented in 1996 with either a metal-semiconductor, or a junction, or an MOS gate region[9]. It was proposed that the sub-surface P^+ region be formed by using ion-implantation with appropriate energy to locate the junction below the surface so as to create an un-depleted N-type channel region. An obvious alternative approach is to grow an N-type epitaxial layer over the P^+ regions implanted into a substrate to create the N-type channel region. Experimental results on the lateral channel, vertical (planar) MESFET/JFET structures are discussed here. The MOS gate structures are discussed in subsequent chapters.

The first planar MESFET structures were successfully fabricated at PSRC in 1996-97 by performing 380 keV boron ion-implants to form the sub-surface P^+ region[19]. Devices were fabricated from both 6H-SiC and 4H-SiC as the starting material with doping concentration of about 2 x 10^{16} cm^{-3}. The energy for the boron implants was chosen to locate the center of the P^+ region at about 0.5 microns below the surface[20]. An additional N-type nitrogen implant was used in the channel region to enhance its doping to produce normally-on devices. Contact to the sub-surface P^+ region was made by additional boron implants at lower energy in selective regions within the cell structure. The devices were found to exhibit a low specific on-resistance of about 12 mΩ-cm^2 but had poor gate–drain breakdown voltage of about 50 volts.

The planar JFET structures that were also fabricated at PSRC in 1996-97 using the same process conditions described above had much better blocking capability[21]. In these devices, the P^+ gate region was formed by using a shallow 10 keV boron implant. The 4H-SiC planar JFETs fabricated using a N-type channel implant with a dose of 1-2 x 10^{13} cm^{-2} exhibited a specific on-resistance of only 11-14 mΩ-cm^2. The devices were able to block a drain bias of 1100 volts with a negative gate bias of 40 volts. Excellent gate controlled pentode-like characteristics with drain current saturation was observed in the devices upto the breakdown voltage with a drain current density of 250 A/cm^2.

Lateral channel, vertical JFET structures, fabricated by epitaxial growth of an N-type layer over a P$^+$ region implanted into the drift region, have also been reported by several groups. The first such devices were reported[11] in 1999 by using epitaxial layers capable of supporting 600 V and 1200 V. By growing a epitaxial layer with doping concentration of about 2 x 10^{16} cm^{-3} with a thickness of 2.5 microns over the sub-surface P$^+$ region, the devices were able to support a gate bias of more than 10 volts above a channel pinch-off voltage of 40 volts. This was sufficient to allow blocking a drain bias of 1200 volts. The specific on-resistance for the devices was 18, 25, and 40 mΩ-cm^2 for devices capable of supporting 550, 800, and 950 volts, respectively. The performance of these structures was subsequently improved[22] to a blocking voltage of 1800 volts with a specific on-resistance of 24.5 mΩ-cm^2.

7.8 Summary

The physics of operation of the normally-on vertical JFET/MESFET structure has been described in this chapter. It is demonstrated that a potential barrier forms in the channel under reverse gate bias which prevents the flow of drain current. The devices exhibit mixed triode-pentode like characteristics. The channel for these devices can be oriented along the vertical drain current flow path by using a trench gate architecture. Alternately, a lateral channel can be formed by introducing a sub-surface P$^+$ region below a planar gate design. The planar devices have been successfully made using either a deep P$^+$ ion implantation step into the N-drift region or by growth of an N-type epitaxial layer above a previously implanted P$^+$ region in the N-type drift region. Both methods have resulted in devices with low specific on-resistance and good gate controlled current saturation capability with relatively low reverse gate bias voltages. These devices are suitable for utilization in the *Baliga-Pair* configuration discussed in the next chapter.

References

[1]   J. Nishizawa, T. Terasaki, and J. Shibata, "Field Effect Transistor versus Analog Transistor (Static Induction Transistor)", IEEE Transactions on Electron Devices, Vol. ED22, pp. 185-197, 1975.

[2]   B. J. Baliga, "Modern Power Devices", John Wiley and Sons, 1987.

[3]   B. J. Baliga, "Evolution of MOS-Bipolar Power Semiconductor Technology", Proceedings of the IEEE, Vol. 74, pp. 409-418, 1988.

[4]   B. J. Baliga, "Silicon Carbide Switching Device with Rectifying Gate", U. S. Patent 5,396,085, Issued March 7, 1995.

[5]   B. J. Baliga, "Power Semiconductor Devices", PWS Publishing Company, 1996.

[6]   P. Plotka and B. Wilamowski, "Interpretation of Exponential type Drain Characteristics of the Static Induction Transistor", Solid State Electronics, Vol. 23, pp. 693-694, 1980.

[7]   B. J. Baliga, "A Power Junction Gate Field Effect Transistor Structure with High Blocking Gain", IEEE Transactions on Electron Devices, Vol. 27, pp. 368-373, 1980.

[8]   X. C. Kun, "Calculation of Amplification Factor of Static Induction Transistors", IEE Proceedings, Vol. 131, pp. 87-93, 1984.

[9]   B. J. Baliga, "High Voltage Junction Gate Field Effect Transistor with Recessed Gates", IEEE Transactions on Electron Devices, Vol. 29, pp. 1560-1570, 1982.

[10]   B. J. Baliga, "Silicon Carbide Semiconductor Devices having Buried Silicon Carbide Conduction Barrier Layers Therein", U. S. Patent 5,543,637, Issued August 6, 1996.

[11]   H. Mitlehner, et al, "Dynamic Characteristics of High Voltage 4H-SiC Vertical JFETs", IEEE International Symposium on Power Semiconductor Devices and ICs, Abstract 11.1, pp. 339-342, 1999.

[12]   P. M. Shenoy and B. J. Baliga, "High Voltage P+ Polysilicon/N- 6H-SiC Heterojunction Diodes", PSRC Technical Report TR-96-050, 1996.

[13]   P. M. Shenoy and B. J. Baliga, "High Voltage P+ Polysilicon/N- 6H-SiC Heterojunction Diodes", Electronics Letters, Vol. 33, pp. 1086-1087, 1997.

[14]   B. Vijay, K. Makeshwar, P. M. Shenoy, and B. J. Baliga, "Analysis of a High Voltage Heterojunction Gate SiC Field Effect Transistor", PSRC Technical Report TR-96-049, 1996.

[15]   P. M. Shenoy, V. Bantval, M. Kothandaraman, and B. J. Baliga, "A Novel P+ Polysilicon/N- SiC Heterojunction Trench Gate Vertical FET",

IEEE International Symposium on Power Semiconductor Devices and ICs, pp. 365-368, 1997.

[16] R. N. Gupta, H.R. Chang, E. Hanna, and C. Bui, "A 600 V SiC Trench JFET". Silicon Carbide and Related Materials – 2001, Material Science Forum, Vol. 389-393, pp. 1219-1222, 2002.

[17] L. Zhu and T. P. Chow, "Design and Processing of High Voltage 4H-SiC Trench Junction Field Effect Transistor", Silicon Carbide and Related Materials – 2001, Material Science Forum, Vol. 389-393, pp. 1231-1234, 2002.

[18] H. Onose, et al, "2 kV 4H-SiC Junction FETs", Silicon Carbide and Related Materials – 2001, Material Science Forum, Vol. 389-393, pp. 1227-1230, 2002.

[19] P. Shenoy and B. J. Baliga, "The Planar Lateral Channel SiC MESFET", PSRC Technical Report TR-97-038, 1997.

[20] M. S. Janson, et al, "Range distributions of Implanted Ions in Silicon Carbide", Silicon Carbide and Related Materials – 2001, Material Science Forum, Vol. 389-393, pp. 779-782, 2002.

[21] P. Shenoy and B. J. Baliga, "A Planar Lateral Channel SiC Vertical High Power JFET", PSRC Technical Report TR-97-036, 1997.

[22] P. Friedrichs, et al, "Static and Dynamic Characteristics of 4H-SiC JFETs Designed for Different Blocking Categories", Silicon Carbide and Related Materials – 1999, Material Science Forum, Vol. 338-342, pp. 1243-1246, 2000.

Chapter 8

The Baliga-Pair Configuration

The advent of semiconductor based power electronics began with the availability of silicon bipolar devices in the 1950s. The development of bipolar power junction transistors for medium power applications and silicon gate turn-off thyristors for high power applications enabled significant enhancement in system performance by replacement of vacuum tubes. However, the low current gain (< 10) for structures designed to support high voltages resulted in large, expensive control circuits. The concurrent development of CMOS technology in the 1960s enabled the power semiconductor industry to introduce power MOSFETs in the 1970s. These devices rapidly captured the market for low power (< 100 watts) systems where the operating voltages are relatively low (< 100 volts). The metal-oxide-semiconductor (MOS) gate structure in these devices could be driven using small pulses of current with essentially negligible steady state currents. The low input power requirements enabled the subsequent integration of the control circuit resulting in a huge improvement in the size and cost of the system.

The relatively high on-resistance of silicon MOSFETs was a significant drawback for application in higher power systems that worked with high operating voltages (>200 volts). However, in the 1980s, a conceptual break-through enabled combining the physics of the MOS and Bipolar structures to create a new class of devices[1]. The commercial introduction of the Insulated Gate Bipolar Transistor (IGBT) with its high input impedance MOS-gate structure and high output power handling capability revolutionized medium and high power system design. As in the case of the power MOSFET, the IGBT could be controlled by using a small integrated drive circuit resulting in compact, low cost solutions for consumer, industrial, medical, and automotive applications. This has set the bench-mark that any new proposed technology must be able to surmount in order to be considered an attractive replacement for the existing silicon MOS-gated devices.

199

As had occurred during the evolution of silicon technology, the ability to produce a high quality interface between silicon carbide and a suitable gate dielectric material has been a significant challenge[2]. In addition to a larger density of charge in the oxide and at the interface that causes threshold voltage shift, the inversion layer mobility was found to be very low when compared with silicon. To compound the problem, the conventional silicon power MOSFET structure could not provide the full benefits of the high breakdown field strength of the silicon carbide material because of reliability and rupture problems associated with the enhanced electric field in the gate oxide.

In order to overcome these problems, it was proposed[3] that a normally-on silicon carbide high voltage JFET/MESFET be used together with a low voltage silicon MOSFET to create a compound switch with the desired features for a high quality power switch. In an analogy to the *Darlington-Pair* configuration[4] commonly used for power control applications, it was suggested[5] that the proposed combination of devices be named the *Baliga-Pair* configuration. The same circuit configuration has also been subsequently called the Cascode circuit[6] with the acknowledgement that it was first disclosed in a textbook in 1996[7]. The idea was also presented in 1996 at the Conference on Silicon Carbide and Related Devices[8] held in Kyoto, Japan.

This chapter discusses the operating principles of the Baliga-Pair configuration. It is demonstrated that a silicon power MOSFET with low breakdown voltage rating (and hence low specific on-resistance) can be used to control a high voltage, normally-on silicon carbide JFET/MESFET structure. This enables supporting large voltages within the silicon carbide FET while allowing control of the composite switch with signals applied to the MOS gate electrode of the silicon MOSFET. The same type of simple, low cost, integrated control circuits used for silicon power MOSFETs and IGBTs can therefore be utilized for the Baliga-Pair configuration. Since both devices in the configuration are unipolar devices, the Baliga-Pair has very fast switching speed and excellent safe-operating-area. In concert with the low on-resistance for both the FETs, the fast switching speed results in very low overall power dissipation in applications[9]. In addition, this configuration contains an excellent fly-back diode allowing replacement of not only the IGBT but also the fly-back rectifier that is usually connected across it in H-bridge power circuits. From this stand-point, it is preferable to use the silicon carbide MESFET structure due to lower on state voltage drop of the Schottky diode.

8.1 The *Baliga-Pair* Configuration

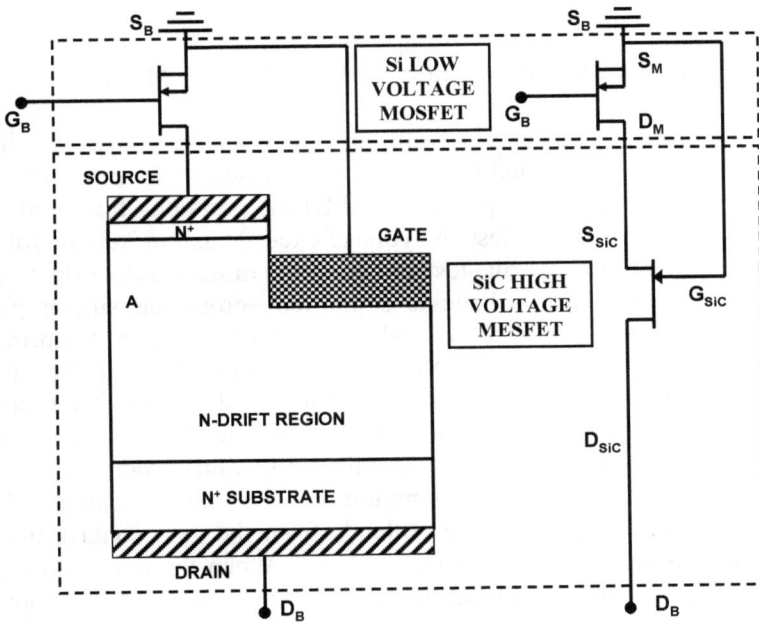

Fig. 8.1 The *Baliga-Pair* Power Switch Configuration.

The Baliga-Pair configuration consists of a low breakdown voltage silicon MOSFET and a normally-on, high voltage silicon carbide JFET/MESFET connected together as shown in Fig. 8.1. Any of the trench gate or planar gate JFET/MESFET structures discussed in the previous chapter can be used to provide the high blocking voltage capability. It is important that the silicon carbide FET structure be designed for normally-on operation with a low specific on-resistance. It is also necessary for the silicon carbide FET to be able to block the drain bias voltage with a gate bias less than the breakdown voltage of the silicon power MOSFET.

The silicon power MOSFET can be either a planar DMOS structure or a trench-gate UMOS structure to provide low specific on-resistance. The source of the silicon carbide FET is connected to the drain of the silicon power MOSFET. Note that the gate of the silicon carbide FET is connected directly to the reference or ground terminal. The path formed between the drain and the gate contact of the silicon carbide FET creates the fly-back diode. The composite switch is controlled by the signal applied to the gate of the silicon power MOSFET.

8.1.1 Baliga-Pair Configuration: Voltage Blocking Mode

The composite switch can block current flow when the gate of the silicon power MOSFET is shorted to ground by the external drive circuit. With zero gate bias, the silicon power MOSFET supports any bias applied to its drain terminal (D_M) unless the voltage exceeds its breakdown voltage. Consequently, at lower voltages applied to the drain terminal (D_B) of the composite switch, the voltage is supported across the silicon power MOSFET because the silicon carbide JFET is operating in its normally-on mode. However, as the voltage at the drain (D_M) of the silicon MOSFET increases, an equal positive voltage develops at the source (S_{SiC}) of the silicon carbide FET. Since the gate (G_{SiC}) of the silicon carbide FET is connected to the ground terminal, this produces a reverse bias across the gate-source junction of the silicon carbide FET. Consequently, a depletion layer extends from the gate contact/junction into the channel of the silicon carbide FET. When the depletion region pinches off the channel at location A, further increase in the bias applied to the drain (D_B) of the composite switch is supported across the silicon carbide FET. Since the potential at the source of the silicon carbide FET is then isolated from the drain bias applied to the silicon carbide FET, the voltage across the silicon power MOSFET is also clamped to a value close to the pinch-off voltage of the silicon carbide FET. This feature enables utilization of a silicon power MOSFET with a low breakdown voltage. Such silicon power MOSFETs have very low specific on-resistance with a mature technology available for their production. From this point of view, it is desirable to utilize silicon power MOSFETs with breakdown voltages of below 50 volts.

If the Baliga-Pair is designed to support a drain bias of 3000 volts, the ability to utilize a silicon power MOSFET with a breakdown voltage of 30 volts requires designing the silicon carbide FET so that the channel is pinched-off at a gate bias of below 20 volts. Thus, the blocking gain of the silicon carbide FET should be in excess of 150. This is feasible for both the trench-gate and planar gate architectures for silicon carbide FETs discussed in the previous chapter. As pointed out in that chapter, much larger blocking gains could be achieved with the planar MESFET structure making it an attractive choice for use in the Baliga-Pair configuration.

8.1.2 Baliga-Pair Configuration: Forward Conduction Mode

The composite switch shown in Fig. 8.1 can be turned-on by application of a positive gate bias to the gate (G_B). If the gate bias is well above the threshold voltage of the silicon power MOSFET, it operates with a low on-resistance. Under these conditions, any voltage applied to the drain terminal (D_B) produces current flow through the normally-on silicon carbide FET and the silicon MOSFET. Due to the low specific on-resistance of both structures, the total on-resistance of the Baliga-Pair configuration is also very small:

$$R_{on}(Baliga - Pair) = R_{on}(SiliconMOSFET) + R_{on}(SiCFET) \quad [8.1]$$

Depending up on the size of the two devices, an on-resistance of less than 10 milli-Ohms is feasible even when the switch is designed to support 3000 volts. This indicates that the Baliga-Pair configuration will have an on-state voltage drop of about 1 volt with a nominal on-state current density of 100 A/cm^2 flowing through the devices. This is well below typical values of around 4 volts for an IGBT designed to support such high voltages.

8.1.3 Baliga-Pair Configuration: Current Saturation Mode

One of the reasons for the success of the silicon power MOSFET and IGBT in power electronics applications is the gate controlled current saturation capability of these devices. This feature enables controlling the rate of rise of current is power circuits by tailoring the input gate voltage waveform rather than by utilizing snubbers that are required for devices like gate turn-off thyristors. In addition, current saturation is essential for survival of short-circuit conditions where the device must limit the current.

The current saturation capability is inherent in the Baliga-Pair configuration. If the gate voltage applied to the Baliga-Pair configuration is close to the threshold voltage of the silicon MOSFET, it will enter its current saturation mode when the drain bias increases. This produces a constant current through both the silicon MOSFET and the silicon carbide FET while the drain bias applied to the composite switch increases. At lower drain bias voltages applied to the drain terminal (D_B), the voltage is supported across the silicon power MOSFET. As this voltage increases, the channel in the silicon carbide FET gets pinched-off and further voltage is then supported by the silicon carbide FET. Under these bias conditions, both the devices sustain current flow while supporting voltage. The level of the current flowing through the devices is controlled by the applied gate bias. In this sense, the Baliga-Pair behaves like a silicon power MOSFET from the point of view of the external circuit on both the input and output side. This feature makes the configuration attractive for use in power electronic systems because the existing circuit topologies can be used. The safe-operating-area of the composite switch is mainly determined by the silicon carbide FET because it supports a majority of the applied drain voltage. The excellent breakdown strength, thermal conductivity, and wide band gap of silicon carbide ensure good safe-operating-area for the FET structures.

8.1.4 Baliga-Pair Configuration: Switching Characteristics

The transition between the on and off modes for the Baliga-Pair configuration is controlled by the applied gate bias. During turn-on and turn-off, the gate bias must charge and discharge the capacitance of the silicon power MOSFET. Since silicon power MOSFETs are extensively used for high frequency power conversion, their input capacitance and gate charge have been optimized by the industry[10]. The switching speed of the Baliga-Pair is consequently very high because of the availability of silicon power MOSFETs designed for high frequency applications. The main limitations to the switching speed of the Baliga-Pair will be related to parasitic inductances in the package that could produce high voltage spikes.

8.1.5 Baliga-Pair Configuration: Fly-Back Diode

The Baliga-Pair contains an inherent high quality fly-back rectifier. When the drain bias is reversed to a negative value, the gate-drain contact/junction of the silicon carbide FET becomes forward biased. Since the gate of the silicon carbide FET is directly connected to the ground terminal, current can flow through this path when the drain voltage is negative in polarity. From this stand-point, it is preferable to use a metal-semiconductor contact for the gate rather than a P-N junction. The Schottky gate contact provides for a lower on-state voltage drop by proper choice of the work-function for the gate contact. In addition, the Schottky contact has no significant reverse recovery current. This greatly reduces switching losses in both the rectifier and the FETs[9]. Thus, the Baliga-Pair configuration replaces not just the power switch (such as the IGBT or GTO) in applications but also the power rectifier that is normally used across the switch.

8.2 Baliga-Pair: Simulation Results

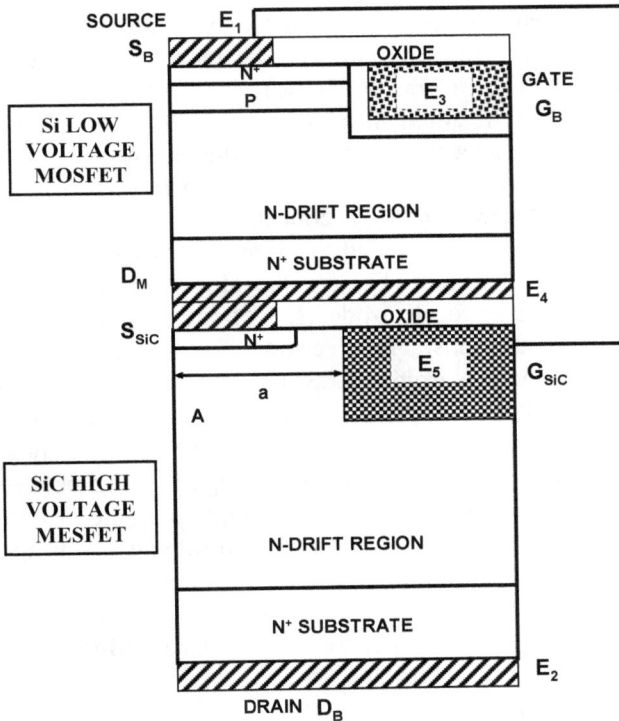

Fig. 8.2 Concatenated Silicon MOSFET and 4H-SiC MESFET Model.

In order to demonstrate the operation of the Baliga-Pair configuration, two-dimensional numerical simulations were performed by concatenating a silicon power MOSFET with a silicon carbide MESFET as illustrated in Fig. 8.2. This allowed observation of the potential and current distribution within both devices during all modes of operation. The silicon carbide MESFET had a drift region with doping concentration of 1×10^{16} cm^{-3} and thickness of 20 microns designed to support 3000 volts. A trench gate region with a depth of 1 micron was chosen based upon the simulation results in the previous chapter. The spacing between the gate regions (dimension 'a' in the figure) was chosen as 0.6 microns because this provided a good trade-off between normally-on operation with low specific on-resistance and a good blocking gain. A work function of 4.5 eV was used for the gate contact corresponding to a barrier height of 0.8 eV.

The silicon power MOSFET had a trench gate design to reduce its on-resistance and make the structure compatible with the trench gate silicon carbide FET for simulations. The gate oxide for the MOSFET was chosen as 500 angstroms with a peak P-base doping concentration of 3×10^{17} cm^{-3}. This resulted in a threshold voltage of 4.5 volts for the MOSFET which is acceptable for high voltage power switches that can be driven using 10 volt gate signals. The drift region for the silicon power MOSFET had a doping concentration of 1.5×10^{16} cm^{-3} and thickness of 2.5 microns designed to support 30 volts.

The gate electrode (E_5) for the silicon carbide MESFET was connected to the source electrode (E_1) of the silicon power MOSFET during all the simulations. The electrode (E_4) served as both the source contact to the silicon carbide MESFET and the drain contact for the silicon power MOSFET. This electrode was treated as a floating electrode (zero current boundary conditions) whose potential was monitored to provide insight into the voltage sharing between the two devices. The high voltage DC bias was applied to the drain (E_2) of the silicon carbide MESFET while the input control signal was applied to the gate electrode (E_3) of the silicon power MOSFET.

8.2.1 Baliga-Pair Simulations: Voltage Blocking Mode

To operate the Baliga-Pair configuration in the voltage blocking mode, the gate electrode was held at zero bias and a drain bias was applied to the drain (E_2) of the silicon carbide MESFET structure. The composite switch was able to support 3000 volts with the potential on the drain electrode (E_4) of the silicon MOSFET remaining below 30 volts. The potential distribution can be observed in Fig. 8.3 within both the transistors. The potential contours in the figure are mapped using an interval of 50 volts. It can be seen that most of the applied drain voltage is supported by the silicon carbide MESFET structure. The channel region adjacent to the Schottky gate contact of the silicon carbide

MESFET has a low potential because the channel is depleted at drain bias of only 5 volts. No potential contours are visible within the silicon MOSFET because it is supporting less than 50 volts.

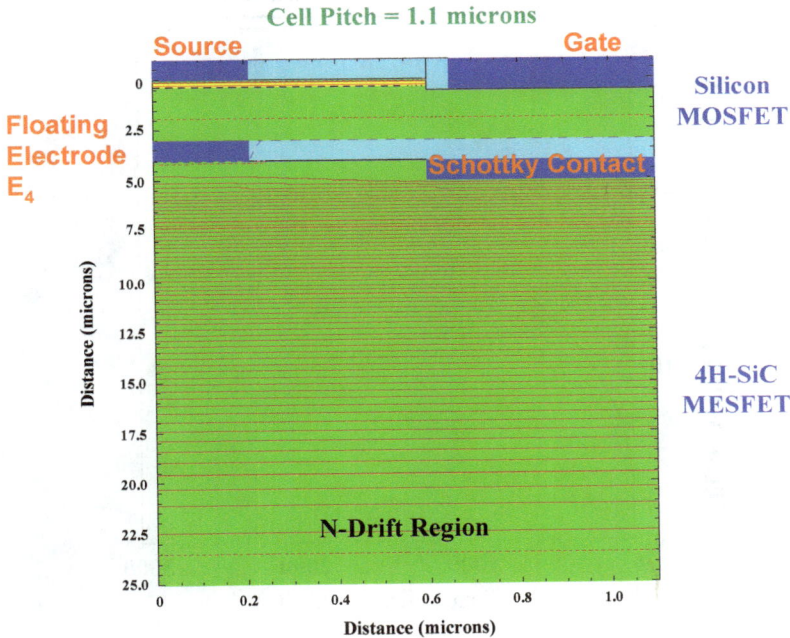

Fig. 8.3 Potential distribution within the Baliga-Pair Configuration:

$(V_{GS} = 0V; V_{DS} = 3000V)$.

The voltage developed across the silicon power MOSFET can be monitored by examining the potential at the floating electrode (E_4). As the applied drain voltage increases, the potential at this electrode also increases as shown in Fig. 8.4. The voltage at the floating electrode increases to 26.5 volts when the applied voltage to the drain of the composite switch reaches 3000 volts. Consequently, the maximum drain voltage experienced by the silicon power MOSFET is maintained below 30 volts for a composite switch capable of supporting 3000 volts. This is an important feature of the Baliga-Pair configuration because it allows the use of silicon power MOSFETs with relatively low breakdown voltages (< 50 volts). Such silicon power MOSFETs are commercially available with very low on-resistance.

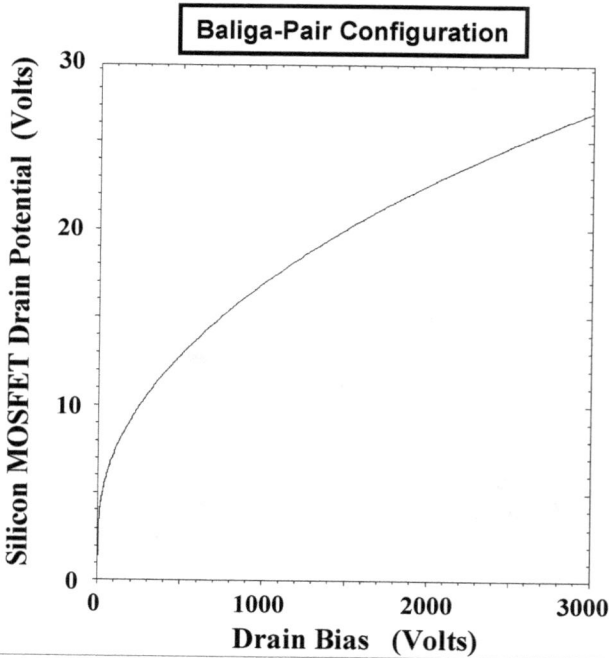

Fig. 8.4 Voltage at the Drain of the Silicon MOSFET in the Baliga-Pair Configuration in the Blocking Mode..

8.2.2 Baliga-Pair Simulations: Forward Conduction Mode

Current flow through the Baliga-Pair configuration is obtained by application of a positive gate bias to the silicon power MOSFET. When the gate bias is well above the threshold voltage of the silicon power MOSFET, it can carry a high current density with very low on-state voltage drop due to its low specific on-resistance. Under these conditions, the voltage difference between the source and gate of the silicon carbide MESFET becomes very small. Consequently, the channel of the silicon carbide MESFET remains un-depleted allowing current flow with low on-resistance.

The current distribution within the Baliga-Pair configuration is shown in Fig. 8.5 for the case of a gate bias of 10 volts and a drain bias of 1 volt. Only the upper portion of the silicon carbide MESFET is displayed in this figure because the current is uniformly distributed through the lower portion of the device. It can be seen that the current flows via the channel of the silicon power MOSFET and becomes uniformly distributed in its drift region. The current is then constricted through the channel of the silicon carbide MESFET before becoming uniformly distributed within its drift region.

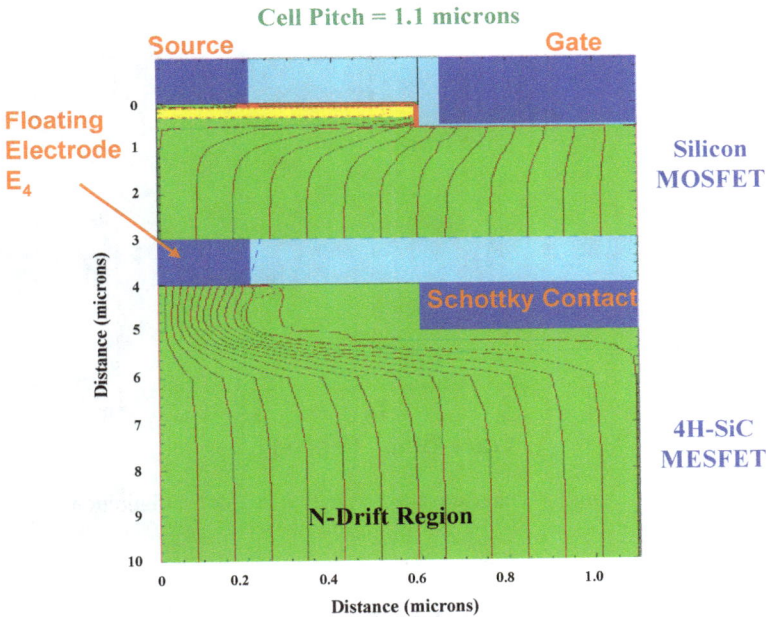

Fig. 8.5 On-State Current Distribution within the Baliga-Pair Configuration.

The transfer characteristic for the Baliga-Pair configuration was obtained by increasing the gate bias applied to the silicon power MOSFET while maintaining a bias of 1 volt at the drain of the silicon carbide MESFET. From the transfer curve shown in Fig. 8.6, it can be seen that a gate bias of 10 volts is sufficient to operate the Baliga-Pair with a low net on-resistance. At this gate bias, the current density flowing through the structures was 570 A/cm^2 corresponding to a specific on-resistance of 1.9 mΩ-cm^2. The silicon carbide MESFET contributes about 90 percent of this on-resistance with the balance contributed by the silicon power MOSFET.

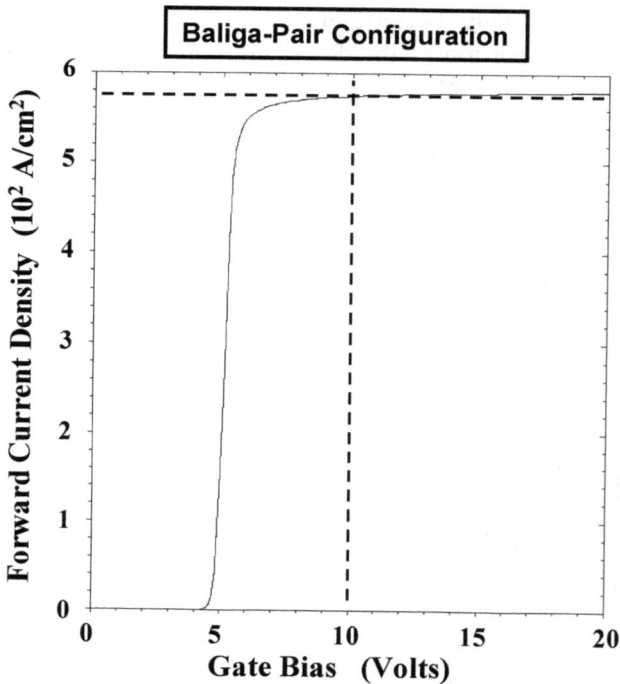

Fig. 8.6 Transfer Characteristics for the Baliga-Pair Configuration.

8.2.3 Baliga-Pair Simulations: Current Saturation Mode

As discussed earlier, the ability to limit the current flow by current saturation within a power switch is an attractive attribute from an applications stand-point. The extension of current saturation to high voltages is very desirable because this provides a broad safe-operating-area for the switching loci during circuit operation. These features were analyzed for the Baliga-Pair configuration by application of various gate bias voltages and examining the resulting output characteristics as shown in Fig. 8.7. It can be seen that the composite switch exhibits an excellent current saturation capability to very high drain bias voltages with very high output resistance until the gate bias voltage exceeds 9 volts. Beyond this gate bias voltage, current compression is observed – a phenomenon commonly observed in transistors at very high current densities[11]. Note that the current density has reached a magnitude of 2 x 10^4 A/cm^2 at a gate bias of 9 volts. Although such high current density can be a problem from a power dissipation point of view, these results demonstrate that the Baliga-Pair has a very broad safe-operating-area within which inductive load switching can be performed.

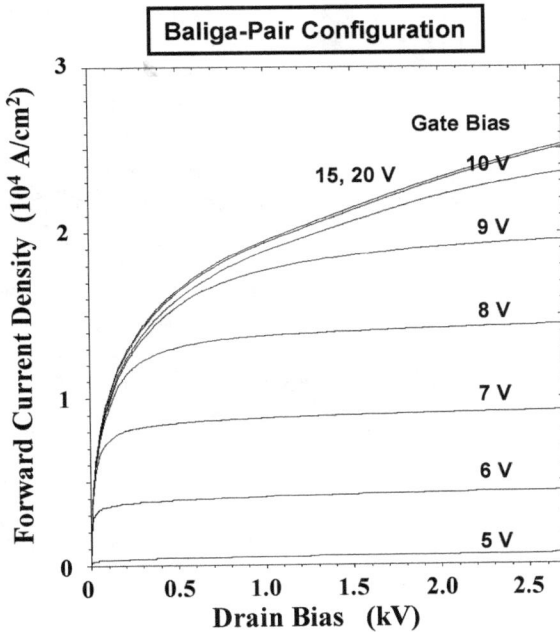

Fig. 8.7 Output Characteristics for the Baliga-Pair Configuration.

8.2.4 Baliga-Pair Simulations: Switching Characteristics

The switching performance of the Baliga-Pair configuration was examined under inductive load conditions. The composite switch was first biased into its blocking state with a drain bias of 2000 volts by keeping the gate bias of the silicon MOSFET at zero. The gate was then biased using a constant input current to charge the gate capacitance. This method is commonly used for characterization of power MOSFETs[12].

The gate voltage at the silicon power MOSFET increases in response to the stimulus, as shown in Fig. 8.8. When the gate voltage exceeds the threshold voltage at time t_1, the drain current begins to rise. The simulations were designed to maintain the drain current density at a constant value once it reached a magnitude of 200 A/cm^2. At this time t_2, the drain voltage falls rapidly until time t_3 resulting in a high dV/dt at the output terminals of the FETs. The plateau in the gate voltage until time t_4 is associated with the charging of the Miller capacitance of the silicon MOSFET.

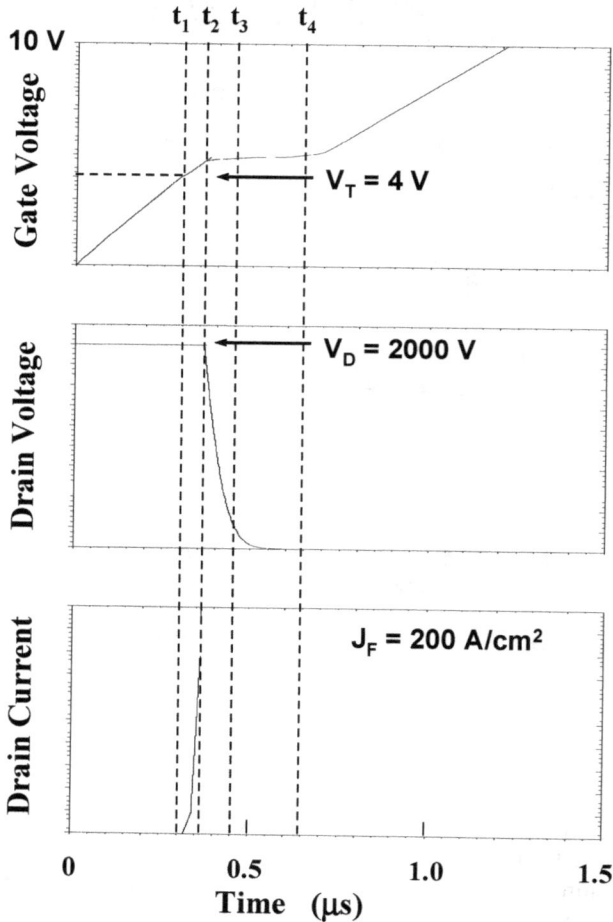

Fig. 8.8 Switching Performance of the Baliga-Pair Configuration.

During the switching event for turning on the Baliga-Pair, high power dissipation occurs during the transition time $(t_2 - t_1)$ for the drain current and the transition time $(t_3 - t_2)$ for the drain voltage. The switching energy associated with these events is given by:

$$E_{off} = 0.5 * V_D * I_D * (t_3 - t_1)$$ [8.2]

For the simulated case with V_D = 2000 volts and J_D = 200 A/cm^2, the calculated switching energy is 30 mJ/cm^2. This would produce a power dissipation of 30 W/cm^2 at a typical operating frequency of 1 kHz. In comparison, the on-state power dissipation is given by:

$$P_{on} = R_{DSonsp} * J_D^2$$ [8.3]

For the simulated Baliga-Pair configuration with specific on-resistance of 2 mΩ-cm^2, the conduction power loss is found to be 80 W/cm^2. If the turn-off switching losses are assumed to be comparable to the turn-on power loss computed above, the total power loss is 140 W/cm^2, which is acceptable from the point of view of cooling the devices with available heat sink technology. This demonstrates that the Baliga-Pair is capable of providing excellent switching performance for high voltage power systems.

It is also interesting to observe the behavior of the drain potential of the silicon power MOSFET during the switching transient. As can be seen in Fig. 8.9, the drain voltage remains close to its value during operation in the blocking mode until time t_2 and then reduces to the on-state voltage drop of the silicon power MOSFET. It is worth pointing out that the drain voltage of the silicon power MOSFET ramps down during the entire time interval from t_2 to t_4 while the drain voltage on the silicon carbide MESFET ramps down in a much shorter time interval from t_2 to t_3. Since most of the voltage is supported by the silicon carbide MESFET, this is a favorable outcome that results in smaller turn-off energy and switching losses.

In addition, the current flow through the gate of MESFET is shown in Fig. 8.9 during the switching transient. This current remains at approximately a constant value during the time interval from t_2 to t_3 when the drain voltage of the MESFET is ramping down. This indicates that the Miller and output capacitances of the MESFET are being charged through the gate electrode of the MESFET. This is responsible for the faster reduction of the drain voltage for the silicon carbide MESFET without being limited by the performance of the silicon

MOSFET. These observations indicate that the Baliga-Pair configuration
has an excellent switching behavior for power electronics applications.

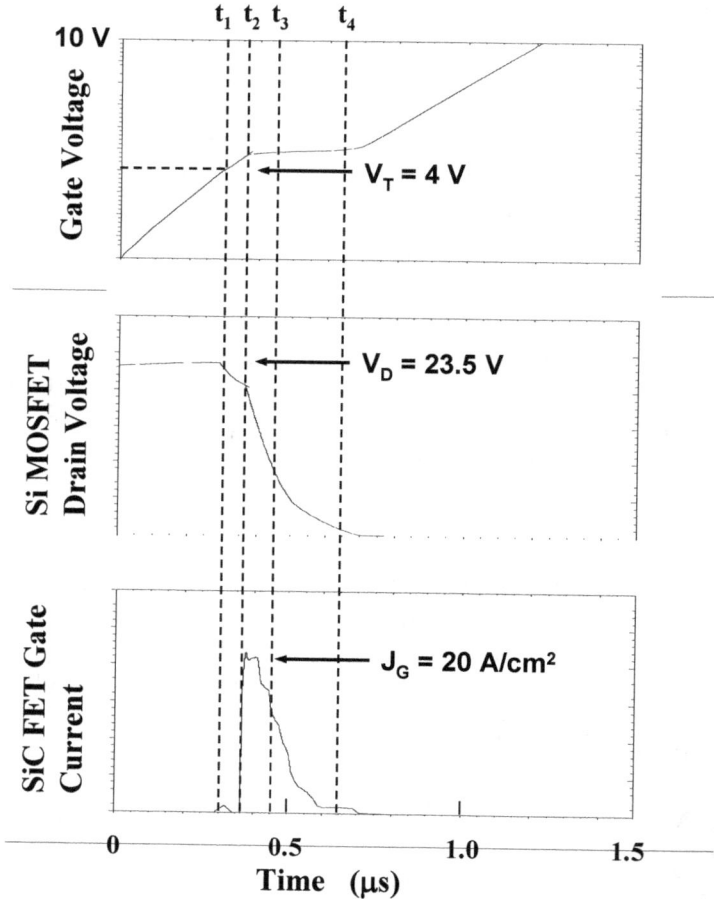

Fig. 8.9 Silicon Power MOSFET Drain Potential during Switching of the
Baliga-Pair Configuration.

8.2.5 Baliga-Pair Simulations: Fly-Back Diode

As mentioned previously, the Baliga-Pair configuration contains an
inherent fly-back rectifier that can be useful in H-bridge and other power
electronic circuit applications. The performance of this rectifier was

examined by doing simulations with a negative bias applied to the drain terminal of the silicon carbide MESFET while connecting the gate of the silicon power MOSFET to the source terminal. The resulting current flow path is shown in Fig. 8.10 for the case of a drain bias of -1 volt. It can be seen that the current flows between the drain of the silicon carbide MESFET and its gate electrode. No current is observed through the body diode of the silicon power MOSFET. This is favorable from the point of view of avoiding the reverse recovery of the silicon MOSFET's body diode[13]. The observed current path through the Schottky gate contact of the silicon carbide MESFET ensures that the fly-back rectifier in the Baliga-Pair operates in the unipolar mode.

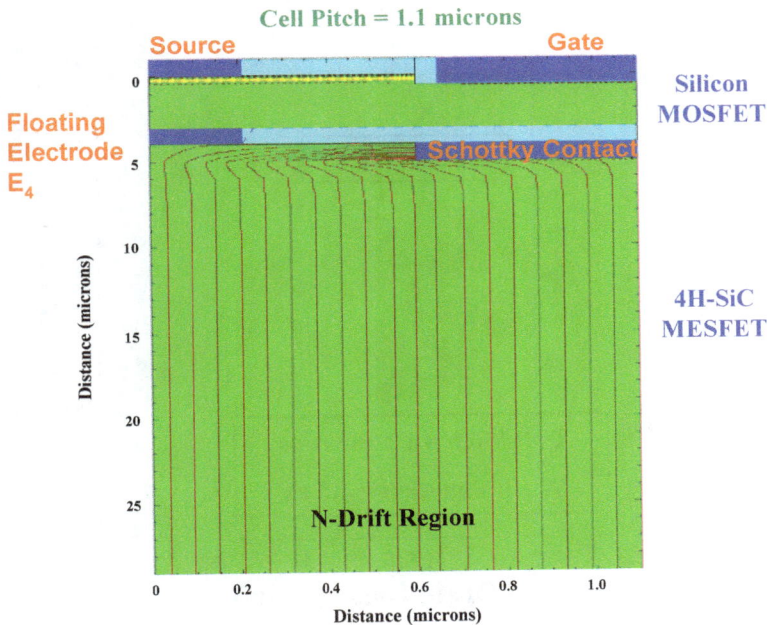

Fig. 8.10 Current Distribution within the Baliga-Pair Configuration with Negative Bias applied to the Drain Terminal.

The current conduction characteristics of the fly-back diode are of interest because they dictate the power losses when the rectifier is carrying the load current in an H-bridge circuit. The current conduction capability of the inherent fly-back rectifier in the Baliga-Pair configuration was determined by performing simulations with a negative voltage applied to the drain of the silicon carbide MESFET. The gate of the silicon MOSFET was held at zero volts during these simulations. The resulting characteristic is shown in Fig. 8.11 at room temperature. Excellent characteristics are observed with an on-state voltage drop of only 0.7 volts. This indicates that a larger work-function could be used for the gate of the silicon carbide MESFET to reduce its leakage current if necessary.

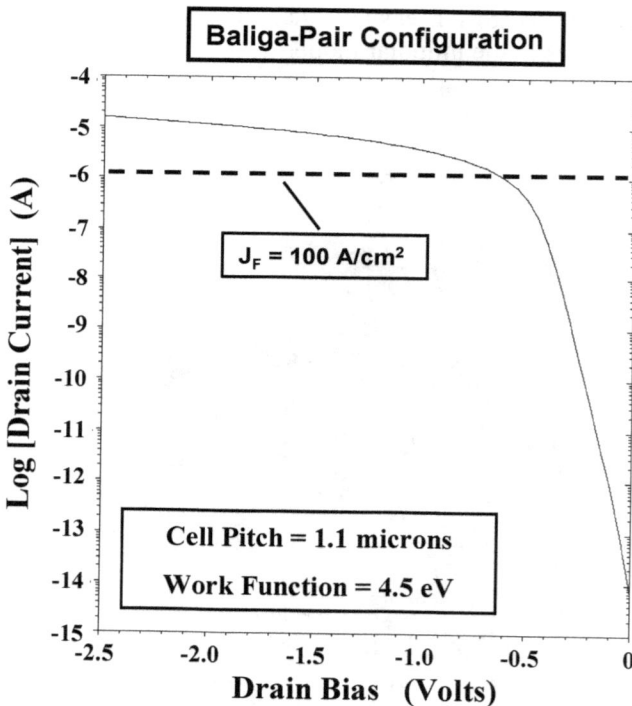

Fig. 8.11 Forward Characteristics of the Fly-Back Diode in the Baliga-Pair Configuration.

The simulations discussed above indicate that the Baliga-Pair configuration provides an excellent power switch for high voltage power system applications. In terms of packaging, the two transistors within the Baliga-Pair configuration replace the IGBT and the fly-back rectifier resulting in a similar level of complexity and cost.

8.3 Baliga-Pair Configuration: Experimental Results

The attractive features of the Baliga-Pair configuration have been acknowledged by several research groups. The same configuration was referred to as the Cascode arrangement in 1999[6] with the acknowledgement that it was originally proposed and published in 1996[7]. The switching behavior of the Cascode arrangement was reported[14] by using a planar silicon carbide JFET structure with a 50 volt silicon power MOSFET. The authors compared the performance of silicon carbide JFETs with the gate formed using the buried sub-surface P^+ region with a JFET fabricated using a gate formed on the upper surface (similar to the structure discussed in the previous chapter). It was found that the buried gate device had inferior switching performance due to the high resistance in the buried P^+ regions. The turn-off time was limited by the R-C charging time-constant for the buried P^+ regions. In addition, the slow response of this JFET structure resulted in the silicon power MOSFET being driven into avalanche breakdown. These problems were not observed for the surface gate device. In a subsequent paper[15], the authors stated "SiCED favors a combination of a silicon switch and a vertical, normally-on SiC junction field effect transistor". Their analysis concluded that the Baliga-Pair configuration is useful upto at least the 4.5 kV range. The excellent performance of the Baliga-Pair circuit has also been confirmed by using numerical simulations and compared with the performance of a silicon carbide MOSFET structure[16]. The authors, who called this a Cascade Configuration, found that the turn-off time for the Baliga-pair configuration was half that for the silicon carbide MOSFET due to the smaller Miller capacitance.

8.4 Summary

It has been demonstrated that the Baliga-pair configuration provides an ideal power switch for high voltage power system applications. It can be used in the same manner as IGBTs packaged with anti-parallel rectifiers without alterations of the gate control techniques. The packaging of this combination of two FETs is similar to that for the two chips currently used because the Baliga-Pair contains an inherent fly-back rectifier. Until the development of reliable silicon carbide power MOSFETs with low specific on-resistance, the Baliga-Pair offers a commercially viable near term option for high power electronic systems.

References

[1] B. J. Baliga, "The Evolution of MOS-Bipolar Power Semiconductor Technology", Proceedings of the IEEE, Vol. 74, pp. 409-418, 1988.

[2] B. J. Baliga, "Critical Nature of Oxide/Interface Quality for SiC Power Devices", Microelectronics Engineering, Vol. 28, pp. 177-184, 1995.

[3] B. J. Baliga, "Silicon Carbide Switching Device with Rectifying Gate", U. S. Patent 5,396,085, Issued March 7, 1995.

[4] S. Darlington, "Semiconductor Signal Translating Device", U. S. Patent 2,663,806, Issued December 22, 1953.

[5] P. M. McLarty, Private Communication, 1995.

[6] P. Friedrichs, et al, "Static and Dynamic Characteristics of 4H-SiC JFETs Designed for Different Blocking Categories", Silicon Carbide and Related Materials – 1999, Material Science Forum, Vol. 338-342, pp. 1243-1246, 2000.

[7] B. J. Baliga, "Power Semiconductor Devices", pp. 418-420, PWS Publishing Company, 1996.

[8] B. J. Baliga, "Prospects for Development of SiC Power Devices", Silicon Carbide and Related Materials – 1995, Institute of Physics Conference Series, Vol. 142, pp. 1-6, 1996.

[9] B. J. Baliga, "Power Semiconductor Devices for Variable-Frequency Drives", Proceeding of the IEEE, Vol. 82, pp. 1112-1122, 1994.

[10] B. J. Baliga and D. Alok, "Paradigm Shift in Planar Power MOSFET Technology", Power Electronics Technology Magazine, pg. 24-32, November 2003.

[11] B. J. Baliga, "Silicon RF Power MOSFETs", World Scientific Publishers, 2005.

[12] B. J. Baliga, "Power Semiconductor Devices", pp. 387-391, PWS Publishing Company, 1996.

[13] B. J. Baliga and J. P. Walden, "Improving the Reverse Recovery of Power MOSFET Integral Diodes by Electron Irradiation", Solid State Electronics, Vol. 26, pp. 1133-1141, 1983.

[14] H. Mitlehner, et al, "Dynamic Characteristics of High Voltage 4H-SiC Vertical JFETs", IEEE International Symposium on Power Semiconductor Devices and ICs, Abstract 11.1, pp. 339-342, 1999.

[15] P. Friedrichs, et al, "Application-Oriented Unipolar Switching SiC Devices", Silicon Carbide and Related Materials – 2001, Material Science Forum, Vol. 389-393, pp. 1185-1190, 2002.

[16] A. Mihaila, et al, "Static and Dynamic Behavior of SiC JFET/Si MOSFET Cascade Configuration for High-Performance Power

Switches", Silicon Carbide and Related Materials – 2001, Material Science Forum, Vol. 389-393, pp. 1239-1242, 2002.

Chapter 9

Planar Power MOSFETs

The planar power MOSFET was the first commercially successful unipolar switch developed using silicon technology[1] once the issues related to the metal-oxide-semiconductor interface had been resolved for CMOS technology. In order to reduce cost, the channel in these devices was created by the double diffusion (or D-MOS) process. In the DMOS process, the P-base and N^+ source regions are formed by ion implantations masked by a common edge defined by a refractory polysilicon gate electrode. A drive-in cycle is used after each ion implantation step to move the P-N junction in the lateral direction under the gate electrode. The separation between the N^+/P-base junction and the P-base/N-drift junction under the gate electrode defines the channel. Consequently, the channel length can be reduced to sub-micron dimensions without the need for high resolution lithography. This approach served the industry quite well from the 1970s into the 1990s with planar power MOSFETs still available for power electronic applications. In the 1990s, the industry borrowed the trench technology originally developed for DRAMs to introduce the UMOSFET structure for commercial applications[2]. This was important for the reduction of the channel and JFET components in the planar MOSFET structure designed for lower (< 30 volts) voltage applications. This UMOSFET structure has also been explored in silicon carbide as described in a subsequent chapter.

In silicon power MOSFETs, the on-resistance becomes dominated by the resistance of the drift region when the breakdown voltage exceeds 200 volts. At high breakdown voltages, the specific on-resistance for these devices becomes greater than 10^{-2} Ohm-cm^2, as shown in Fig. 3.6, leading to an on-state voltage drop of more than 1 volt at a typical on-state current density of 100 A/cm^2. For this reason, the Insulated Gate Bipolar Transistor (IGBT) was developed in the 1980s to serve medium and high power systems[2]. The superior performance of the

IGBT in high voltage applications relegated the silicon MOSFETs to applications with operating voltages below 100 volts. Novel silicon structures that utilize the charge-coupling concept have allowed extending the breakdown voltage of power MOSFETs to the 600 volt range[3]. However, their specific on-resistance is still quite large limiting their use to high operating frequencies where switching losses dominate.

In principle, the much lower resistance of the drift region in silicon carbide should enable development of power MOSFETs with very high breakdown voltages. These devices offer not only fast switching speed but also superior safe operating area when compared to high voltage silicon IGBTs. This allows reduction of both the switching loss and conduction loss components in power circuits[4]. Unfortunately, the power MOSFET structures developed in silicon cannot be directly utilized to form high performance silicon carbide devices. Firstly, the lack of significant diffusion of dopants in silicon carbide prevents the use of the silicon DMOS process. Secondly, a high electric field occurs in the gate oxide of the silicon carbide MOSFET exceeding its rupture strength leading to catastrophic failure of devices in the blocking mode at high voltages. Thirdly, when compared with silicon, the smaller band offset between the conduction band of silicon carbide and silicon dioxide can produce injection of hot carriers into the oxide leading to instability during operation. In addition, the quality of the oxide-semiconductor interface for silicon carbide must be improved to allow good control over the threshold voltage and the channel mobility.

This chapter begins with a review of the basic principles of operation of the planar MOSFET structure. A planar MOSFET structure formed by staggering the ion implantation of the P-base and N^+ source regions to create the channel is then described. Next, the problem of exacerbation of reach-through breakdown in silicon carbide is considered. Based upon fundamental considerations, the difference between the threshold voltage for silicon and silicon carbide structures is analyzed. The relatively high doping concentrations and large channel lengths required to prevent reach-through are shown to be serious limitations to obtaining low specific on-resistance in these devices. In addition, the much larger electric field in silicon carbide is shown to lead to a high electric field in the gate oxide. Structures designed to reduce the electric field at the gate oxide by using a shielding region are therefore essential to realization of practical silicon carbide MOSFETs even after the MOS interface quality is improved. These structures are described and analyzed in the next chapter. The results of the analysis of the planar silicon carbide MOSFET structures by using two-dimensional numerical simulations are described in this chapter followed by the description of experimental results on relevant structures to define the state of the development effort on these devices.

9.1 Planar Power MOSFET Structure

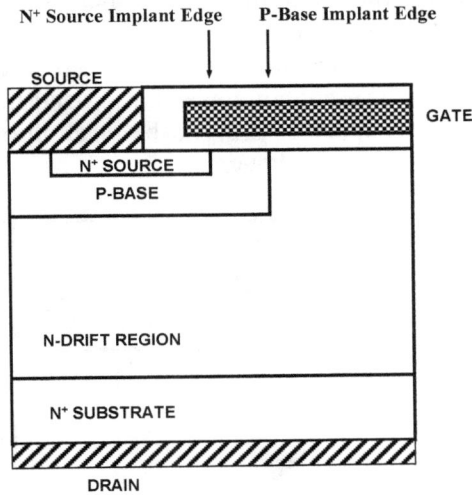

Fig. 9.1 The Planar Power MOSFET Structure.

The basic structure of the planar power MOSFET is shown in Fig. 9.1 together with the location of the ion implantation edges. In recognition of the low diffusion coefficients for dopants in silicon carbide, it was proposed[5] that the P-base and N[+] source ion implantations be staggered by using photoresist masks rather that be defined by the gate edge. It was also suggested that the P-base region be first amorphized to obtain a more uniform dopant distribution to suppress reach-through. This approach with staggered P-base and N[+] source implants was subsequently taken by several groups to fabricate high voltage devices[6]. These devices have been called DIMOSFETs because of the Double-Implant process used for their fabrication. The measured performance of these structures is discussed at the end of the chapter.

9.1.1 Planar Power MOSFET Structure: Blocking Characteristics

In the forward blocking mode of the planar power MOSFET, the voltage is supported by a depletion region formed on both sides of the P-base/N-drift junction. The maximum blocking voltage can be determined by the electric field at this junction becoming equal to the critical electric field for breakdown if the parasitic $N^+/P/N$ bipolar transistor is completely suppressed. This suppression is accomplished by short-circuiting the N^+ source and P-base regions using the source metal as shown on the upper left hand side of the cross-section. However, a large leakage current can occur when the depletion region in the P-base region reaches-through to the N^+ source region. The doping concentration and thickness of the P-base region must be designed to prevent the reach-through phenomenon from limiting the breakdown voltage.

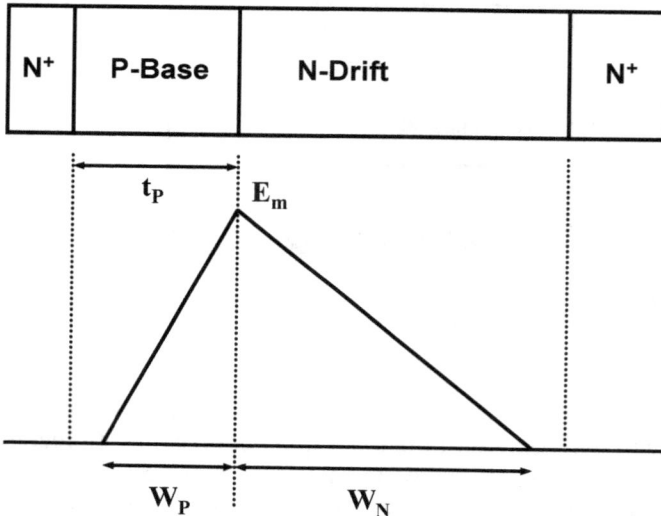

Fig. 9.2 Reach-Through in a Power MOSFET Structure.

The applied drain voltage is supported by the N-drift region and the P-base region with a triangular electric field distribution as shown in Fig. 9.2 if the doping is uniform on both sides. The maximum electric field occurs at the P-base/N-drift junction. The depletion width on the P-base side is related to the maximum electric field by:

$$W_P = \frac{\varepsilon_S E_m}{qN_A}$$ [9.1]

where N_A is the doping concentration in the P-base region. The minimum P-base thickness required to prevent reach-though limited breakdown can be obtained by assuming that the maximum electric field at the P-base/N-drift junction reaches the critical electric field for breakdown when the P-base region is completely depleted:

$$t_P = \frac{\varepsilon_S E_C}{qN_A}$$ [9.2]

where E_C is the critical electric field for breakdown in the semiconductor.

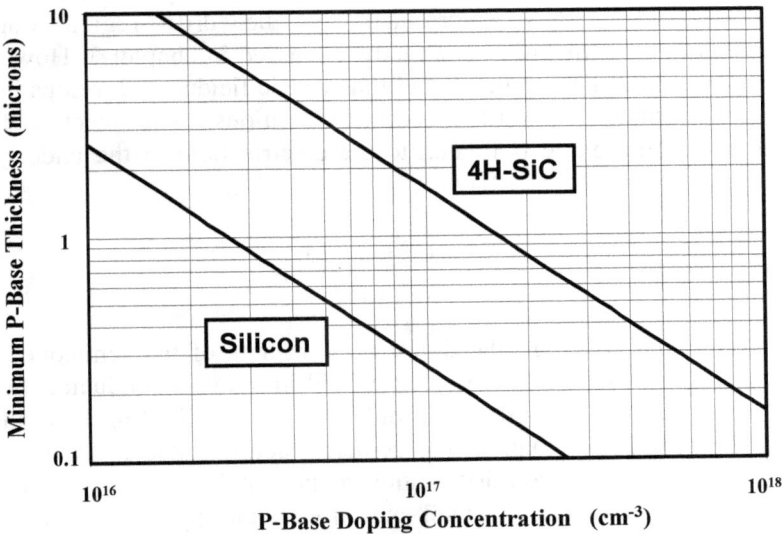

Fig. 9.3 Comparison of Minimum P-Base Thickness to prevent Reach-Through Breakdown in 4H-SiC and Silicon.

The calculated minimum P-base thickness for 4H-SiC power MOSFETs is compared with that for silicon in Fig. 9.3. At any given P-base doping concentration, the thickness for 4H-SiC is about six times larger than for silicon. This implies that the minimum channel length required for silicon carbide devices is much larger than for silicon devices resulting in a big increase in the on-resistance. The enhancement of the on-resistance is compounded by the lower channel inversion layer mobility observed for silicon carbide.

The minimum thickness of the P-base region required to prevent reach-through breakdown decreases with increasing doping concentration as shown in Fig. 9.3. The typical P-base doping concentration for silicon power MOSFETs is 1×10^{17} cm^{-3} to obtain a threshold voltage between 1 and 2 volts for a gate oxide thickness of 500 to 1000 angstroms. At this doping level, the P-base thickness can be reduced to 0.5 microns without reach-through limiting the breakdown voltage. In contrast, for 4H-SiC, it is necessary to increase the P-base doping concentration to above 4×10^{17} cm^{-3} to prevent reach-through with a 0.5 micron P-base thickness. This higher doping concentration increases the threshold voltage as discussed in the next section.

The maximum blocking voltage capability of the power MOSFET structure is determined by the drift region doping concentration and thickness as already discussed in chapter 3. However, in the power MOSFET structure, a high electric field also develops in the gate oxide under forward blocking conditions. The electric field developed in the oxide is related to the electric field in the underlying semiconductor by Gausses Law:

$$E_{Oxide} = \left(\frac{\varepsilon_{Semi}}{\varepsilon_{Oxide}} \right) . E_{Semi} \qquad [9.3]$$

where ε_{Semi} and ε_{Oxide} are the dielectric constants of the semiconductor and the oxide and E_{Semi} is the electric field in the semiconductor. In the case of both silicon and silicon carbide, the electric field in the oxide is about 3 times larger than in the semiconductor. Since the maximum electric field in the silicon drift region remains below 3×10^5 V/cm, the electric field in the oxide does not exceed its reliability limit of about 3×10^6 V/cm. However, for 4H-SiC, the electric field in the oxide reaches a value of 9×10^6 V/cm when the field in the semiconductor reaches its breakdown strength. This value not only exceeds the reliability limit but can cause rupture of the oxide leading to catastrophic breakdown. It is therefore important to monitor the electric field in the gate oxide when designing and modeling the silicon carbide MOSFET structures. Novel structures[7] that shield the gate oxide from high electric field have also been proposed and demonstrated to resolve this problem. These structures will be discussed in the next chapter.

9.1.2 Planar Power MOSFET Structure: Forward Conduction

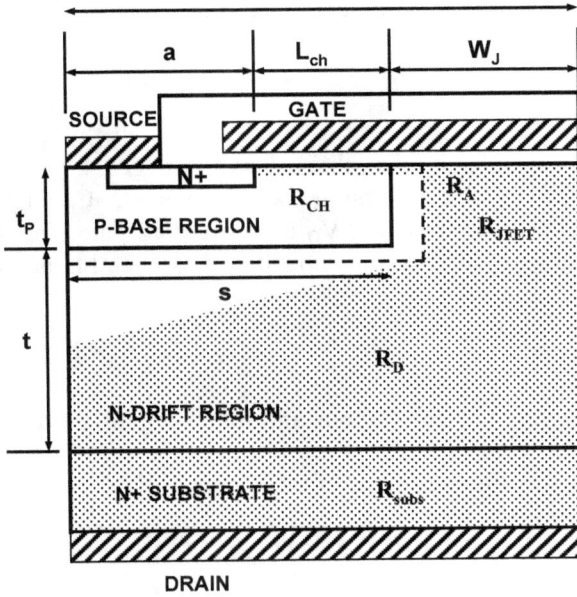

Fig. 9.4 Current Flow Path in the Planar 4H-SiC Power MOSFET.

Current flow between the drain and source can be induced by creating an inversion layer channel on the surface of the P-base region. The current path is illustrated in Fig. 9.4 by the dotted area. The current flows through the inversion layer channel formed due to the applied gate bias into the JFET region via the accumulation layer formed above it under the gate oxide. It then spreads into the N-drift region at a 45 degree angle and becomes uniform through the rest of the structure. The total on-resistance for the planar power MOSFET structure is determined by the resistance of these components in the current path:

$$R_{on,sp} = R_{CH} + R_A + R_{JFET} + R_D + R_{subs} \qquad [9.4]$$

where R_{CH} is the channel resistance, R_A is the accumulation region resistance, R_{JFET} is the resistance of the JFET region, R_D is the resistance of the drift region after taking into account current spreading from the JFET region, and R_{subs} is the resistance of the N^+ substrate. These resistances can be analytically modeled by using the current flow pattern indicated by the shaded regions in Fig. 9.4. In this figure, the depletion region boundaries have also been shown using dashed lines.

The specific channel resistance is given by:

$$R_{CH} = \frac{(L_{CH} \cdot p)}{\mu_{inv} C_{ox} (V_G - V_T)} \qquad [9.5]$$

where L_{CH} is the channel length as shown in Fig. 9.4, μ_{inv} is the mobility for electrons in the inversion layer channel, C_{ox} is the specific capacitance of the gate oxide, V_G is the applied gate bias, and V_T is the threshold voltage. The specific capacitance can be obtained using:

$$C_{ox} = \frac{\varepsilon_{ox}}{t_{ox}} \qquad [9.6]$$

where ε_{ox} is the dielectric constant for the gate oxide and t_{ox} is its thickness.

The specific resistance of the accumulation region is given by:

$$R_A = \frac{K(W_J - W_P)p}{\mu_a C_{ox} (V_G - V_T)} \qquad [9.7]$$

where μ_a is the mobility for electrons in the accumulation layer, C_{ox} is the specific capacitance of the gate oxide, V_G is the applied gate bias, and V_T is the threshold voltage. The factor K is used to account for two-dimensional current spreading from the channel into the JFET region with a typical value of 0.6 for silicon devices. In this equation, W_P is the zero-bias depletion width at the P-base/N-drift junction. It can be determined using:

$$W_P = \sqrt{\frac{2\varepsilon_S V_{biP}}{qN_D}} \qquad [9.8]$$

where the built-in potential V_{biP} for the P-N junction is typically 3.3 volts for 4H-SiC.

The specific JFET region resistance is given by:

$$R_{JFET} = \rho_D . t_P \left(\frac{p}{W_J - W_P} \right) \qquad [9.9]$$

where t_p is the depth of the P-base region.

The drift region spreading resistance can be obtained by using:

$$R_D = \rho_D . p . \ln\left(\frac{p}{W_J - W_P}\right) + \rho_D . (t - s - W_P)$$ [9.10]

where t is the thickness of the drift region below the P-base region and s is the width of the P-base region.

The contribution to the resistance from the N^+ substrate is given by:

$$R_{subs} = \rho_{subs} . t_{subs}$$ [9.11]

where ρ_{subs} and t_{subs} are the resistivity and thickness of the substrate, respectively. A typical value for this contribution is 4×10^{-4} Ω-cm^2.

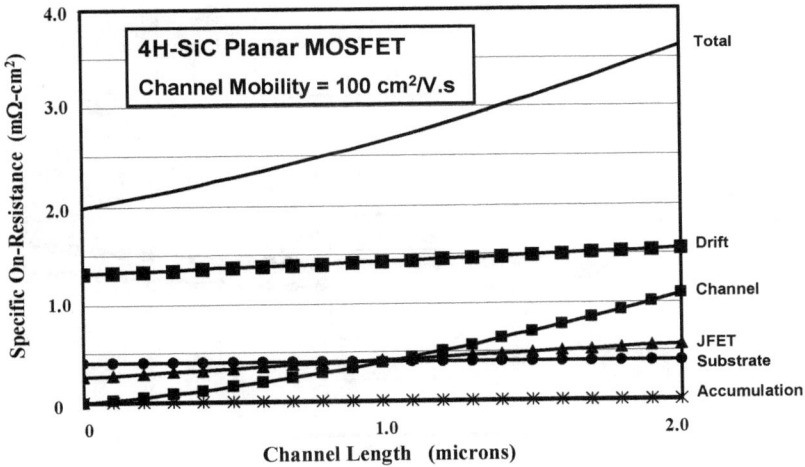

Fig. 9.5 On-Resistance Components for a 4H-SiC Planar MOSFET.

The specific on-resistances of 4H-SiC planar MOSFETs with a drift region doping concentration of 1×10^{16} cm^{-3} and thickness of 20 microns were modeled using the above analytical expressions. In all the devices, a gate oxide thickness of 0.1 microns was used. The dimension 'a' in the structure shown in Fig. 9.4 was kept at 1 micron. This value can be achieved using 0.5 micron design rules. The accumulation mobility was assumed to be twice the inversion layer mobility and a K-factor of 0.1 was used to provide correlation with the simulations. The P-base was assumed to have a depth of 1 micron and the effective gate drive voltage ($V_G - V_T$) was assumed to be 20 volts. The various components of the on-resistance are shown in Fig. 9.5 when a channel inversion layer mobility of 100 cm^2/Vs was used. This magnitude of inversion layer mobility has been experimentally observed[8]. It can be seen that the drift region resistance is dominant under these conditions even for channel lengths of 2 microns. At a channel length of 1 micron, a total specific on-resistance of 2.7 $m\Omega$-cm^2 is obtained, which is about three times the ideal specific on-resistance. Although this is a good value for a device with the drift regions parameters used for the modeling, the breakdown voltage of the structure was determined to be well below its ideal parallel-plane case (3000 volts), as shown later under device simulations, due to the reach-through problem.

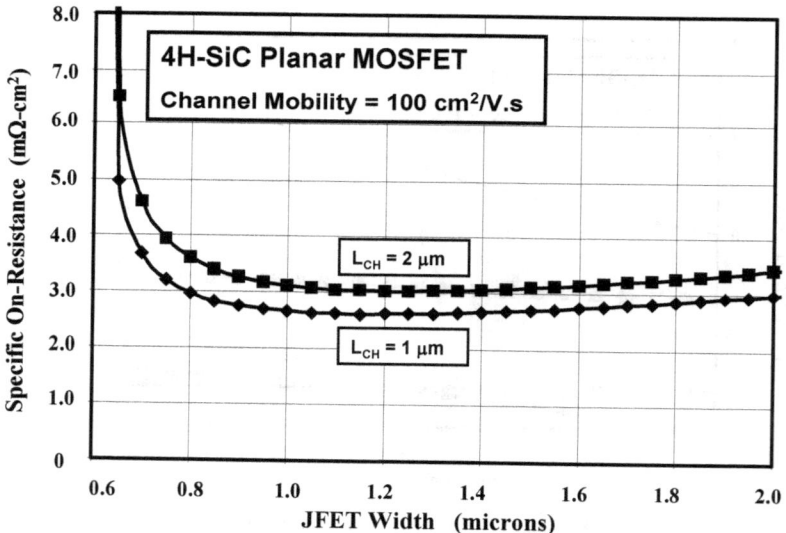

Fig. 9.6 Impact of JFET Width on the On-Resistance for a 4H-SiC Planar MOSFET.

The JFET width in the planar MOSFET structure must be optimized not only to obtain the lowest specific on-resistance but to also control the electric field at the gate oxide interface. The impact of changes to the JFET width on the specific on-resistance is shown in Fig. 9.6 for a structure with the same cell parameters that were given above. It can be observed that the specific on-resistance becomes very large when the JFET width is made smaller than about 0.7 microns. This occurs because the zero bias depletion width for 4H-SiC P-N junctions is about 0.6 microns. When the JFET width approaches this value, the current becomes severely constricted resulting in a high specific on-resistance. As the JFET width is increased, the contribution from its resistance decreases rapidly. However, the increase in cell pitch enhances the specific resistance contributions from the channel resulting in an increase in the total specific on-resistance. A minimum specific on-resistance is observed to occur at a JFET width of about 1 micron.

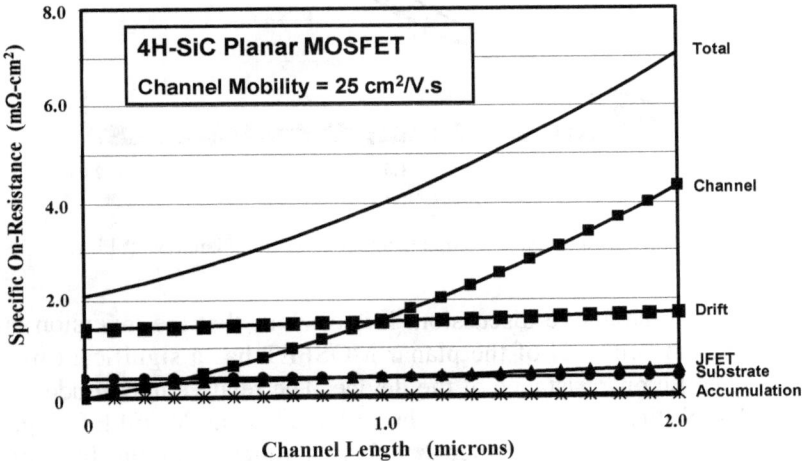

Fig. 9.7 On-Resistance Components for a 4H-SiC Planar MOSFET.

The above results are based upon using a channel mobility of 100 cm^2/Vs. However, most groups working on silicon carbide have reported much lower magnitudes for the inversion layer mobility in 4H-SiC structures[9]. Taking this into consideration, it is interesting to examine the impact of reducing the inversion layer mobility as depicted in Fig. 9.7 and 9.8 for the case of mobility values of 25 and 2.5 cm^2/Vs. With an inversion layer mobility of 25 cm^2/Vs, the channel contribution becomes comparable to the contribution from the drift region. This results in approximately doubling of the specific on-resistance to a value of 4 mΩ-cm^2 for a channel length of 1 micron. When the inversion layer mobility is reduced to 2.5 cm^2/Vs, the channel resistance becomes dominant as shown in Fig. 9.8, with an increase in the specific on-resistance to a value of 18 mΩ-cm^2 for a channel length of 1 micron.

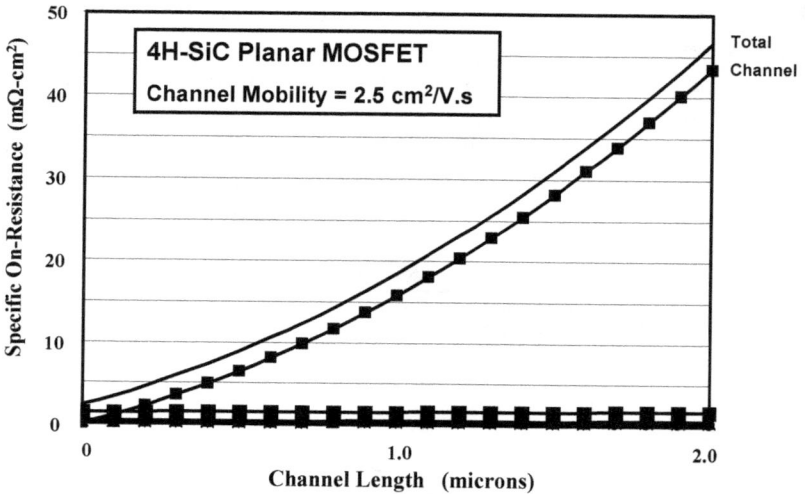

Fig. 9.8 On-Resistance Components for a 4H-SiC Planar MOSFET.

From the above discussion, it is obvious that the inversion layer mobility in the channel of the planar MOSFET has a significant impact on the specific on-resistance of the device. The relative magnitude of the channel mobility on the performance of the planar MOSFET depends upon the drift region resistance which is a function of the breakdown voltage of the device. In order to provide a better understanding of this, consider a 4H-SiC planar MOSFET with a JFET region whose doping concentration is held at 1×10^{16} cm^{-3} while the properties of the underlying drift region are adjusted to obtain the desired breakdown voltage. This approach for designing planar MOSFETs with enhanced doping in the JFET region is common practice in silicon technology[2]. The other cell parameters are maintained at the following values: parameter 'a' of 1 micron, P-base depth of 1 micron, gate oxide thickness of 0.1 microns, and channel length of 1 micron.

The specific on-resistance for the planar 4H-SiC MOSFETs is plotted in Fig. 9.9 as a function of the breakdown voltage for several cases of channel mobility. In performing this modeling, it is important to recognize that the thickness of the drift region (parameter 't' in Fig. 9.4) can become smaller than the cell parameter 's' at lower breakdown voltages (below 700 volts). Under these conditions, the current does not distribute at a 45 degree angle into the drift region from the JFET region.

Instead, the current flows from a cross-sectional width of $(W_J - W_P)$ to a cross-section of $(t + W_J - W_P)$. The drift region resistance for these cases can be modeled using:

$$R_D = \rho_D \cdot p \cdot \ln\left(\frac{t + W_J - W_P}{W_J - W_P}\right) \qquad [9.12]$$

In addition, the resistance contributed by the JFET region should be modeled using:

$$R_{JFET} = \rho_{JFET} \cdot t_P \left(\frac{p}{W_J - W_P}\right) \qquad [9.13]$$

where ρ_{JFET} is the resistivity of the JFET region, which is different from the resistivity of the underlying drift region.

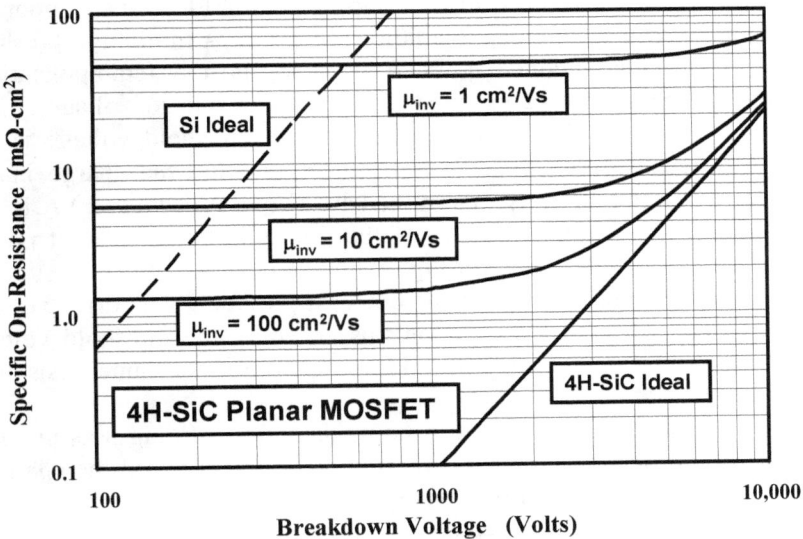

Fig. 9.9 Specific On-Resistance for 4H-SiC Planar MOSFETs.

From Fig. 9.9, it can be seen that the specific on-resistance of 4H-SiC planar MOSFETs approaches the ideal specific on-resistance when the breakdown voltage exceeds 5000 volts if a channel mobility of 100 cm^2/Vs is achieved. Even at this relatively high inversion layer mobility for silicon carbide, the channel resistance limits the performance of the planar MOSFET when the breakdown voltage falls below 1000 volts. For a breakdown voltage of 1000 volts, the anticipated improvement in specific on-resistance over silicon devices is then about 100 times (as opposed to the 2000x improvement in the specific on-resistance of the drift region). When the inversion layer mobility is reduced to 10 cm^2/Vs, the degradation in performance extends to much larger breakdown voltages. This highlights the importance of developing process technology to achieve high inversion layer mobility in 4H-SiC structures. In comparison with silicon, whose ideal specific on-resistance is shown by the dashed line in Fig. 9.9, the 4H-SiC planar MOSFET can surpass the performance by an order of magnitude at breakdown voltages above 1000 volts if a channel mobility of at least 10 cm^2/Vs is obtained.

9.1.3 Planar Power MOSFET Structure: Threshold Voltage

The threshold voltage of the power MOSFET is an important design parameter from an application stand-point. A minimum threshold voltage must be maintained at above 1 volt for most system applications to provide immunity against inadvertent turn-on due to voltage spikes arising from noise. At the same time, a high threshold voltage is not desirable because the voltage available for creating the charge in the channel inversion layer is determined by ($V_G - V_T$) where V_G is the applied gate bias voltage and V_T is the threshold voltage. Most power electronic systems designed for high voltage operation (the most suitable application area for silicon carbide devices) provide a gate drive voltage of only upto 10 volts. Based upon this criterion, the threshold voltage should be kept below 3 volts in order to obtain a low channel resistance contribution.

The threshold voltage can be modeled by defining it as the gate bias at which on-set of *strong inversion* begins to occur in the channel. This voltage can be determined using[10]:

$$V_{TH} = \frac{\sqrt{4\varepsilon_S kTN_A \ln(N_A/n_i)}}{C_{ox}} + \frac{2kT}{q}\ln\left(\frac{N_A}{n_i}\right) \qquad [9.14]$$

where N_A is the doping concentration of the P-base region, k is Boltzmann's constant, and T is the absolute temperature. The presence of positive fixed oxide charge shifts the threshold voltage in the negative direction by:

$$\Delta V_{TH} = \frac{Q_F}{C_{ox}}$$ [9.15]

A further shift of the threshold voltage in the negative direction by 1 volt can be achieved by using heavily doped N-type polysilicon as the gate electrode as routinely done for silicon power MOSFETs.

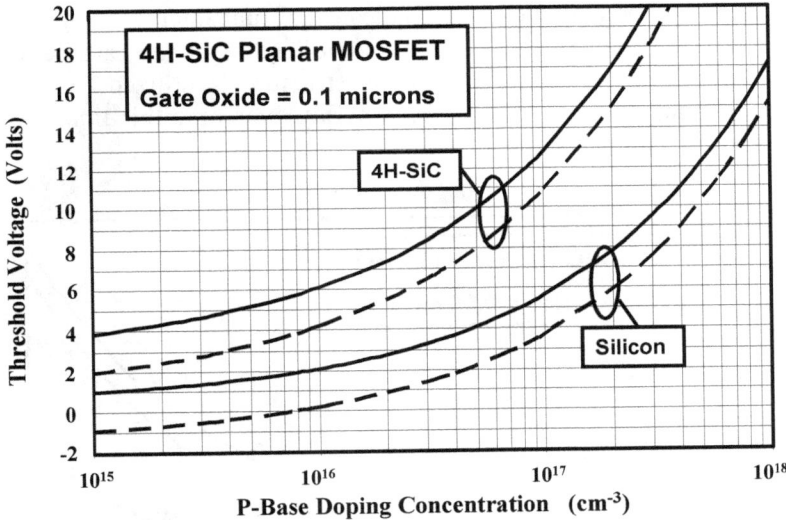

Fig. 9.10 Threshold Voltage of 4H-SiC Planar MOSFETs.

(Dashed lines include the impact of N^+ Polysilicon Gate and an Oxide Fixed Charge of 2×10^{11} cm^{-2})

The analytically calculated threshold voltage for 4H-SiC planar MOSFETs is shown in Fig. 9.10 for the case of a gate oxide thickness of 0.1 microns. The results obtained for a silicon power MOSFET with the same gate oxide thickness is also provided in this figure for comparison. In the case of silicon devices, a threshold voltage of about 3 volts is obtained for a P-base doping concentration of 1×10^{17} cm^{-3}. At this doping concentration, the depletion width in the P-base region for silicon devices is less than 0.5 microns, as shown earlier in Fig. 9.3, even when the electric field in the semiconductor approaches the critical electric field for breakdown. This allows the design of silicon power MOSFETs with channel lengths of below 0.5 microns without encountering reach-through breakdown limitations. In contrast, a P-base doping concentration of about 3×10^{17} cm^{-3} is required in 4H-SiC (see Fig. 9.3) to keep the depletion width in the P-base region below 1 micron when the electric field in the semiconductor approaches the critical electric field for breakdown. At this doping concentration, the threshold voltage for the 4H-SiC MOSFET approaches 20 volts. The much larger threshold voltage for silicon carbide is physically related to its larger band gap as well as the higher P-base doping concentration required to suppress reach-through breakdown. This indicates a fundamental problem for achieving reasonable levels of threshold voltage in silicon carbide power MOSFETs if the conventional silicon structure is utilized. Innovative structures that shield the P-base region[7] can overcome this limitation as discussed in the next chapter.

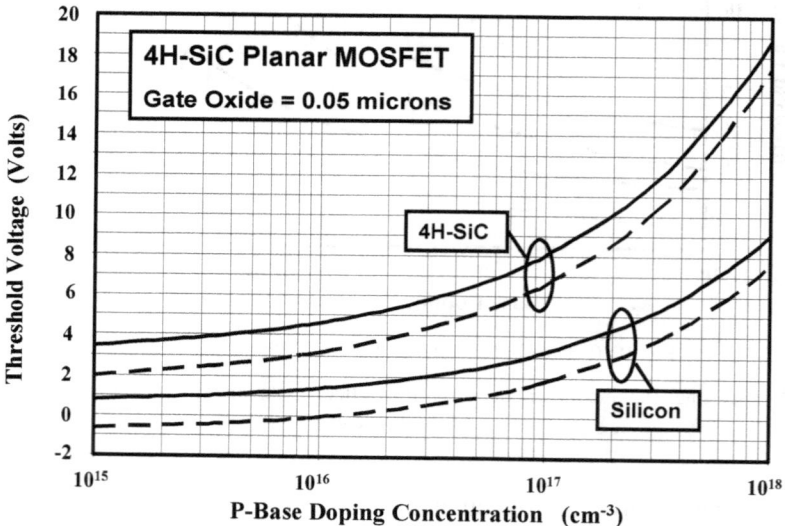

Fig. 9.11 Threshold Voltage of 4H-SiC Planar MOSFETs.

(Dashed lines include the impact of N^{+} Polysilicon Gate and an Oxide Fixed Charge of 2×10^{11} cm^{-2})

The threshold voltage for a MOSFET can be reduced by decreasing the gate oxide thickness. If the gate oxide thickness is reduced to 0.05 microns, the threshold voltage for the silicon power MOSFET decreases to just below 2 volts as shown in Fig. 9.11. This is common practice for the design of low voltage silicon power MOSFETs that are often driven with logic-level (5 volt) gate signals. In the case of the 4H-SiC MOSFET, the threshold voltage is reduced to 10 volts which is still too large from an applications stand-point. A further reduction of the threshold voltage is obtained by reducing the gate oxide thickness to 0.025 microns as shown in Fig. 9.12. For this gate oxide thickness, the threshold voltage for the silicon MOSFET drops below 1 volt indicating the need to increase the P-base doping concentration to 2×10^{17} cm^{-3}. For the case of 4H-SiC MOSFETs, a threshold voltage now becomes about 6 volts which is marginally acceptable. However, operation of very high voltage power MOSFETs with such thin gate oxides may create manufacturing and reliability issues when taking into account the high electric fields under the gate oxide in the semiconductor.

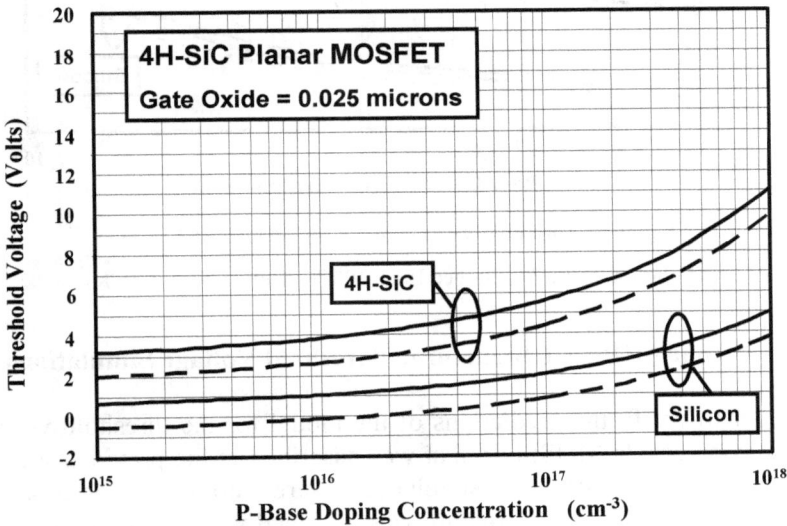

Fig. 9.12 Threshold Voltage of 4H-SiC Planar MOSFETs.

(Dashed lines include the impact of N$^+$ Polysilicon Gate and an Oxide Fixed Charge of 2×10^{11} cm^{-2})

The threshold voltage of power MOSFETs decreases with increasing temperature. This variation is shown in Fig. 9.13 for the case of 4H-SiC and silicon MOSFETs using a gate oxide thickness of 0.025 microns. It can be seen that there is a relatively small shift in the threshold voltage with temperature. Consequently, this effect is not an important consideration during the design of 4H-SiC power MOSFETs.

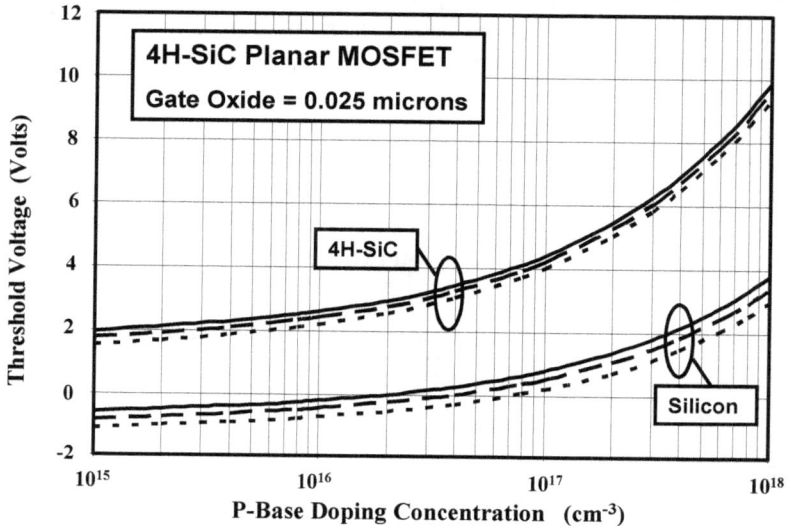

Fig. 9.13 Threshold Voltage of 4H-SiC Planar MOSFETs.

(Solid Line: 300 °K; Dashed Line: 400 °K; Dotted Line: 500 °K)

9.1.4 Planar MOSFET Structure: Threshold Voltage Simulations

In order to verify the predictions of the model for the threshold voltage for silicon carbide MOSFETs that was described in the previous section, two-dimensional numerical simulations were performed for a lateral MOSFET structure. This structure was designed with a long channel (10 microns) and uniform P-base doping concentration, as shown in Fig. 9.14, to allow extraction of the threshold voltage and the channel inversion layer mobility. Simulations performed with this structure using different P-base doping concentrations and gate oxide thickness values enables extraction of the dependence of the threshold voltage on these parameters.

The transfer characteristics obtained for the case of a gate oxide thickness of 0.05 microns, are shown in Fig. 9.15, for various P-base doping concentrations. The threshold voltage for various P-base doping concentrations was extracted by using a drain current level indicated by

the dashed line in the figure. The extracted threshold voltage is compared
with the predictions of the analytical model in Fig. 9.16.

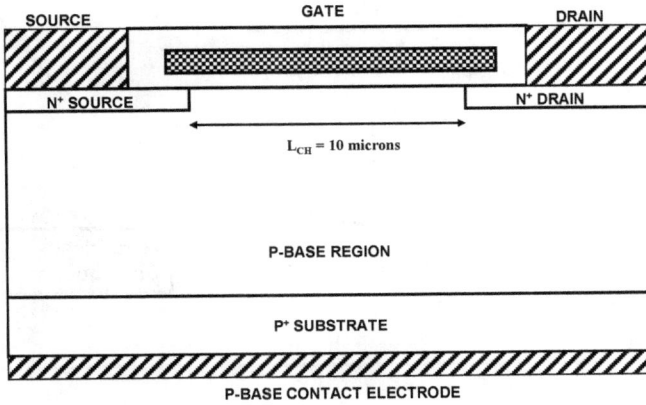

Fig. 9.14 The Planar Lateral 4H-SiC MOSFET Structure.

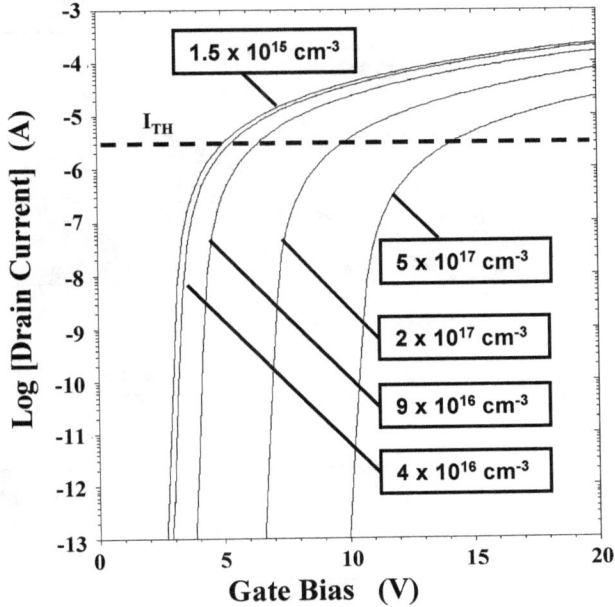

Fig. 9.15 Transfer Characteristics of the Planar Lateral 4H-SiC MOSFET Structure.

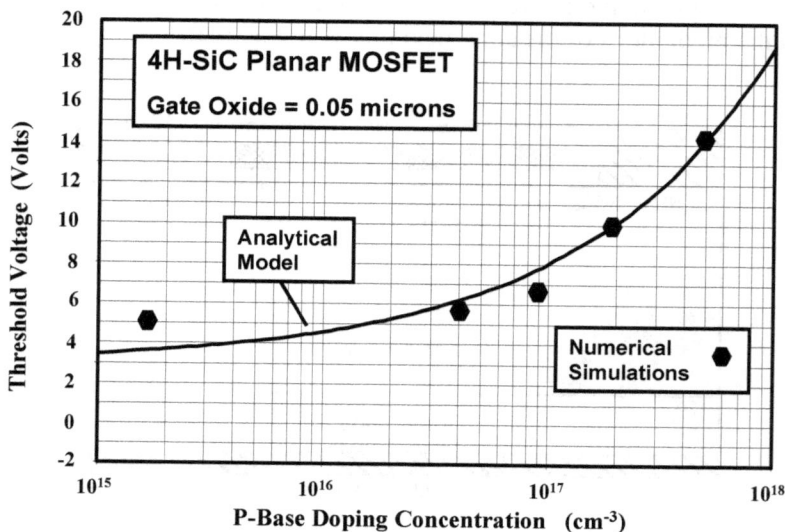

Fig. 9.16 Threshold Voltage of 4H-SiC Planar MOSFETs.

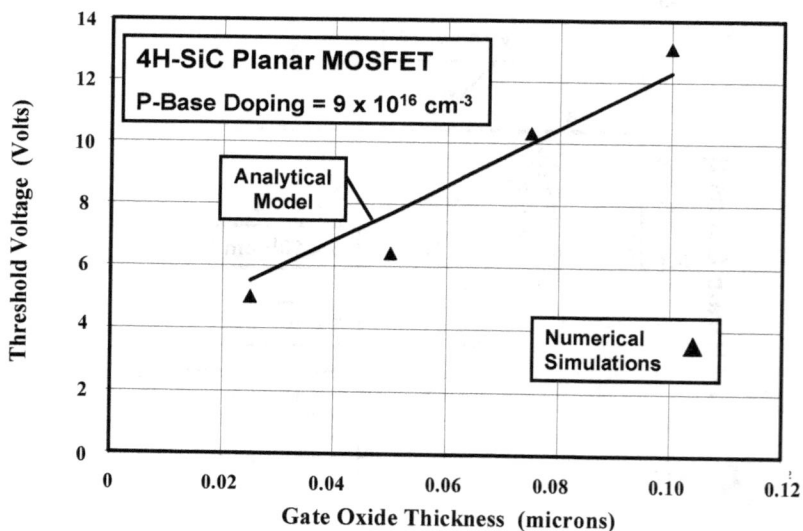

Fig. 9.17 Threshold Voltage of 4H-SiC Planar MOSFETs.

From Fig. 9.16, it can be seen that there is an excellent agreement between the threshold voltages predicted by the analytical model and the numerical simulations over a broad range of P-base doping concentrations. Note that no adjustment was made for the work function of the gate electrode or fixed oxide charge in both the simulations and in the analytical model. Numerical simulation of the lateral 4H-SiC MOSFET were also performed with various gate oxide thicknesses while maintaining a P-base doping concentration of 9 x 10^16 cm^-3. The results of the threshold voltage extracted from these simulations are compared with those obtained using the analytical model in Fig. 9.17. Once again, the predictions of the analytical model are in good agreement with the simulations indicating that this model can be utilized for the design of silicon carbide MOSFETs.

9.1.5 Silicon Carbide MOSFET Structure: Hot Electron Injection

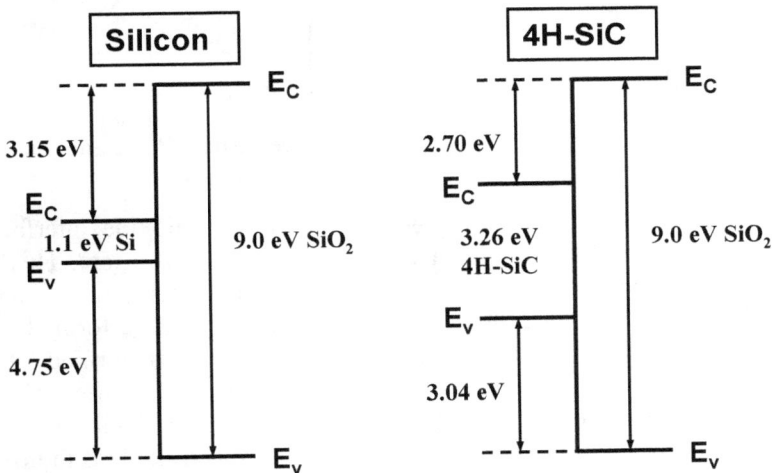

Fig. 9.18 Energy Band Offsets between Oxide and Semiconductors.

In any MOSFET structure, the injection of electrons into the gate oxide can occur when the electrons gain sufficient energy in the semiconductor to surmount the potential barrier between the semiconductor and the oxide. The energy band offsets between the semiconductor and silicon dioxide are shown in Fig. 9.18 for silicon and 4H-SiC for comparison purposes. Due to the larger band gap of 4H-SiC, the band offset between the conduction band edges for the semiconductor and silicon dioxide is significantly smaller than for the case of silicon.

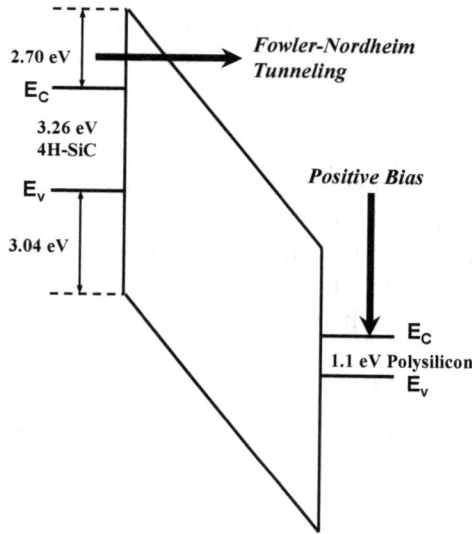

Fig. 9.19 Energy Band Diagram for the Polysilicon/Oxide/4H-SiC Structure.

The band diagram for the 4H-SiC silicon dioxide interface is shown in Fig. 9.19 when a positive bias is applied to the gate. This bias condition is typical for the on-state mode of operation in power MOSFETs. In this illustration, the gate was assumed to be formed using polysilicon. It can be seen that a narrow barrier is formed at the semiconductor-oxide interface which can be penetrated by the tunneling of electrons from the conduction band into the oxide. This produces a Fowler-Nordheim tunneling current that injects electrons into the oxide. This current has been observed in measurements reported on both 6H-SiC[11] and 4H-SiC[12] MOS-capacitors. The trapping of these electrons within the gate oxide can cause shifts in the threshold voltage of the MOSFET leading to reliability problems.

9.2 Planar Power MOSFET Structure: Numerical Simulations

In order to gain insight into the operation of the planar 4H-SiC power MOSFET structure, two-dimensional numerical simulations were performed with a drift region doping concentration of 1×10^{16} cm^{-3} and thickness of 20 microns corresponding to a parallel-plane breakdown voltage of 3000 volts. The baseline device had a gate oxide thickness of 0.1 microns, which is typical for devices that have been fabricated and reported in the literature. The P-base region had a depth of 1 micron for all the structures with a doping concentration of 9×10^{16} cm^{-3}. The JFET width for the baseline device was 1 micron.

9.2.1 Planar Power MOSFET Structure: Blocking Characteristics

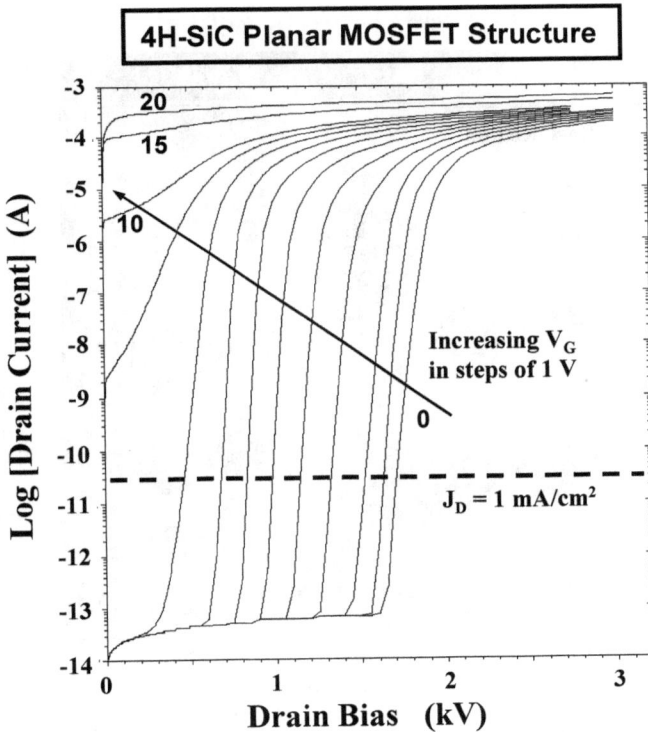

Fig. 9.20 Blocking Characteristics of the Planar 4H-SiC MOSFET Structure.

The blocking capability of the planar 4H-SiC power MOSFET was investigated by maintaining zero gate bias while increasing the drain voltage. It was found that the drain current begins to flow at a drain bias of about 1600 volts and then increases at an exponential rate as shown in

Fig. 9.20. In this figure, the drain current is also shown when various positive gate bias voltages are applied to the planar 4H-SiC power MOSFET structure. The exponential increase in the drain current with increasing drain bias voltage occurs for gate bias voltages of upto 8 volts. Thus, the exponential current-voltage relationship holds true as long as the gate bias is below the threshold voltage for the device. (The threshold voltage for this structure is about 10 volts as shown later in this section.) This behavior is associated with current flow due to the reach-through phenomenon.

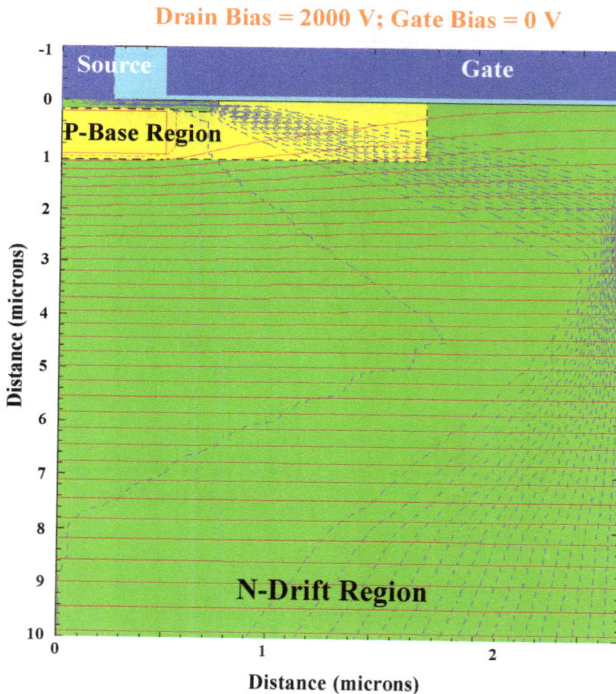

Fig. 9.21 Reach-Through Phenomenon in the Planar 4H-SiC Power MOSFET.

Reach-through breakdown in the planar 4H-SiC power MOSFET is induced when the depletion region extends through the P-base region. In order to illustrate this effect, the potential and current distributions within the planar 4H-SiC planar MOSFET structure are shown in Fig. 9.21 when a drain bias of 2000 volts is applied with zero gate bias. It can be seen that the current flow (indicated by the dashed lines) occurs from the N^+ source region into the drift region through the P-base region. If the device had undergone breakdown due to impact ionization, the current flow would have occurred into the contact for the P-base region located on the left-hand edge of the structure and not through the N^+ source region.

Drain Bias = 2000 V

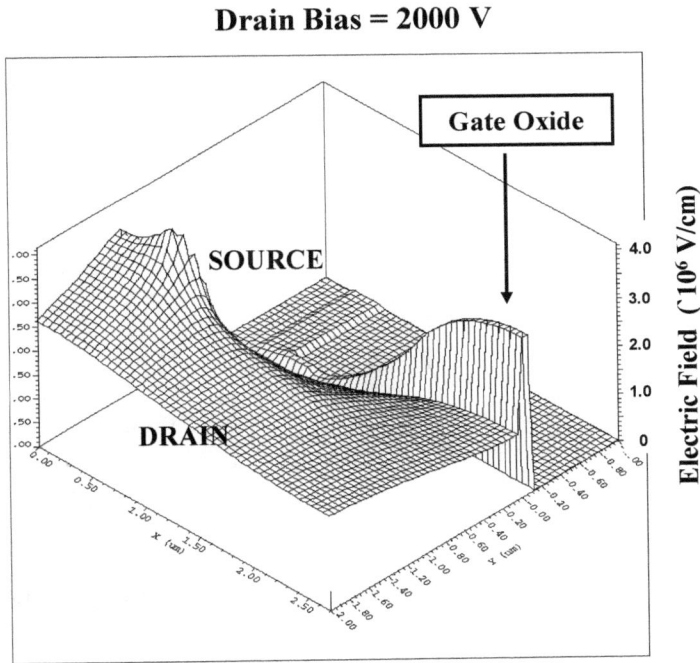

Fig. 9.22 Electric Field Distribution in the Planar 4H-SiC MOSFET Structure.

The above discussion illustrates one of the drawbacks of the planar power MOSFET structure when applied to silicon carbide. A second problem with this structure is associated with the high electric field generated at the gate oxide interface. This problem can be illustrated by examination of the electric field distribution within the planar 4H-SiC power MOSFET structure under voltage blocking conditions. A three-dimensional view of the electric field distribution is shown in Fig. 9.22 for a drain bias of 2000 volts including the portion of the structure containing the gate oxide. The field distribution shown in this figure indicates that the electric field at the gate oxide is slightly suppressed by the presence of the JFET region. However, in spite of this field reduction, it can be seen that the electric field in the oxide has reached a magnitude of close to 3×10^6 V/cm. Such larger electric fields in the oxide can lead to reliability problems due to injection of hot electrons into the oxide.

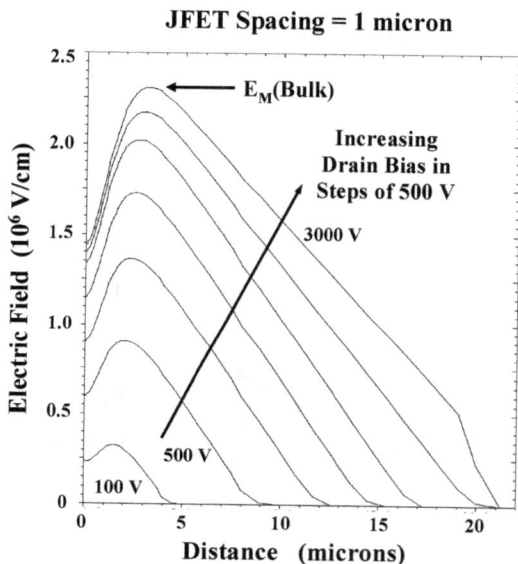

Fig. 9.23 Electric Field Distribution in the Planar 4H-SiC MOSFET Structure.

In the planar 4H-SiC power MOSFET structure, the largest electric field at the gate oxide interface occurs at the center of the JFET region. The behavior of this electric field with increasing drain bias is shown in Fig. 9.23 for the case of a JFET width of 1 micron. From this figure, it is apparent that there is some shielding of the gate oxide by the JFET region due to the formation of a potential barrier when this region becomes depleted at lower drain voltages. The maximum electric field occurs inside the semiconductor at a depth of about 3 microns. The electric field at the gate oxide interface is reduced to 60 percent of the bulk electric field. However, this is insufficient to keep the electric field in the gate oxide in the reliable operating regime. As previously discussed, the specific on-resistance for the planar MOSFET structure can be reduced by increasing the JFET width. However, the suppression of the electric field at the gate oxide is almost completely lost when this is done as illustrated in Fig. 9.24 for the case of a JFET width of 3.3 microns.

JFET Spacing = 3.3 microns

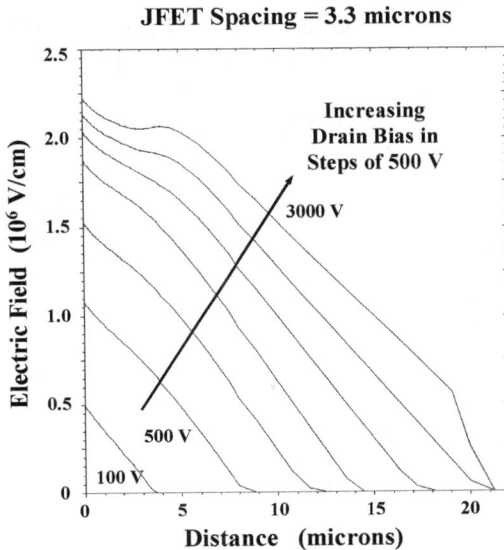

Fig. 9.24 Electric Field Distribution in the Planar 4H-SiC MOSFET Structure.

9.2.2 Planar Power MOSFET Structure: On-State Characteristics

The on-state voltage drop for the planar 4H-SiC power MOSFET structure is determined by its specific on-resistance. The on-resistance for the structure was extracted by performing numerical simulations with a gate bias of 20 volts. A relatively high gate bias was required because of the large threshold voltage of these devices as already discussed in previous sections of this chapter. As expected, the specific on-resistance was dependent upon the JFET region width and the channel inversion layer mobility.

The current distribution within the planar 4H-SiC power MOSFET structure is shown in Fig. 9.25 for the case of JFET widths of 1 and 8.3 microns. It can be seen that the current is constricted in the JFET region when its width is 1 micron. This phenomenon makes the JFET resistance contribution equal to the channel resistance contribution when the channel mobility is assumed to be 100 cm^2/Vs. When the JFET width is increased, the JFET resistance decreases significantly and becomes a small fraction of the overall resistance of the device. However, the increase in cell pitch required to accommodate the larger JFET region produces an increase in the specific on-resistance of the structure. The current flow-lines shown in the lower part of Fig. 9.25 demonstrate that the current flow in the JFET region is occurring primarily near the edge of the channel. This justifies the use of a small value for the parameter 'K' in the analytical model for the accumulation region resistance.

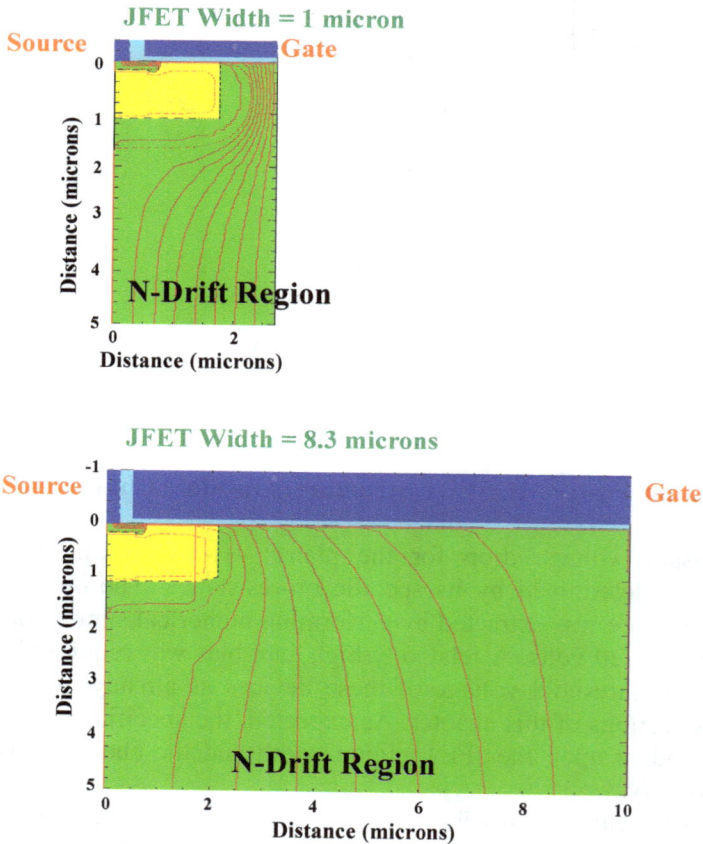

Fig. 9.25 On-State Current Distribution in the Planar 4H-SiC Power MOSFETs.

The simulations of the specific on-resistance were done for each of the structures with different values for the inversion layer mobility.

The specific on-resistance was obtained by using a drain bias of 1 volt and sweeping the gate bias to 20 volts to obtain the transfer characteristic. A drain bias of 1 volt was used for these simulations because the on-state voltage drop for these structures is typically in this range. A typical transfer characteristic obtained using this method is shown in Fig. 9.26 for the device with JFET width of 1 micron. From the transfer characteristics, it can be seen that the threshold voltage that determines the on-resistance is approximately 10 volts. This value is close to that calculated using the analytical model discussed in the previous section for the chosen P-base doping concentration of 9×10^{16} cm^{-3} and gate oxide thickness of 0.1 microns. The impact of changing the channel inversion layer mobility is also shown in this figure. As expected, there is a reduction in the drain current as the channel mobility is reduced.

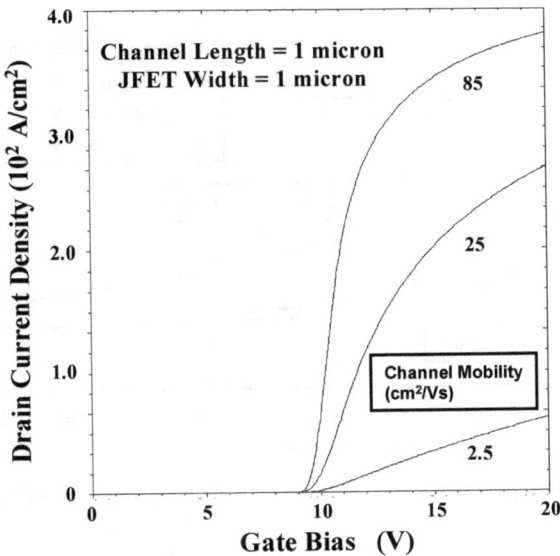

Fig. 9.26 Transfer Characteristics of the 4H-SiC Planar MOSFETs.

The specific on-resistances obtained from the on-state simulations are compared in Fig. 9.27 with the analytically calculated values obtained by using the model developed in the previous section. Three cases of JFET region width have been taken into consideration.

For the simulation cases, the channel mobility values were extracted from the *i-v* characteristics obtained for the lateral 4H-SiC MOSFET structure that was previously described (see Fig. 9.14) by using the same interface degradation factor as used during the simulation of the vertical 4H-SiC planar MOSFET structure. It can be seen that the predictions of the analytical model are in excellent agreement with the values obtained from the simulations in all cases. These results provide confidence in the ability to use the analytical model during the design of the planar 4H-SiC power MOSFET structure. From the figure, it can be concluded that a channel mobility of more than 50 cm^2/Vs is satisfactory for achieving a low specific on-resistance for a 2 kV planar 4H-SiC power MOSFET structure with JFET width of 1 micron. In addition, it can be concluded that there is no benefit to increasing the JFET width beyond 1 micron.

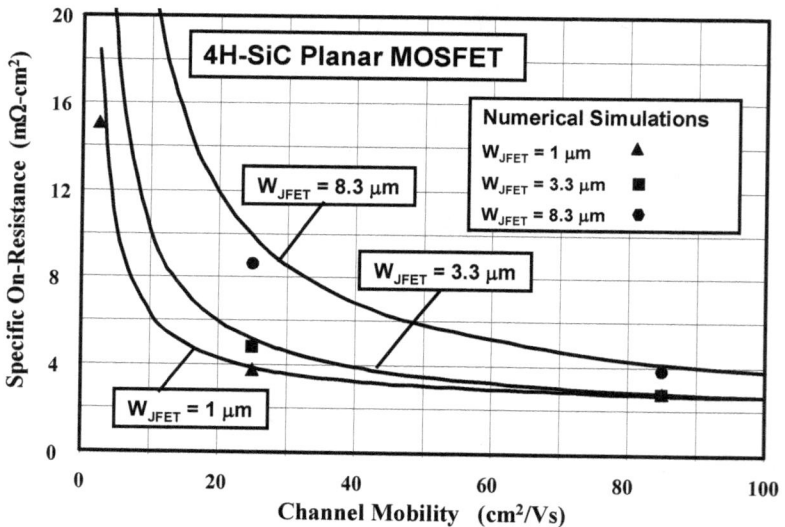

Fig. 9.27 Specific On-Resistance for the 4H-SiC Planar MOSFETs.

9.2.3 Planar Power MOSFET Structure: Output Characteristics

The output characteristics of the power MOSFET defines the region within which the device can operate during the switching transients in applications. The current-voltage loci during the transient must be chosen to remain within the boundary defined by the output characteristics. For a power MOSFET, one of the boundaries is defined by the on-resistance and a second one by its breakdown voltage. These boundaries limit the maximum on-state current that can be tolerated by the device after taking into consideration the thermal impedance of the package and the maximum off-state voltage that can be impressed upon the device without excessive power dissipation due to leakage current. The gate controlled current saturation in power MOSFETs defines its third boundary. Here, the magnitude of the output resistance becomes important with a larger value preferred for providing greater control over the drain current by the applied gate bias voltage.

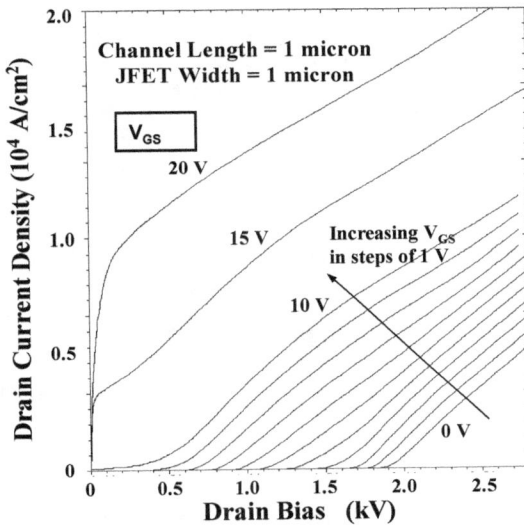

Fig. 9.28 Output Characteristics of a 4H-SiC Planar MOSFET.

The output characteristics for the 4H-SiC planar power MOSFET were analyzed by biasing the gate at various values and sweeping the drain voltage upto 2800 volts. The characteristics obtained for the structure with JFET width of 1 micron are shown in Fig. 9.28. It can be observed that significant current flow occurs when the drain bias exceeds 1900 volts for a zero gate bias voltage. This current flow is due to the reach-through breakdown problem discussed in the previous sections. The reach-through current also degrades the output resistance of the device at all positive gate bias voltages. Even further degradation of the output resistance and reduction of the maximum blocking voltage is observed when the JFET width is increased as shown in Fig. 9.29. Consequently, the output characteristics observed for the 4H-SiC planar MOSFET structure are poor when compared with those described for the Baliga-Pair configuration in the previous chapter.

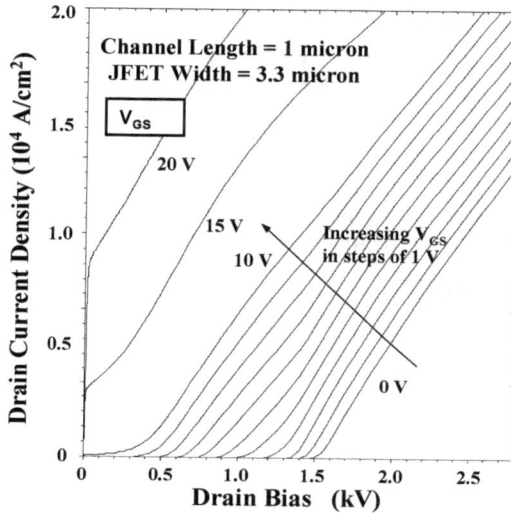

Fig. 9.29 Output Characteristics of a 4H-SiC Planar MOSFET.

9.3 Planar Power MOSFET: Experimental Results

The need to obtain a high quality interface between silicon carbide and the gate dielectric was identified as a challenging endeavor from the inception of interest in the development of unipolar transistors from this semiconductor material[13]. Initially, it was impossible to fabricate silicon carbide power MOSFETs with specific on-resistances below those reported for silicon devices until the interface was sufficiently improved to obtain adequate inversion layer mobility. Further, early work on planar MOSFETs did not sufficiently take into consideration the fundamental issues that have been discussed in the previous sections of this chapter dealing with the reach-through problem and the high electric field in the oxide. These issues are discussed here in the historical context of developing silicon carbide planar power MOSFET structures.

9.3.1 Silicon Carbide MOSFET: Inversion Layer Mobility

Early investigations of the interface between P-type silicon carbide and thermally grown oxide indicated a high density of interface states and positive charge in the oxide. Significant improvement in the interface quality was achieved for 6H-SiC by anneal the thermally grown oxide in a NO ambient[14] leading to a peak inversion layer mobility of 70 cm^2/Vs. However, the inversion layer mobility for 4H-SiC was reported[15] to be less than 1 cm^2/Vs.

In 1998, a break-through was achieved at PSRC[16] with the use of deposited oxides on 4H-SiC leading to a reported[17] record high inversion layer mobility of 165 cm^2/Vs. A detailed study[18] of the process steps responsible for producing the improved interface resulting in the high channel mobility was also undertaken at PSRC. This work produced the first observation of phonon scattering limited inversion layer mobility (which decreased with increasing temperature), and demonstrated that a wet oxide anneal of the deposited oxide was the critical step for producing the high inversion layer mobility. The process proposed and demonstrated at PSRC was subsequently reproduced by the sponsors[19] as well by other research groups[20]. These results indicate that it is possible to achieve sufficiently high inversion layer mobility in 4H-SiC power MOSFETs to obtain low specific on-resistance in high voltage structures.

The low inversion layer mobility observed in 4H-SiC MOSFETs fabricated using thermally grown gate oxides has been traced to the trapping of the carriers at interface states. Hall effect measurements performed on lateral MOSFETs have demonstrated that the effective Hall mobility for carriers in the inversion layer is high when the trapping effect is taken into consideration[21]. Due to the trapping effect, as recently as 2001, the measured effective mobility in 4H-SiC MOSFETs has been reported[22] to be less than 10 cm^2/V-s when thermal oxidation is used to form the gate oxide. In these devices, the effective mobility was found to increase with temperature which is a signature of trap dominated current conduction in the channel. By using NO ambient anneals of the thermally grown gate oxide, an improvement in the effective channel mobility to 30-35 cm^2/V-s was reported[23] in 2001.

In conclusion, process technology has been demonstrated to obtain sufficiently high inversion layer mobility on the surface of P-type 4H-SiC by using deposited silicon dioxide films. The quality of interface between thermally grown oxide and 4H-SiC remains inferior with significant progress being made towards its improvement.

9.3.2 The 4H-SiC Planar MOSFET: Experimental Results

As already mentioned earlier in the chapter, in recognition of the low diffusion coefficients for dopants in silicon carbide, it was proposed[5] that the P-base and N^+ source ion implantations for the planar silicon carbide power MOSFET be staggered by using photoresist masks rather that be defined by the gate edge as conventionally done for silicon devices. This idea was first successfully demonstrated[6] for 6H-SiC devices in 1997 under the DI-MOSFET moniker. The authors fabricated the devices using multiple energy boron implants to form a P-base region with a box profile at a depth of 1 micron. Devices were designed with channel length of 2 and 5 microns, and the JFET region width (W_J in Fig. 9.4) was varied from 2.5 to 15 microns. The channel inversion layer mobility was found to be about 20 cm^2/Vs for these 6H-SiC MOSFETs fabricated using a thermally grown oxide. The best specific on-resistance of 130 $m\Omega$-cm^2 was obtained for the smallest JFET width and channel length. The breakdown voltage of the device was found to be 760 volts as limited by reach-through as well as gate oxide rupture leading to catastrophic failure.

The DI-MOSFET structure was fabricated[9] from 4H-SiC more recently by using 25 micron thick epitaxial layers with doping concentration of 3 x 10^{15} cm^{-3}. In spite of using the previously discussed deposited oxide process[8], an inversion layer mobility of only 14 cm^2/Vs was observed. Due to the low channel mobility, the devices exhibited a relatively large specific on-resistance of 55 $m\Omega$-cm^2 even when a high gate bias of 25 volts was applied. The device was able to support nearly 2000 volts at zero gate bias. The authors associated the blocking voltage to be limited by open-base bipolar breakdown, namely, the reach-through problem discussed earlier in this chapter. This work was extended[24] to achieve a breakdown voltage of 2400 volts with a specific on-resistance of 42 $m\Omega$-cm^2 in large area (0.1 cm^2) device structures; as well as devices[25] with a breakdown voltage of 10 kV with a specific on-resistance of 236 $m\Omega$-cm^2.

9.4 Summary

The extension of the silicon power DMOSFET architecture to silicon carbide has been critically examined in this chapter. It has been shown that the direct applicability of the structure to silicon carbide is hindered by the following issues:

(a) The lack of any significant diffusion of impurities in silicon carbide requires the staggering of the P-base and N^+ source ion implants with photoresist masking layers, which limits the minimum achievable channel length;

(b) The higher threshold voltage for silicon carbide when compared with silicon for the same doping concentration exacerbates the reach-through problem within the P-base region limiting the blocking voltage capability;

(c) The higher threshold voltage for silicon carbide MOSFETs increases the specific on-resistance due to the enhanced channel contribution;

(d) A high electric field develops at the gate oxide that can lead to catastrophic failure of the silicon carbide device well before the electric field in the semiconductor can reach the critical electric field for breakdown;

(e) The poor inversion layer mobility obtained with thermally grown oxide can limit the specific on-resistance to well above the capability of the drift region.

These limitations of the conventional DMOSFET structure can be overcome by shielding the P-base region as discussed in the next chapter.

References

[1]   D. A. Grant and J. Gowar, "Power MOSFETs: Theory and Applications", John Wiley and Sons, 1989.

[2]   B. J. Baliga, "Power Semiconductor Devices", PWS Publishing Company, 1996.

[3]   L. Lorenz, G. Deboy, A. Knapp, and M. Marz, "COOLMOS – A New Milestone in High Voltage Power MOS", IEEE International Symposium on Power Semiconductor Devices and ICs, Abstract 1.1, pp. 3-10, 1999.

[4]   B.J. Baliga, "Power Semiconductor Devices for Variable Frequency Drives", Proceedings of the IEEE, Vol. 82, pp. 1112-1122, 1994.

[5]   B. J. Baliga and M. Bhatnagar, "Method of Fabricating Silicon Carbide Field Effect Transistor", U. S. Patent 5,322,802, Issued June 21, 1994.

[6]   J. N. Shenoy, J. A. Cooper, and M. R. Melloch, "High Voltage Double-Implanted Power MOSFETs in 6H-SiC", IEEE Electron Device Letters, Vol. 18, pp. 93-95, 1997.

[7]   B. J. Baliga, "Silicon Carbide Semiconductor Devices having Buried Silicon Carbide Conduction Barrier Layers Therein", U. S. Patent 5,543,637, Issued August 6, 1996.

[8]   S. Sridevan and B. J. Baliga, "Inversion Layer Mobility in SiC MOSFETs", Silicon Carbide and Related Materials – 1997, Material Science Forum, Vol. 264-268, pp. 997-1000, 1998.

[9]   S-H Ryu, et al, "Design and Process Issues for Silicon Carbide Power DiMOSFETs", Material Research Society Symposium Proceeding, Vol. 640, pp. H4.5.1-H4.5.6, 2001.

[10] B. J. Baliga, "Power Semiconductor Devices", Chapter 7, pp. 357-362, PWS Publishing Company, 1996.

[11] D. Alok, P. McLarty, and B. J. Baliga, "Electrical Properties of Thermal Oxide Grown on N-type 6H-Silicon Carbide", Applied Physics Letters, Vol. 64, pp. 2845-2846, 1994.

[12] J. B. Casady, et al, "4H-SiC Power Devices: Comparative Overview of UMOS, DMOS, and GTO Device Structures", Material Society Research Symposium Proceedings, Vol. 483, pp. 27-38, 1998.

[13] B. J. Baliga, "Impact of SiC on Power Devices", Proceedings of the 4th International Conference on Amorphous and Crystalline Silicon Carbide", pp. 305-313, 1991.

[14] L. Lipkin and J. W. Palmour, "Improved Oxidation Procedures for Reduced SiO_2/SiC Defects", J. Electronic Materials, Vol. 25, pp. 909-915, 1996.

[15] R. Schorner, et al, "Significantly improved performance of MOSFETs on Silicon Carbide using the 15R-SiC Polytype", IEEE Electron Device Letters, Vol. 20, pp. 241-244, 1999.

[16] S. Sridevan and B. J. Baliga, "Lateral N-channel Inversion Mode 4H-SiC MOSFETs", PSRC Technical Report TR-97-019, 1997.

[17] S. Sridevan and B. J. Baliga, "Lateral N-channel Inversion Mode 4H-SiC MOSFETs", IEEE Electron Device Letters, Vol. 19, pp. 228-230, 1998.

[18] S. Sridevan and B. J. Baliga, "Phonon Scattering Limited Mobility in SiC Inversion Layers", PSRC Technical Report TR-98-03, 1998.

[19] D. Alok, E. Arnold, and R. Egloff, "Process Dependence of Inversion Layer Mobility in 4H-SiC Devices", Silicon Carbide and Related Materials – 1999, Material Science Forum, Vol. 338-342, pp. 1077-1080, 2000.

[20] K. Chatty, et al, "Hall Measurements of Inversion and Accumulation-Mode 4H-SiC MOSFETs", Silicon Carbide and Related Materials – 2002, Material Science Forum, Vol. 389-393, pp. 1041-1044, 2002.

[21] N. S. Saks, S. S. Mani, and K. Agarwal, Applied Physics Letters, Vol. 76, pp. 2250-2251, 2000.

[22] S.Harada, et al, "Temperature Dependence of the Channel Mobility and Threshold Voltage in 4H and 6H-SiC MOSFETs", Material Society Research Symposium Proceedings, Vol. 640, pp. H5.37.1-H5.37.6, 2001.

[23] G. Y. Chung, et al, "Improved Inversion Channel Mobility for 4H-SiC MOSFETs following High Temperature Anneals in Nitric Oxide", IEEE Electron Device Letters, Vol. 22, pp. 176-178, 2001.

[24] S-H. Ryu, et al, "Large-Area 93.3 mm x 3.3 mm) Power MOSFETs in 4H-SiC", Silicon Carbide and Related Materials – 2001, Material Science Forum, Vol. 389-393, pp. 1195-1198, 2002.

[25] S-H. Ryu, et al, "Development of 10 kV 4H-SiC Power DMOSFETs", Silicon Carbide and Related Materials – 2003, Material Science Forum, Vol. 457-460, pp. 1385-1388, 2004.

Chapter 10

Shielded Planar MOSFETs

In the previous chapter, it was demonstrated that the conventional silicon planar power DMOSFET structure is not satisfactory for utilization in silicon carbide due to a reach-through problem, a relatively high threshold voltage, a high electric field developed in the gate oxide, and the relatively low inversion layer mobility in the channel. All of these issues can be addressed in a satisfactory manner by shielding the channel from the high electric field developed in the drift region. The concept of shielding of the channel region was first proposed[1] at PSRC in the early 1990's with a U.S. patent issued in 1996. The shielding was accomplished by formation of either a P-type region under the channel or by creating a high resistivity conduction barrier region under the channel. The issue of low inversion layer channel mobility was addressed by utilizing an accumulation mode of operation. As discussed in this chapter, the accumulation mode is also attractive from the point of view of reducing the threshold voltage so as to enhance the channel conductivity.

This chapter begins with a review of the basic principles of operation of the shielded planar MOSFET structure. The impact of shielding on ameliorating the reach-through breakdown in silicon carbide is described. This shielding approach has more recently been used very effectively for improvement of even silicon low voltage planar power MOSFETs[2]. Based upon fundamental considerations, the difference between the threshold voltage for inversion and accumulation mode silicon carbide structures is then analyzed. The results of the analysis of the planar silicon carbide MOSFET structures by using two-dimensional numerical simulations are described in this chapter. It is shown that the shielding concept enables reduction of the electric field developed in the gate oxide as well leading to the possibility to fully utilize the breakdown field strength of the underlying semiconductor drift region. Experimental results on relevant structures are discussed at the end of this chapter to define the state of the development effort on these devices. The shielded accumulation-mode MOSFET structures (named the ACCUFET) discussed in this chapter have the most promising characteristics for the development of monolithic power switches from silicon carbide.

10.1 Shielded Planar MOSFET Structure

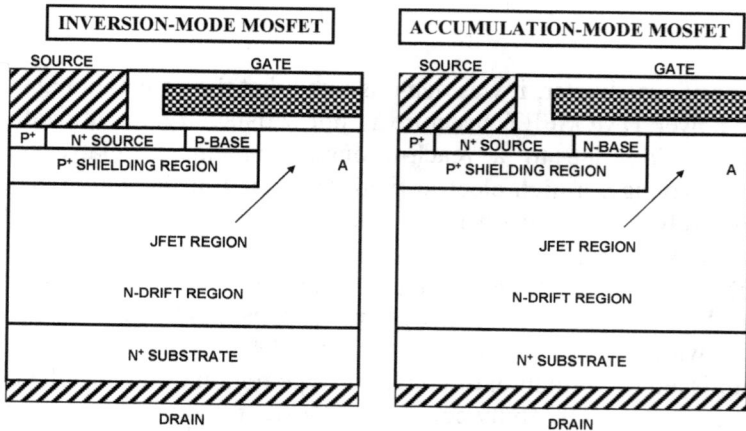

Fig. 10.1 Shielded Planar MOSFET Structures.

The basic structure of the shielded planar MOSFET is shown in Fig. 10.1 with either an inversion layer channel or an accumulation layer channel. In the case of the structure with the inversion layer channel, the P^+ shielding region extends under both the N^+ source region as well as under the P-base region. It could also extend beyond the edge of the P-base region. In the case of the structure with the accumulation layer channel, the P^+ shielding region extends under the N^+ source region and the N-base region located under the gate. This N-base region can be formed using an uncompensated portion of the N-type drift region or it can be created by adding N-type dopants with ion implantation to control its thickness and doping concentration.

The gap between the P^+ shielding regions is optimized to obtain a low specific on-resistance while simultaneously shielding the gate oxide interface from the high electric field in the drift region. In both structures shown above, a potential barrier is formed at location A after the JFET region becomes depleted by the applied drain bias in the blocking mode. This barrier prevents the electric field from becoming large at the gate oxide interface. When a positive bias is applied to the gate electrode, an inversion layer or accumulation layer channel is formed in the structures enabling the conduction of drain current with a low specific on-resistance.

10.1.1 Shielded Planar MOSFET Structure: Blocking Mode

In the forward blocking mode of the shielded planar MOSFET structure, the voltage is supported by a depletion region formed on both sides of the P^+ region/N-drift junction. The maximum blocking voltage can be determined by the electric field at this junction becoming equal to the critical electric field for breakdown if the parasitic $N^+/P/N$ bipolar transistor is completely suppressed. This suppression is accomplished by short-circuiting the N^+ source and P^+ regions using the source metal as shown on the upper left hand side of the cross-section. This short circuit can be accomplished at a location orthogonal to the cell cross-section if desired to reduce the cell pitch while optimizing the specific on-resistance. If the doping concentration of the P^+ region is high, the reach-through breakdown problem discussed in the previous chapter is completely eliminated. In addition, the high doping concentration in the P^+ region promotes the depletion of the JFET region at lower drain voltages providing enhanced shielding of the channel and gate oxide.

With the shielding provided by the P^+ region, the minimum P-base thickness for 4H-SiC power MOSFETs is no longer constrained by the reach-through limitation. This enables reducing the channel length below the values associated with any particular doping concentration of the P-base region in Fig. 9.3. In addition, the opportunity to reduce the P-base doping concentration enables decreasing the threshold voltage. The smaller channel length and threshold voltage provide the benefits of reducing the channel resistance contribution.

In the case of the accumulation mode planar MOSFET structure, the presence of the sub-surface P^+ shielding region under the N-base region provides the potential required for *completely depleting* the N-base region if its doping concentration and thickness are appropriately chosen. This enables *normally-off operation* of the accumulation mode planar MOSFET with zero gate bias. It is worth pointing out that this mode of operation is fundamentally different that that of buried channel MOS devices[3]. Buried channel devices contain an un-depleted N-type channel region that provides a current path for the drain current at zero gate bias. This region must be depleted by a negative gate bias creating a normally-on device structure.

In the accumulation-mode, planar MOSFET structure, the depletion of the N-base region is accompanied by the formation of the potential barrier for the flow of electrons through the channel. The channel potential barrier does not have to have a large magnitude because the depletion of the JFET region screens the channel from the drain bias as well.

The maximum blocking voltage capability of the shielded planar MOSFET structure is determined by the drift region doping concentration and thickness as already discussed in chapter 3. However, to fully utilize the high breakdown electric field strength available in silicon carbide, it is important to screen the gate oxide from the high field within the semiconductor. In the shielded planar MOSFET structure, this is achieved by the formation of a potential barrier at location A by the depletion of the JFET region at a low drain bias voltage.

10.1.2 Shielded Planar MOSFET Structure: Forward Conduction

In the shielded planar MOSFET structure, current flow between the drain and source can be induced by creating an inversion layer channel on the surface of the P-base region or an accumulation layer channel on the surface of the N-base region. The current path is similar to that already shown in Fig. 9.4 by the dotted area. The current flows through the channel formed due to the applied gate bias into the JFET region via the accumulation layer formed above it under the gate oxide. It then spreads into the N-drift region at a 45 degree angle and becomes uniform through the rest of the structure. The total on-resistance for the planar power MOSFET structure is determined by the resistance of these components in the current path:

$$R_{on,sp} = R_{CH} + R_A + R_{JFET} + R_D + R_{subs} \qquad [10.1]$$

where R_{CH} is the channel resistance, R_A is the accumulation region resistance, R_{JFET} is the resistance of the JFET region, R_D is the resistance of the drift region after taking into account current spreading from the JFET region, and R_{subs} is the resistance of the N$^+$ substrate. These resistances can be analytically modeled by using the current flow pattern indicated by the shaded regions in Fig. 9.4.

For the shielded planar MOSFET structure with the P-base region, the specific channel resistance is given by:

$$R_{CH} = \frac{(L_{CH} \cdot p)}{\mu_{inv} C_{ox} (V_G - V_T)}$$

[10.2]

where L_{CH} is the channel length as shown in Fig. 9.4, μ_{inv} is the mobility for electrons in the inversion layer channel, C_{ox} is the specific capacitance of the gate oxide, V_G is the applied gate bias, and V_T is the threshold voltage. For the shielded planar MOSFET structure with the N-base region, the specific channel resistance is given by:

$$R_{CH} = \frac{(L_{CH} \cdot p)}{\mu_a C_{ox} (V_G - V_T)}$$

[10.3]

where μ_a is the mobility for electrons in the accumulation layer channel. As discussed later in this chapter, much larger accumulation layer mobility has been experimentally observed in silicon carbide allowing reduction of the specific on-resistance. In addition, the threshold voltage for the accumulation mode is smaller than for the inversion mode allowing further improvement in the channel resistance contribution. The rest of the resistance components in the shielded planar MOSFET structure can be modeled by using the same equations provided in chapter 9.

The specific on-resistances of 4H-SiC shielded planar MOSFETs with a drift region doping concentration of 1×10^{16} cm^{-3} and thickness of 20 microns were modeled using the analytical expressions. For comparison of the inversion and accumulation mode structures, a gate oxide thickness of 0.1 microns was used. The dimension 'a' in the structure shown in Fig. 9.4 was kept at 1 micron. This value can be achieved using 0.5 micron design rules. The accumulation mobility was assumed to be 200 cm^2/Vs while an inversion layer mobility of 50 cm^2/Vs was used based upon typical values reported in the literature as discussed later in this chapter. A K-factor of 0.1 was used to provide correlation with the simulations. The P$^+$ region was assumed to have a depth of 1 micron and the effective gate drive voltage ($V_G - V_T$) was assumed to be 10 volts for the inversion mode structure versus 15 volts for the accumulation mode structure due to the difference between their threshold voltages. A channel length of 1 micron and a JFET width of 2 microns were used for these structures.

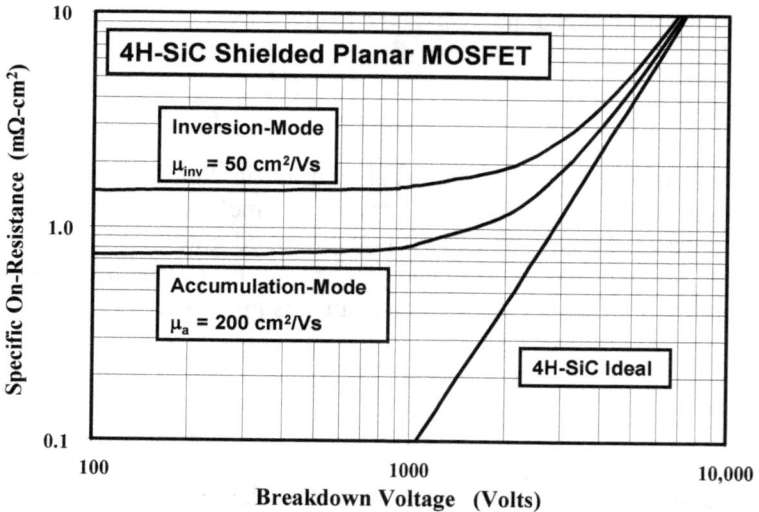

Fig. 10.2 On-Resistance for the 4H-SiC Shielded Planar MOSFETs.

The specific on-resistance for the shielded planar 4H-SiC MOSFETs is plotted in Fig. 10.2 as a function of the breakdown voltage. As discussed in the previous chapter, in performing this modeling, it is important to recognize that the thickness of the drift region (parameter 't' in Fig. 9.4) can become smaller than the cell parameter 's' at lower breakdown voltages (below 700 volts). Under these conditions, the current does not distribute at a 45 degree angle into the drift region from the JFET region. Instead, the current flows from a cross-sectional width of $(W_J - W_P)$ to a cross-section of $(t + W_J - W_P)$.

From Fig. 10.2, it can be seen that specific on-resistance of 4H-SiC shielded planar MOSFETs approaches the ideal specific on-resistance when the breakdown voltage exceeds 5000 volts because the drift region resistance becomes dominant. However, when the breakdown voltage falls below 1000 volts, the channel contribution becomes dominant. In this design regime, the advantage of using an accumulation channel becomes quite apparent. The accumulation-mode device has about 2x lower specific on-resistance. When compared with the planar MOSFET structure without the shielded channel (see Fig. 9.9), the difference in specific on-resistance does not appear to be very significant. However, as discussed in the previous chapter, the unshielded devices were able to support only about half the blocking voltage capability of the drift region due to the reach-through breakdown problem. This limitation is not observed for the shielded planar MOSFET structure as will be shown in the section of this chapter dealing with simulated results.

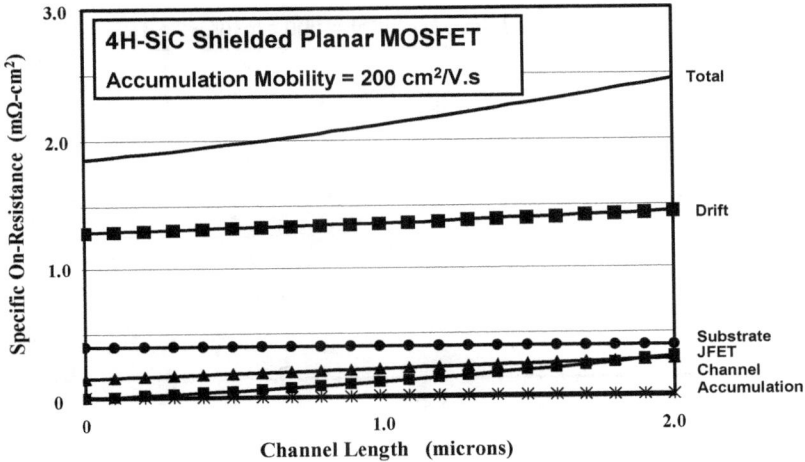

Fig. 10.3 Components of the On-Resistance for the 4H-SiC Shielded Planar MOSFET with Accumulation Channel.

In order to contrast the performance of the accumulation mode planar structure with the performance of the inversion mode structure described in the previous chapter, it is interesting to examine the components of the on-resistance for the accumulation-mode device. The various components of the on-resistance are shown in Fig. 10.3 when a channel accumulation layer mobility of 200 cm^2/Vs was used. This structure has a channel length of 1 micron, a JFET width of 1.5 microns, and a gate oxide thickness of 0.05 microns. A thinner gate oxide was chosen because of the shielding of the gate oxide in this structure. It can be seen that the drift region resistance is dominant under these conditions even for channel lengths of 2 microns. At a channel length of 1 micron, a total specific on-resistance of 2.1 mΩ-cm^2 is obtained, which is about twice the ideal specific on-resistance. This is an excellent value for a device with the drift regions parameters used for the modeling providing a breakdown voltage of 3000 volts. It is worth pointing out that the contribution from the N$^+$ substrate is exceeding that from the channel for this structure. This indicates the need to either reduce the substrate thickness or reduce its resistivity to bring the total specific on-resistance close to that of the drift region.

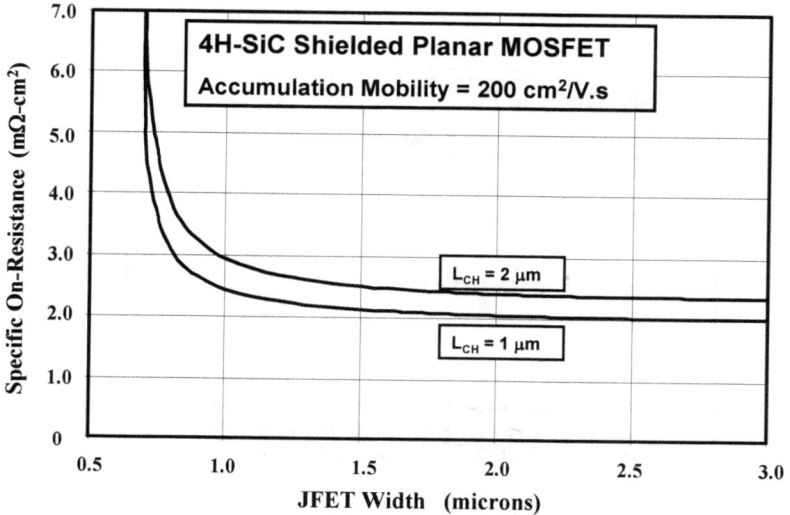

Fig. 10.4 Impact of JFET Width on the On-Resistance for the 4H-SiC Shielded Planar MOSFET with Accumulation Channel.

The JFET width in the planar MOSFET structure must be optimized not only to obtain the lowest specific on-resistance but to also control the electric field at the gate oxide interface. The impact of changes to the JFET width on the specific on-resistance for the accumulation-mode 4H-SiC shielded planar MOSFET structure is shown in Fig. 10.4 for a structure with the same cell parameters that were given above. It can be observed that the specific on-resistance becomes very large when the JFET width is made smaller than about 0.7 microns. This occurs because the zero bias depletion width for 4H-SiC P-N junctions is about 0.6 microns. When the JFET width approaches this value, the current becomes severely constricted resulting in a high specific on-resistance. As the JFET width is increased, the contribution from its resistance decreases rapidly. However, the increase in cell pitch enhances the specific resistance contributions from the channel resulting in a nearly constant total specific on-resistance. A minimum specific on-resistance is observed to occur at a JFET width of above 1.5 microns, which is larger than the value for the planar MOSFETs discussed in the previous chapter. This is due to the larger mobility in the accumulation channel. In addition to the slightly lower specific on-resistance for this structure when compared with the planar MOSFET discussed in the previous chapter (see Fig. 9.6), the specific on-resistance is observed to be less sensitive to the channel length. This provides more latitude with regard to the design rules used for the fabrication of the accumulation mode device.

10.1.3 Shielded Planar MOSFET Structure: Threshold Voltage

As pointed out in the previous chapter, the threshold voltage of the power MOSFET is an important design parameter from an application stand-point. A minimum threshold voltage must be maintained at above 1 volt for most system applications to provide immunity against inadvertent turn-on due to voltage spikes arising from noise. At the same time, a high threshold voltage is not desirable because the voltage available for creating the charge in the channel is determined by ($V_G -$ V_T) where V_G is the applied gate bias voltage and V_T is the threshold voltage. Most power electronic systems designed for high voltage operation (the most suitable application area for silicon carbide devices) provide a gate drive voltage of only upto 10 volts. Based upon this criterion, the threshold voltage should be kept below 3 volts in order to obtain a low channel resistance contribution.

For the inversion-mode shielded planar MOSFET, the threshold voltage can be modeled by defining it as the gate bias at which on-set of *strong inversion* begins to occur in the channel. This voltage can be determined using[4]:

$$V_{TH} = \frac{\sqrt{4\varepsilon_S kTN_A \ln(N_A / n_i)}}{C_{ox}} + \frac{2kT}{q} \ln\left(\frac{N_A}{n_i}\right) \qquad [10.4]$$

where N_A is the doping concentration of the P-base region, k is Boltzmann's constant, and T is the absolute temperature. The presence of positive fixed oxide charge shifts the threshold voltage in the negative direction by:

$$\Delta V_{TH} = \frac{Q_F}{C_{ox}} \qquad [10.5]$$

A further shift of the threshold voltage in the negative direction by 1 volt can be achieved by using heavily doped N-type polysilicon as the gate electrode as routinely done for silicon power MOSFETs.

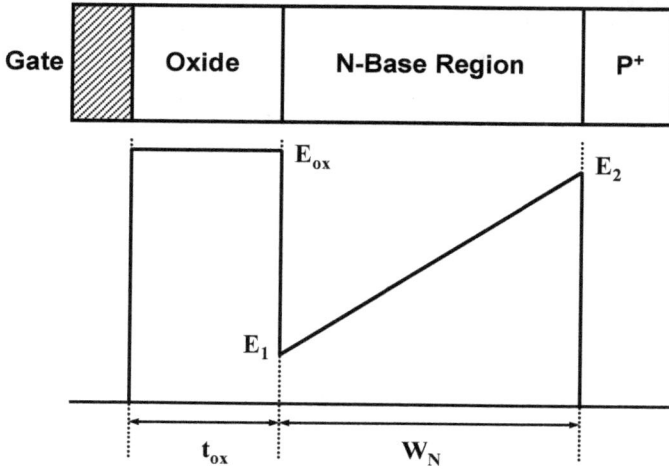

Fig. 10.5 Electric Field Profile in the Gate Region for the Accumulation-Mode MOSFET structure.

The band bending required to create a channel in the accumulation-mode planar MOSFET is much smaller than required for the inversion mode device. This provides the opportunity to reduce the threshold voltage while obtaining the desired normally-off device behavior. A model for the threshold voltage of accumulation-mode MOSFETs has been developed[5] using the electric field profile shown in Fig. 10.5 when the gate is biased at the threshold voltage. In this figure, the electric fields in the semiconductor and oxide are given by:

$$E_1 = \frac{V_{bi}}{W_N} - \frac{qN_D W_N}{2\varepsilon_S} \qquad [10.6]$$

$$E_2 = \frac{V_{bi}}{W_N} + \frac{qN_D W_N}{2\varepsilon_S} \qquad [10.7]$$

$$E_{ox} = \frac{\varepsilon_S}{\varepsilon_{ox}} E_1 \qquad [10.8]$$

Note that this model is based upon neglecting any voltage supported within the P$^+$ region under the assumption that it is very heavily doped. Using these electric fields, the threshold voltage is found to be given by:

$$V_{TH} = \phi_{MS} + \left(\frac{\varepsilon_S V_{bi}}{\varepsilon_{ox} W_N} - \frac{q N_D W_N}{2\varepsilon_{ox}} \right) t_{ox} \qquad [10.9]$$

The first term in this equation accounts for the work function difference between the gate material and the lightly doped N-Base region. The second term represents the effect of the built-in potential of the P$^+$/N junction that depletes the N-base region.

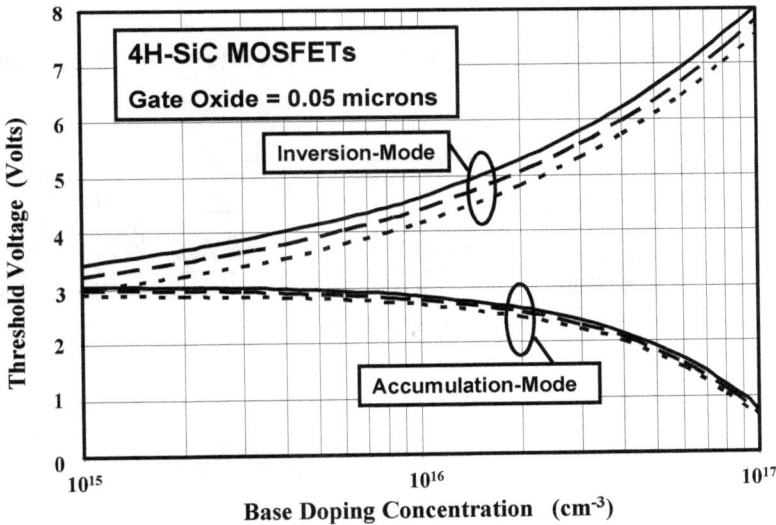

Fig. 10.6 Threshold Voltage of 4H-SiC Accumulation-Mode MOSFETs.

(Solid Line: 300°K; Dash Line: 400°K; Dotted Line: 500°K)

The analytically calculated threshold voltage for 4H-SiC accumulation-mode MOSFETs are provided in Fig. 10.6 for the case of a gate oxide thickness of 0.05 microns, and N-base thickness of 0.2 microns as a function of the N-base doping concentration with the inclusion of a metal-semiconductor work-function difference of 1 volt. For comparison purposes, the threshold voltage for the inversion-mode 4H-SiC MOSFET is also given in this figure for the same gate oxide thickness. A strikingly obvious difference between the structures is a decrease in the threshold voltage for the accumulation-mode structure with increasing doping concentration in the N-base region. This occurs due to the declining influence of the P^+/N junction at the gate oxide interface when the doping concentration of the N-base region is increased. It can also be noted that the temperature dependence of the threshold voltage is smaller for the accumulation-mode structure. Of course, the most important benefit of the accumulation-mode is that lower threshold voltages can be achieved than in the inversion-mode structures. At an N-base doping concentration of 1×10^{16} cm^{-3}, the threshold voltage for the accumulation-mode structure is 2.8 volts. This will be reduced to just below 2 volts in the presence of typical fixed oxide charge in the 4H-SiC/oxide system.

10.1.4 Accumulation-Mode MOSFET Structure: Threshold Voltage Simulations

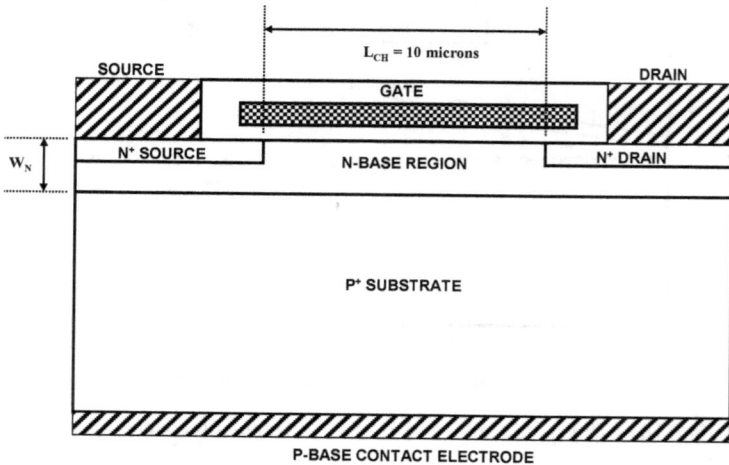

Fig. 10.7 The Planar Lateral Accumulation-Mode 4H-SiC MOSFET Structure.

In order to verify the predictions of the analytical model for the threshold voltage for silicon carbide accumulation-mode MOSFETs that was described in the previous section, two-dimensional numerical simulations were performed for a lateral MOSFET structure. This structure was designed with a long channel (10 microns) and uniform N-base doping concentration, as shown in Fig. 10.7, to allow extraction of the threshold voltage and the channel accumulation layer mobility. Simulations performed with this structure using different N-base doping concentrations and gate oxide thickness values enables extraction of the dependence of the threshold voltage on these parameters. In addition, the impact of changing the thickness of the N-base region was examined.

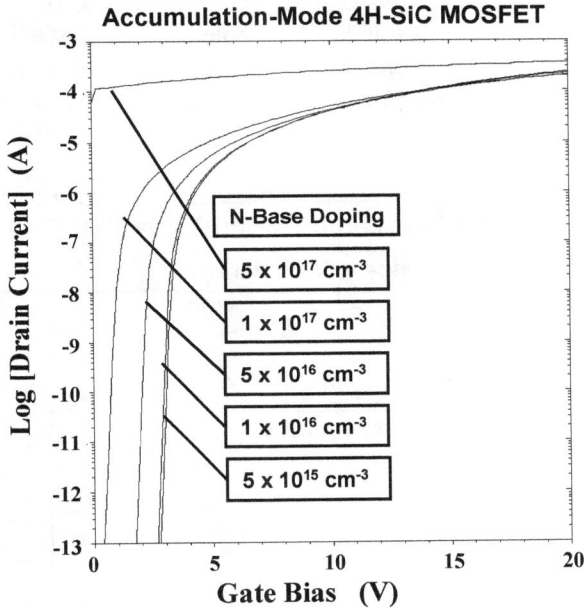

Fig. 10.8 Transfer Characteristics of the Planar Lateral Accumulation-Mode 4H-SiC MOSFET Structure.

The transfer characteristics obtained for the case of a gate oxide thickness of 0.05 microns and N-base thickness of 0.2 microns, are shown in Fig. 10.8, for various N-base doping concentrations. The threshold voltages for the different N-base doping concentrations were extracted by using a low drain current level (10^{-13} A/micron) because this defines the on-set of the formation of the accumulation layer. The extracted threshold voltage is compared with the predictions of the analytical model in Fig. 10.9. A good agreement is observed by using the model to compute the threshold voltage until it approaches a value close to zero. The threshold voltage predicted by the analytical model is larger than that obtained with the simulations at high doping concentration in the N-base region. This occurs because the depletion region and voltage supported by the P$^+$ region were neglected when deriving the analytical model. For the P$^+$ region doping concentration of 1 x 10^{19} cm^{-3} used during the simulations, this assumption results in a deviation of about 0.3 volts in the analytically calculated threshold voltage at a N-base doping concentration of 1 x 10^{17} cm^{-3}. It is worth pointing out that when the doping concentration of the N-base region exceeds 1 x 10^{17} cm^{-3} (for a thickness of 0.2 microns), the built-in potential of the P$^+$/N junction becomes insufficient to completely deplete the N-base region. The threshold voltage then becomes negative indicting the presence of a conductive channel at zero gate bias.

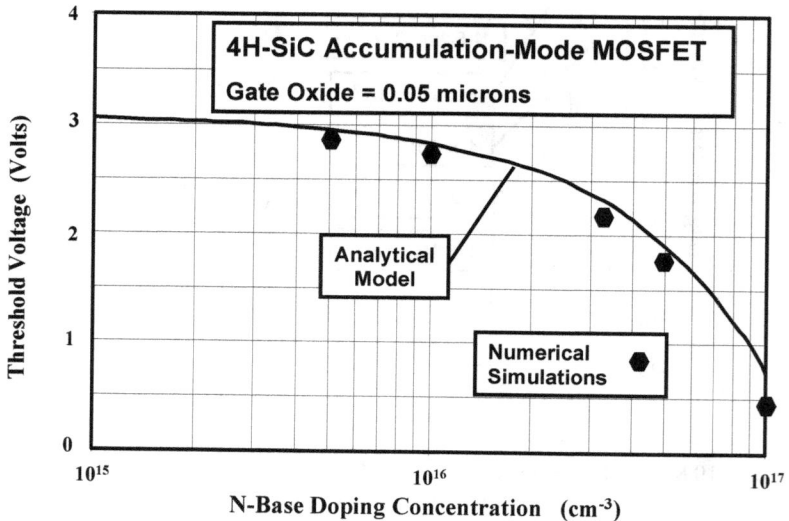

Fig. 10.9 Threshold Voltage of 4H-SiC Accumulation-Mode MOSFETs.

The thickness of the N-base region also has a strong influence on the threshold voltage of the accumulation-mode MOSFET. The results of simulations performed on the lateral accumulation-mode structure with various N-base widths are provided in Fig. 10.10 for the case of a gate oxide thickness of 0.05 microns and an N-base doping concentration of 1 x 10^{16} cm^{-3}. It can be observed that the threshold voltage decreases as the N-base width is increased eventually becoming less than zero for an N-base width of 1 micron. The threshold voltage extracted from these simulations is compared with the predictions of the analytical model in

Fig. 10.11. A good agreement is observed by using the model to compute the threshold voltage for N-base widths of 0.2 and 0.5 microns. However, the threshold voltage predicted by the analytical model is about 0.7 volts larger than that obtained with the simulations for the case of an N-base width of 0.1 microns. This deviation is again due to neglecting the depletion width and voltage supported by the P^+ region when deriving the analytical model. The contribution from this voltage becomes more significant as the width of the N-base region is reduced because it now supports a smaller fraction of the voltage across the gate structure.

The analytically predicted threshold voltage was also found to be larger than that obtained from the numerical simulations by about 0.7 volts when the width of the N-base region was increased to 1 micron. In this case, the N-base region is on the verge of being completely depleted according to the simulations. The influence of majority-carrier distribution tails[6] on the current transport at the gate oxide interface must now be taken into account, which is beyond the scope of the model. It is worth pointing out that a mobility degradation factor was not included during the simulations performed to extract the threshold voltage.

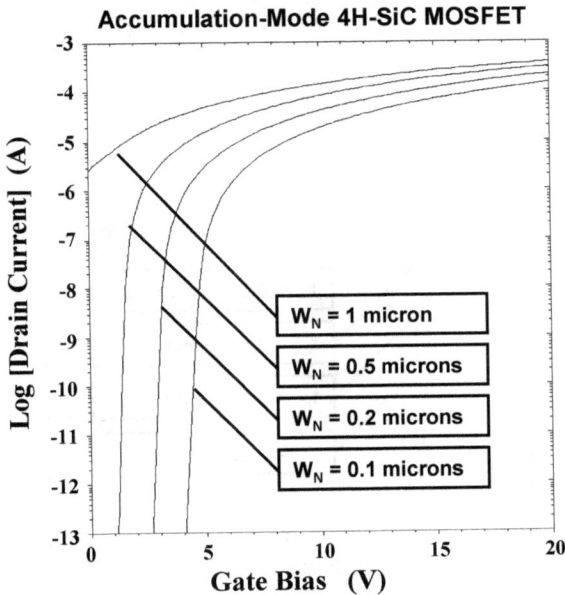

Fig. 10.10 Transfer Characteristics of the Planar Lateral Accumulation-Mode 4H-SiC MOSFET Structure.

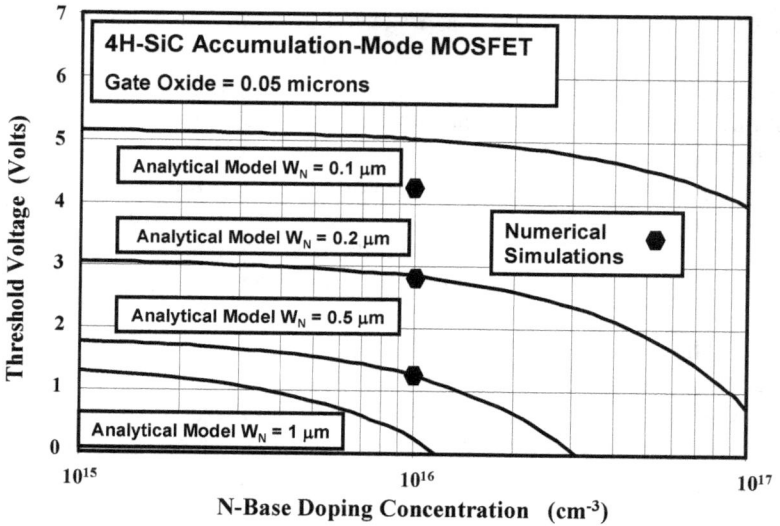

Fig. 10.11 Threshold Voltage of 4H-SiC Accumulation-Mode MOSFETs.

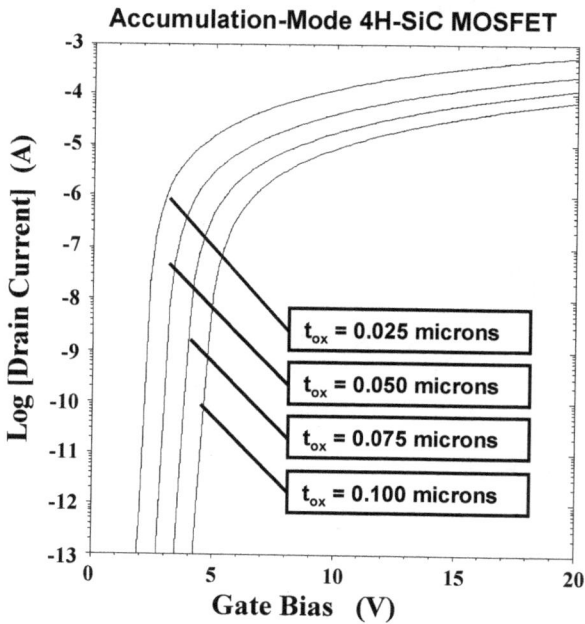

Fig. 10.12 Transfer Characteristics of the Planar Lateral Accumulation-Mode
4H-SiC MOSFET Structure.

As in the case of inversion-mode MOSFETs, the threshold voltage for the accumulation-mode MOSFETs depends upon the gate oxide thickness. This dependence is predicted by the second term for the analytical expression for the threshold voltage in Eq. [10.9]. The results of numerical simulations are shown in Fig. 10.12 for the case of an N-base doping concentration of 8.5 x 10^{15} cm^{-3} and thickness of 0.2 microns. The threshold voltage extracted from these simulations is compared with the analytically predicted values in Fig. 10.13. Once again, the predictions of the analytical model are in good agreement with the simulations indicating that this model can be utilized for the design of silicon carbide MOSFETs.

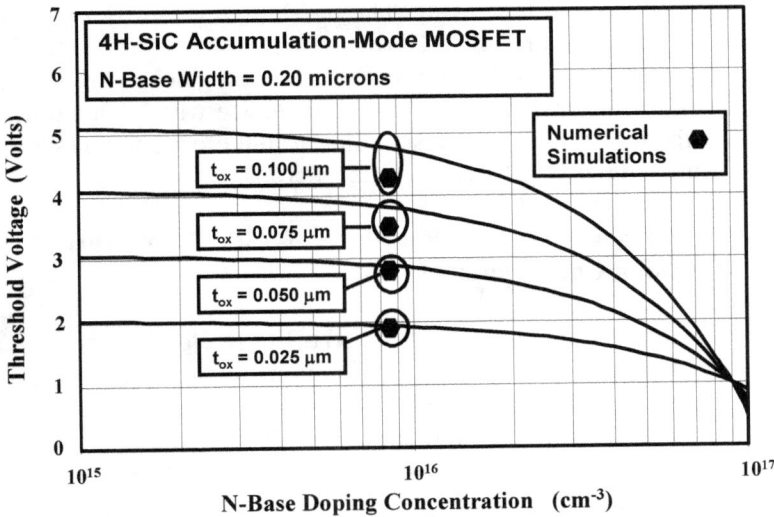

Fig. 10.13 Threshold Voltage of 4H-SiC Accumulation-Mode MOSFETs.

It is interesting to observe that the threshold voltage becomes independent of the gate oxide thickness in Fig. 10.13 for an N-base doping concentration of 9 x 10^{16} cm^{-3}. This behavior occurs due to the cancellation of the two parts in the second term in Eq. [10.9] under the condition:

$$N_D.W_N^2 = \frac{2.\varepsilon_S.V_{bi}}{q}$$ [10.10]

This condition corresponds to the combination of N-base region doping concentration and thickness that is just completely depleted by the built-in potential of the P$^+$/N junction.

10.2 Planar Shielded Inversion-Mode MOSFET Structure: Numerical Simulations

In order to gain insight into the operation of the planar 4H-SiC shielded inversion-mode MOSFET structure, two-dimensional numerical simulations were performed with a drift region doping concentration of 1 x 10^{16} cm^{-3} and thickness of 20 microns corresponding to a parallel-plane breakdown voltage of 3000 volts. The baseline device had a gate oxide thickness of 0.04 microns and a P$^+$ region depth of 1 micron. The P-base region had a thickness of 0.2 microns with a doping concentration of 9 x 10^{16} cm^{-3}. The JFET width for the baseline device was 1.5 microns. The cell pitch for the structure, corresponding to the cross-section shown on the left-hand-side of Fig. 10.1, was 3.25 microns.

10.2.1 Planar Shielded Inversion-Mode MOSFET Structure: Blocking Characteristics

The blocking capability of the planar 4H-SiC shielded inversion-mode MOSFET was investigated by maintaining zero gate bias while increasing the drain voltage. It was found that the drain current remains below 1 x 10^{-13} amperes upto a drain bias of 3000 volts. This behavior is shown in Fig. 10.14 to contrast the characteristics of the shielded device with that shown earlier in Fig. 9.20 for the planar inversion-mode MOSFET without the shielding of the P-base region. Consequently, the shielding of the P-base region is very effective for preventing the reach-through breakdown problem allowing the device to operate upto the full capability of the drift region.

In this figure, the drain current is also shown when various positive gate bias voltages are applied to the planar 4H-SiC inversion-mode MOSFET structure. The on-set of drain current flow is observed when the gate bias exceeds 4 volts. Excellent drain current saturation under gate control is observed at all gate bias voltages upto the blocking voltage capability of the drift region. These characteristics indicate effective suppression of the reach-through phenomenon as well as the parasitic N-P-N bipolar transistor.

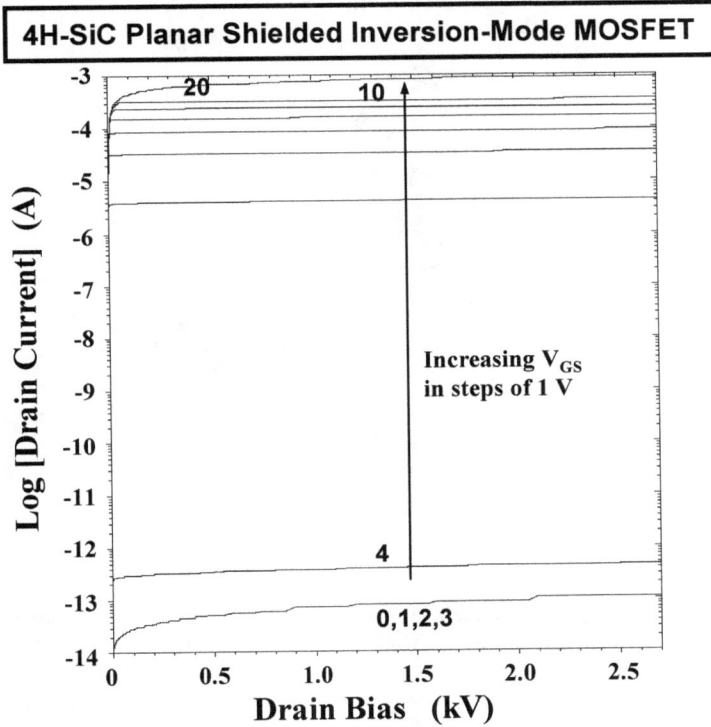

Fig. 10.14 Blocking Characteristics of the Planar 4H-SiC Shielded
Inversion-Mode MOSFET Structure.

In the previous chapter, it was demonstrated that a high electric field is generated at the gate oxide interface in the planar 4H-SiC MOSFET structure when the shielding is ineffective. This problem can be mitigated with the P^+ region to shield the surface from the high electric fields developed in the drift region. A three-dimensional view of the electric field distribution is shown in Fig. 10.15 for a drain bias of 3000 volts with zero gate bias. The field distribution shown in this figure indicates that the electric field at the gate oxide is significantly suppressed by the presence of the P^+ region. At the maximum blocking capability of the drift region, the electric field at the gate oxide interface has reached a magnitude of 1.5×10^6 V/cm while the electric field in the bulk has reached a magnitude of 3×10^6 V/cm. A slight field enhancement is observed at the edge of the P^+ region at a depth of 1 micron below the semiconductor surface.

Drain Bias = 3000 V

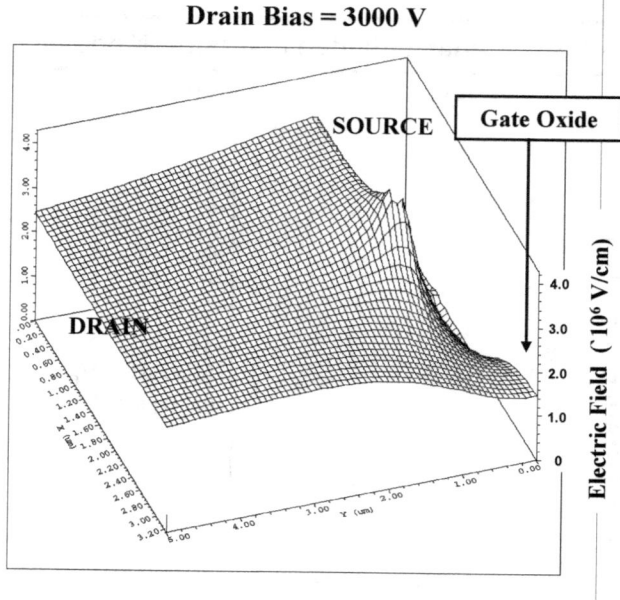

Fig. 10.15 Electric Field Distribution in the Planar 4H-SiC Shielded Inversion-Mode MOSFET Structure.

JFET Spacing = 1.5 microns

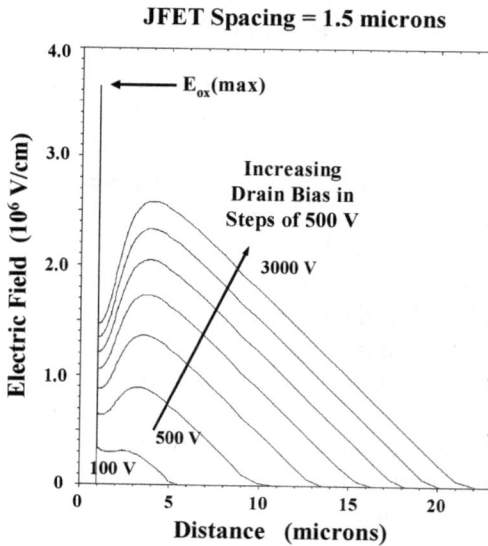

Fig. 10.16 Electric Field Distribution in the Planar 4H-SiC Shielded Inversion-Mode MOSFET Structure.

In the planar 4H-SiC shielded inversion-mode MOSFET structure, the largest electric field at the gate oxide interface occurs at the center of the JFET region. The behavior of this electric field with increasing drain bias is shown in Fig. 10.16 for the case of a JFET width of 1.5 microns. From this figure, it is apparent that there is considerable shielding of the gate oxide by the JFET region due to the formation of a potential barrier when this region becomes depleted at lower drain voltages. The maximum electric field occurs inside the semiconductor at a depth of about 4 microns. The electric field in the oxide reaches a magnitude of 3.5×10^6 V/cm at a drain bias of 3000 volts. The same magnitude for the electric field was observed in the unshielded structure for a smaller JFET width of 1 micron. This difference in behavior is due to the larger doping concentration of the P^+ shielding region when compared with P-base region. The high doping level in the P^+ region promotes the depletion of the JFET region setting up the same magnitude of the potential barrier at a larger JFET width.

10.2.2 Planar Shielded Inversion-Mode MOSFET Structure: On-State Characteristics

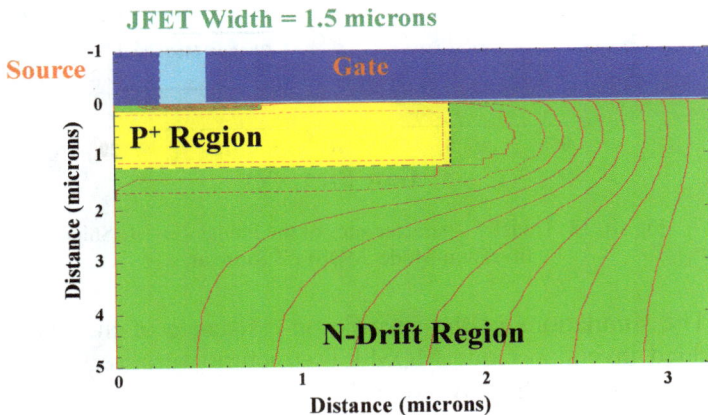

Fig. 10.17 On-State Current Distribution in the Planar 4H-SiC Shielded Inversion-Mode MOSFET Structure.

The on-resistance for the planar 4H-SiC shielded inversion-mode structure was extracted by performing numerical simulations with a gate bias of upto 20 volts. The current distribution within the structure is shown in Fig. 10.17 for the case of JFET width of 1.5 microns. It can be seen that the current is not as severely constricted in the JFET region when compared with the current flow in the unshielded structure with JFET width of 1 micron (see top part of Fig. 9.25). Consequently, for the same magnitude of electric field reduction at the gate oxide interface, a lower contribution to the on-resistance can be obtained for the JFET region with the shielded structure.

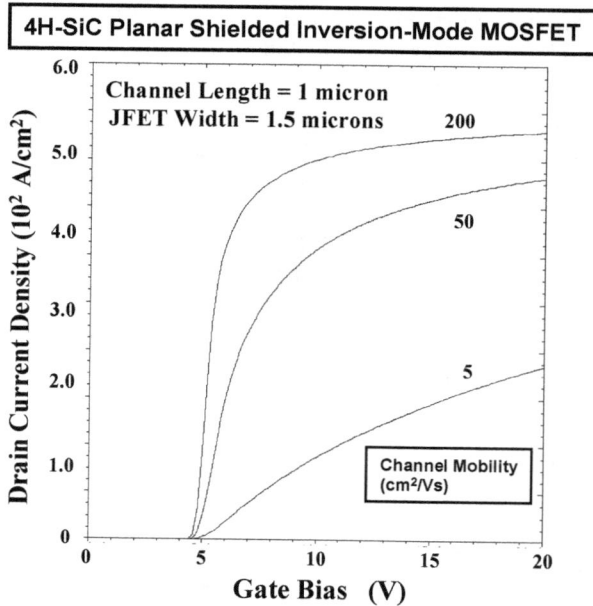

Fig. 10.18 Transfer Characteristics of the Planar 4H-SiC Shielded Inversion-Mode MOSFET Structures.

The simulations of the specific on-resistance of the planar 4H-SiC shielded inversion-mode MOSFET structure were performed with different values for the inversion layer mobility. The specific on-resistance was obtained by using a drain bias of 1 volt and sweeping the gate bias to 20 volts to obtain the transfer characteristic. A drain bias of 1 volt was used for these simulations because the on-state voltage drop for these structures is typically in this range. A typical transfer characteristic obtained using this method is shown in Fig. 10.18 for the device with JFET width of 1.5 microns. From the transfer characteristics, it can be seen that the threshold voltage that determines the on-resistance is approximately 5 volts, which is much lower than for the unshielded device structure (see Fig. 9.26). This allows operating the device in the on-state with a gate bias of 10 volts. The impact of changing the channel inversion layer mobility is also shown in this figure. As expected, there is a reduction in the drain current as the channel mobility is reduced.

When the inversion layer mobility is assumed to be 200 cm^2/Vs, it can be noted from Fig. 10.18 that nearly the same drain current is obtained at a gate bias of 10 volts as at 20 volts. As discussed in the previous chapter, inversion layer effective mobility values as high as 165 cm^2/Vs have been experimentally obtained with even higher Hall mobility values reported in the literature. The simulation results indicate that it is possible to operate the planar 4H-SiC shielded inversion-mode MOSFET structure with a gate bias of 10 volts. The drain current is also observed to be larger than that observed in the unshielded structure. Consequently, the specific on-resistance for the shielded 4H-SiC inversion-mode MOSFET structure is found to be 2.4 $m\Omega$-cm^2 (after including a substrate contribution of 0.4 $m\Omega$-cm^2) compared with 2.8 $m\Omega$-cm^2 obtained for the unshielded structure. This value is in excellent agreement with the value obtained by using the analytical model for the specific on-resistance (see Fig. 10.2). Although the difference between the specific on-resistance for the shielded and unshielded structures is only 17 percent, it is worth pointing out the maximum blocking voltage capability for the shielded structure is twice that observed for the unshielded structure whose blocking capability was limited by the reach-through problem.

10.2.3 Planar Shielded Inversion-Mode MOSFET Structure: Output Characteristics

In a previous section, it was demonstrated that the shielding of the P-base region within the planar MOSFET structure suppresses the reach-through breakdown problem. The shielding effect can also be expected to improve the output characteristics of the power MOSFET structure. In addition, the elimination of the reach-through induced leakage current should improve the output resistance of the device. These factors indicate that the output characteristics for the shielded 4H-SiC inversion-mode MOSFET should be much superior to those observed for the unshielded device structure.

The output characteristics for the shielded 4H-SiC inversion-mode MOSFET structure were obtained by using two-dimensional numerical simulations with sweeping of the drain voltage at various gate

bias voltages. The results obtained for a structure with channel length of
1 micron, gate oxide thickness of 0.04 microns, and a JFET width of 1.5
microns are shown in Fig. 10.19. The device structure exhibits excellent
drain current saturation for gate bias voltages upto 10 volts with a very
high output resistance. The saturated drain current increases as the square
of the gate bias voltage, which is typical for MOSFETs that operate with
channel pinch-off. For the gate bias values of 15 and 20 volts, a
pronounced quasi-saturation region is observed. However, when taking
into consideration the very high drain current density obtained even for a
gate bias of 10 volts, it is not necessary to operate the planar 4H-SiC
shielded inversion-mode MOSFET with gate bias voltages above 10
volts.

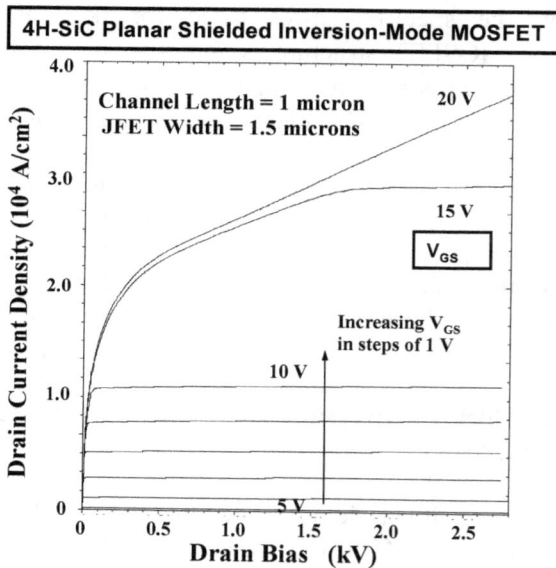

Fig. 10.19 Output Characteristics of a Planar 4H-SiC Shielded
Inversion-Mode MOSFET structure.

10.3 Planar Shielded Accumulation-Mode MOSFET Structure: Numerical Simulations

In order to gain insight into the operation of the planar 4H-SiC shielded
accumulation-mode MOSFET structure, two-dimensional numerical

simulations were performed with a drift region doping concentration of 1 x 10^{16} cm^{-3} and thickness of 20 microns corresponding to a parallel-plane breakdown voltage of 3000 volts. The baseline device had the same cell parameters as used for the inversion-mode structure, namely, a gate oxide thickness of 0.04 microns, a P^{+} region depth of 1 micron, and a JFET width of 1.5 microns.. The N-base region had a thickness of 0.2 microns with a doping concentration of 8.5 x 10^{15} cm^{-3}. The cell pitch for the structure, corresponding to the cross-section shown on the right-hand-side of Fig. 10.1, was 3.25 microns.

10.3.1 Planar Shielded Accumulation-Mode MOSFET Structure: Blocking Characteristics

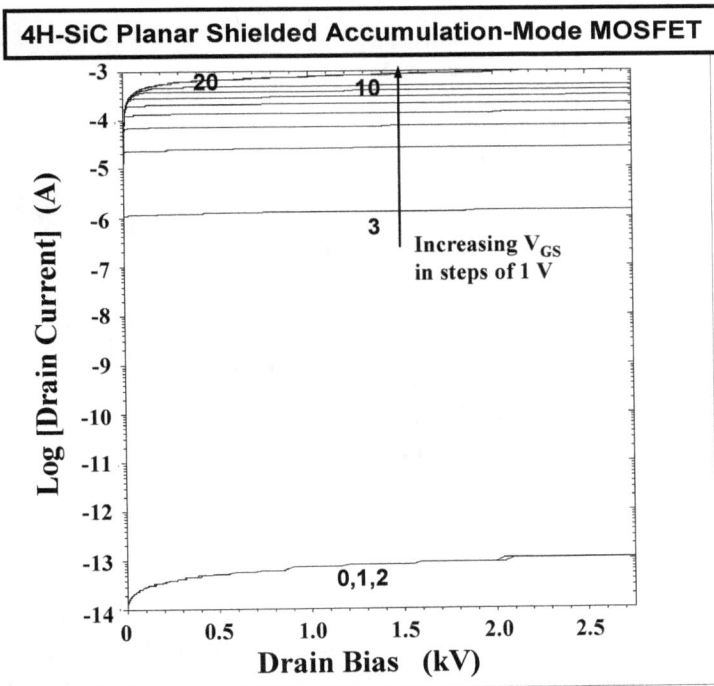

Fig. 10.20 Blocking Characteristics of the Planar 4H-SiC Shielded Accumulation-Mode MOSFET Structure.

The blocking capability of the planar 4H-SiC shielded accumulation-mode MOSFET was investigated by maintaining zero gate bias while increasing the drain voltage. It was found that the drain current remains below 1 x 10^{-13} amperes upto a drain bias of 3000 volts. This behavior, shown in Fig. 10.20, demonstrates that the depletion of the N-base region by the built-in potential of P$^+$/N junction creates a potential barrier for electron transport through the channel. In addition, the shielding of the N-base region by the underlying P$^+$ region is very effective for preventing the reach-through breakdown problem allowing the device to operate upto the full capability of the drift region.

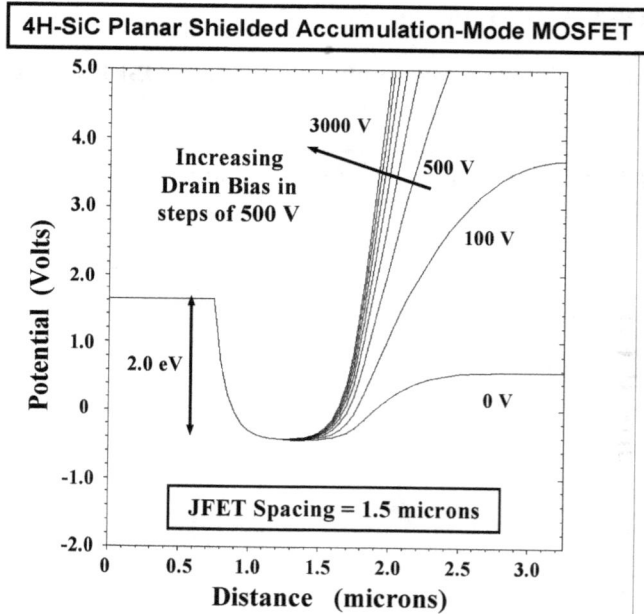

Fig. 10.21 Channel Potential Barrier in the Planar 4H-SiC Shielded
Accumulation-Mode MOSFET Structure.

In order to understand the ability to suppress current flow in the accumulation-mode structure in the absence of the P-base region, it is instructive to examine the potential distribution along the channel. The variation of the potential along the channel is shown in Fig. 10.21 for the planar 4H-SiC shielded accumulation-mode MOSFET described above for various drain bias voltages when the gate bias voltage is held at zero volts. It can be seen that there is a potential barrier in the channel with a magnitude of 2 eV at zero drain bias. This barrier for the transport of electrons from the source to the drain is upheld even when the drain bias is applied upto 3000 volts. This demonstrates the fundamental operating principle for creating a normally-off device with a depleted N-base region formed using the built-in potential of the P^+/N.

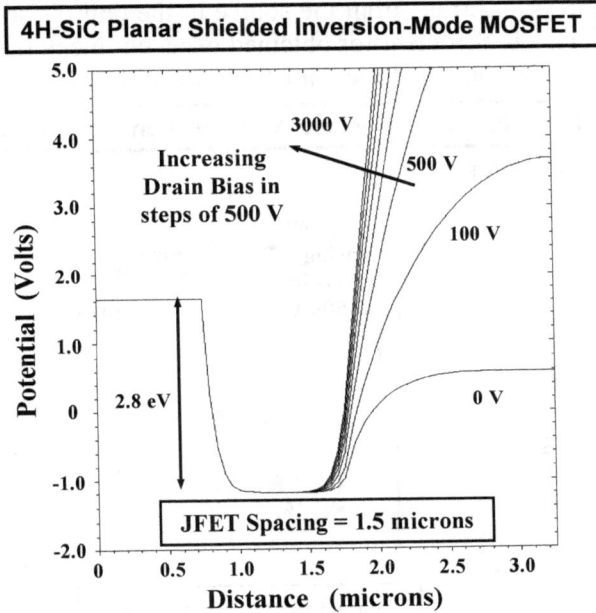

Fig. 10.22 Channel Potential Barrier in the Planar 4H-SiC Shielded Inversion-Mode MOSFET Structure.

For a better perspective on the operation of the accumulation-mode structure, the potential distribution along the channel is provided in Fig. 10.22 for the inversion-mode structure. From this figure, it can be seen that a potential barrier of 2.8 eV in magnitude is created due to the presence of the P-base region. This barrier is also retained intact when the drain bias voltage is increased to 3000 volts. By examining the potential distributions shown in the last two figures, it can be concluded that the accumulation-mode structure operates as well as the inversion-mode structure from the point of view of supporting high drain bias voltages with zero gate bias.

The normally-off feature can be preserved in the planar 4H-SiC shielded accumulation-mode MOSFET structure even when the channel length is reduced to 0.5 microns. The potential distribution along the channel for this structure is shown in Fig. 10.23 for various drain bias voltages. It can be seen that, although the width of the barrier along the channel (x-direction in Fig. 10.23) is smaller than for the structure with the 1 micron channel length, the magnitude of the potential barrier for the transport of electrons from the source to the drain is still maintained at 2.0 eV. This is sufficient for obtained excellent blocking capability in the planar 4H-SiC shielded accumulation-mode MOSFET.

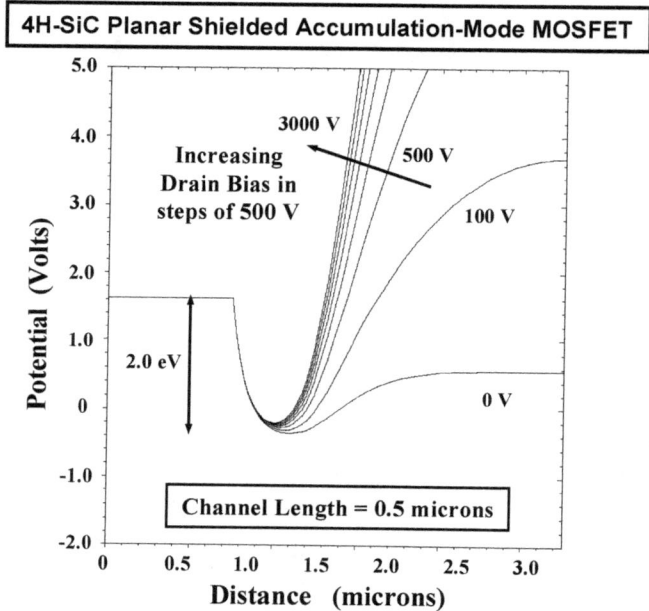

Fig. 10.23 Channel Potential Barrier in the Planar 4H-SiC Shielded Accumulation-Mode MOSFET Structure.

The blocking characteristic for the planar 4H-SiC shielded accumulation-mode MOSFET with channel length of 0.5 microns is shown in Fig. 10.24 together with traces obtained at various positive gate bias voltages. The device exhibits the same leakage current as the device with the longer channel length at zero gate bias. This demonstrates that the blocking voltage capability is preserved in the accumulation-mode structure even when the channel length is shortened to 0.5 microns. It is worth pointing out that an increased current is observed at a gate bias of 2 volts when the channel length is reduced to 0.5 microns. In addition, the traces for higher gate bias voltages are all shifted to larger drain current levels. This is, of course, related to the increase in the transconductance of the MOSFET when the channel length is reduced. The change in the transconductance with channel length is similar to that observed in the conventional inversion-mode MOSFET structure.

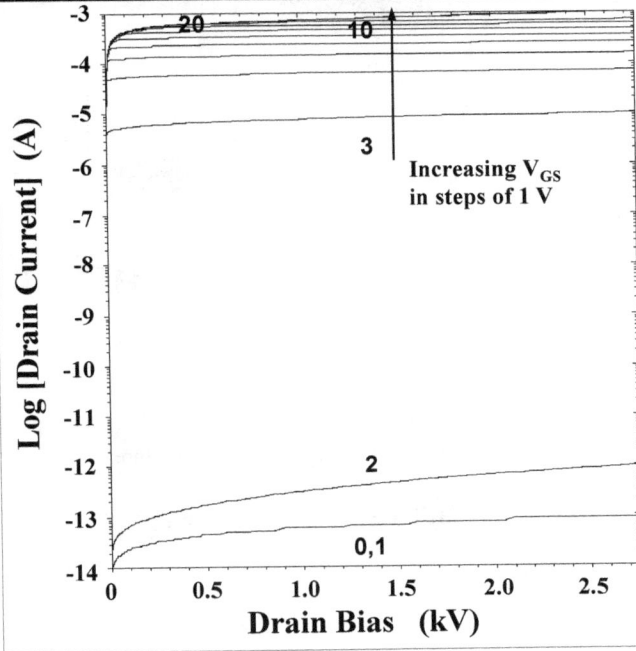

Fig. 10.24 Blocking Characteristics of the Planar 4H-SiC Shielded Accumulation-Mode MOSFET Structure.

As in the case of the inversion-mode structure, the P^+ region in the planar 4H-SiC shielded accumulation-mode MOSFET structure can isolate the surface under the gate oxide from the high electric fields developed in the drift region. In order to demonstrate this, a three-dimensional view of the electric field distribution is shown in Fig. 10.25 for this structure at a drain bias of 3000 volts with zero gate bias. The field distribution shown in this figure indicates that the electric field at the gate oxide is suppressed by the presence of the P^+ region to the same extent as for the inversion-mode structure. At the maximum blocking capability of the drift region, the electric field at the gate oxide interface has reached a magnitude of 1.5×10^6 V/cm while the electric field in the bulk has reached a magnitude of 3×10^6 V/cm.

Drain Bias = 3000 V

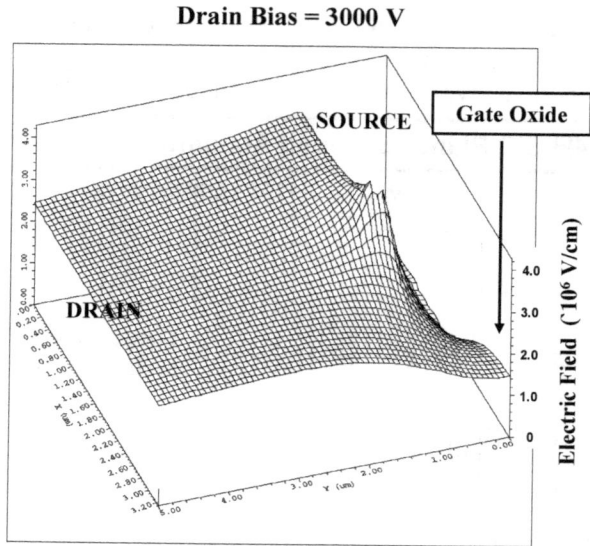

Fig. 10.25 Electric Field Distribution in the Planar 4H-SiC Shielded Accumulation-Mode MOSFET Structure.

JFET Spacing = 1 micron

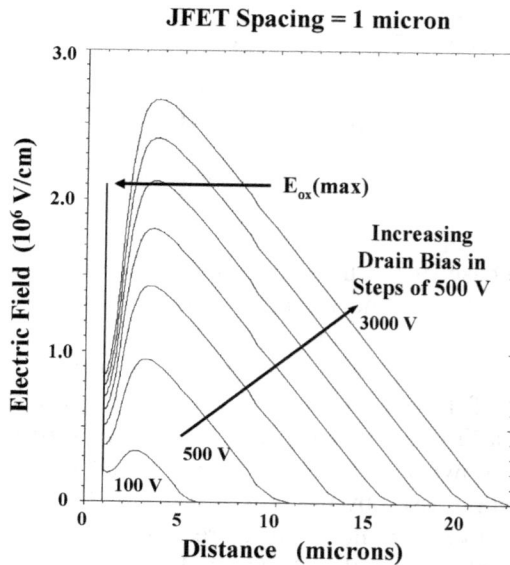

Fig. 10.26 Electric Field Distribution in the Planar 4H-SiC Shielded Accumulation-Mode MOSFET Structure.

In the planar 4H-SiC shielded accumulation-mode MOSFET structure, the largest electric field at the gate oxide interface occurs at the center of the JFET region. The behavior of this electric field with increasing drain bias was found to be identical to the profiles shown in Fig. 10.16 for the inversion-mode structure with the same JFET width of 1.5 microns. A further reduction of the electric field in the gate oxide can be achieved by reducing the width of the JFET region. In order to illustrate this, the electric field distribution is shown in Fig. 10.26 for the case of a planar 4H-SiC shielded accumulation-mode MOSFET structure with JFET width of 1 micron. It can be seen that the maximum electric field at the gate oxide interface is reduced to 8.5 x 10^5 V/cm resulting in a maximum electric field in the gate oxide of only 2.1 x 10^6 V/cm at a drain bias of 3000 volts. This value is sufficiently low for reliable operation of the structure especially due to the planar gate structure, where there are no localized electric field enhancements under the gate electrode.

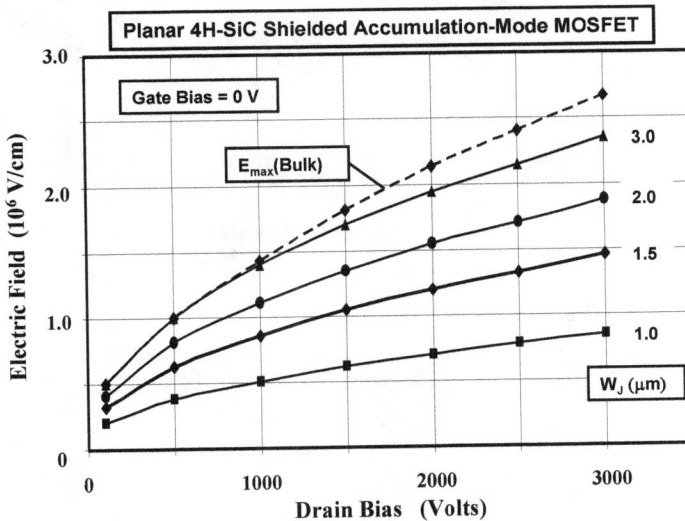

Fig. 10.27 Electric Field Suppression in the Planar 4H-SiC Shielded Accumulation-Mode MOSFET Structure.

The magnitude of the electric field developed in the vicinity of the gate oxide interface in the planar 4H-SiC shielded accumulation-mode MOSFET structure is dependent upon the width of the JFET region. The electric field at the gate oxide interface becomes smaller as the JFET width is reduced as shown in Fig. 10.27. For the structure with a drift region doping concentration of 1×10^{16} cm^{-3}, the optimum spacing lies between 1 and 2 microns. The JFET width cannot be arbitrarily reduced because the resistance of the JFET region becomes very large when the width approaches the zero-bias depletion width of the P$^+$/N junction. For a drift region doping concentration of 1×10^{16} cm^{-3}, the zero-bias depletion width is approximately 0.6 microns. Consequently, using a JFET width of 1.5 microns provides adequate space in the JFET region for current flow as shown in the next section of the chapter.

10.3.2 Planar Shielded Accumulation-Mode MOSFET Structure: On-State Characteristics

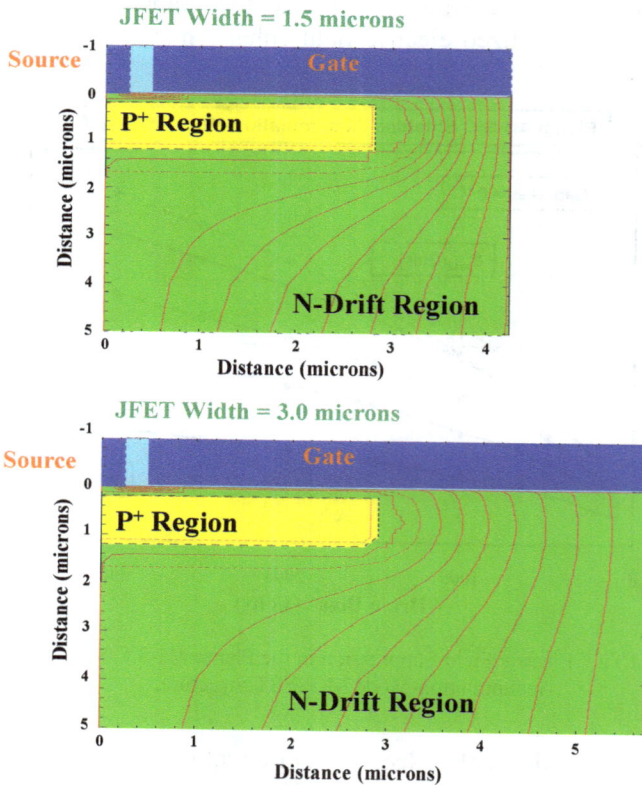

Fig. 10.28 On-State Current Distribution in the Planar 4H-SiC Shielded Accumulation-Mode MOSFET.

The on-resistance for the planar 4H-SiC shielded accumulation-mode structure was extracted by performing numerical simulations with a gate bias of 10 volts. The lower gate bias voltage, when compared with that used in the previous chapter for the unshielded inversion-mode structure, was chosen because of the smaller threshold voltage of these devices. A gate bias of 10 volts is a typical value for the available gate drive circuits used in medium and high power electronic systems. In addition, due to the improved channel resistance contribution, a channel length of 2 microns was also investigated because this provides greater margin when fabricating the devices.

The current distribution within the planar 4H-SiC shielded accumulation-mode MOSFET structure is shown in Fig. 10.28 for the case of JFET widths of 1.5 microns and 3 microns. These structures have a channel length of 2 microns. The absence of the P-type doping is apparent above the P^+ shielding region as is to be expected for an accumulation-mode structure. It can be seen that the current does not flow through the bulk of the N-base region and is confined to a channel formed below the gate oxide. This confirms the depletion of the N-base region by the built-in potential of the P^+/N junction, thus, preventing current flow through the bulk. In Fig. 10.28, the current is observed to flow from the accumulation layer channel into the JFET region and then spread into the drift region at a 45 degree angle. The current flow is not as severely constricted in the JFET region when the JFET width is increased from 1.5 to 3 microns.

Simulations to obtain the specific on-resistance of the planar 4H-SiC shielded accumulation-mode MOSFET structure were performed with different accumulation layer mobility values. The specific on-resistance was obtained by using a drain bias of 1 volt and sweeping the gate bias to 20 volts to obtain the transfer characteristic. A drain bias of 1 volt was used for these simulations because the on-state voltage drop for these structures is typically in this range. The specific on-resistance was computed by using the drain current extracted at a gate bias of 10 volts.

Typical transfer characteristics obtained using this method are shown in Fig. 10.29 for the device with JFET width of 1.5 microns and channel length of 1 micron. From the transfer characteristics, it can be seen that the threshold voltage that determines the on-resistance is approximately 3 volts, which is lower than for the shielded inversion-mode device structure (see Fig. 10.18). The impact of changing the channel accumulation mobility is shown in this figure. The mobility values chosen for these simulations extend to larger values because the accumulation mobility measured in the literature is larger as discussed later in this chapter. As expected, there is a reduction in the drain current as the channel mobility is reduced.

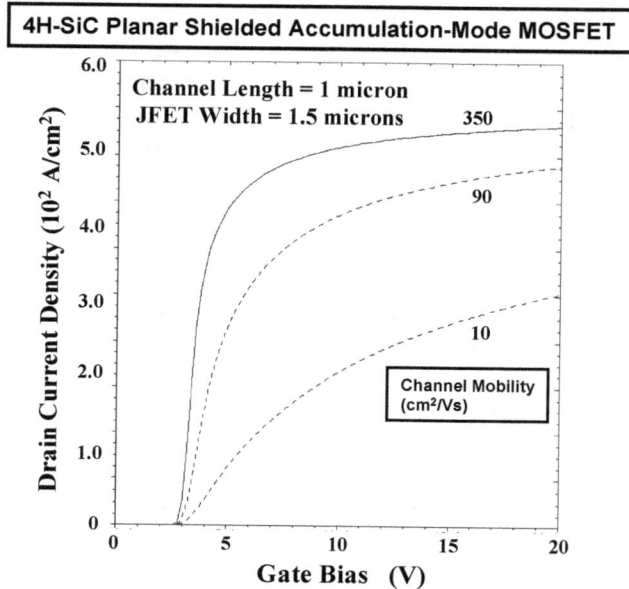

Fig. 10.29 Transfer Characteristics of the Planar 4H-SiC Shielded Accumulation-Mode MOSFETs.

When the accumulation layer mobility is assumed to be 350 cm^2/Vs, it can be noted from Fig. 10.29 that nearly the same drain current is obtained at a gate bias of 10 volts as at 20 volts. This demonstrates that it is possible to operate the planar 4H-SiC shielded accumulation-mode MOSFET structure with a gate bias of 10 volts. The specific on-resistance for the shielded 4H-SiC accumulation-mode MOSFET structure is found to be 2.4 $m\Omega$-cm^2 (after including a substrate contribution of 0.4 $m\Omega$-cm^2) which is the same as that obtained for the inversion-mode structure if an inversion layer mobility of 200 cm^2/Vs is assumed. This is because the channel resistance component for both structures becomes a small fraction of the total on-resistance for the 3 kV devices analyzed here. However, if the inversion layer mobility is low, as more commonly observed in the literature, the accumulation-mode structure will have a lower specific on-resistance. Thus, the accumulation-mode structure provides an alternate approach to obtaining high performance MOSFETs in silicon carbide until consistently high inversion layer mobility can be reproducibly achieved when manufacturing these structures.

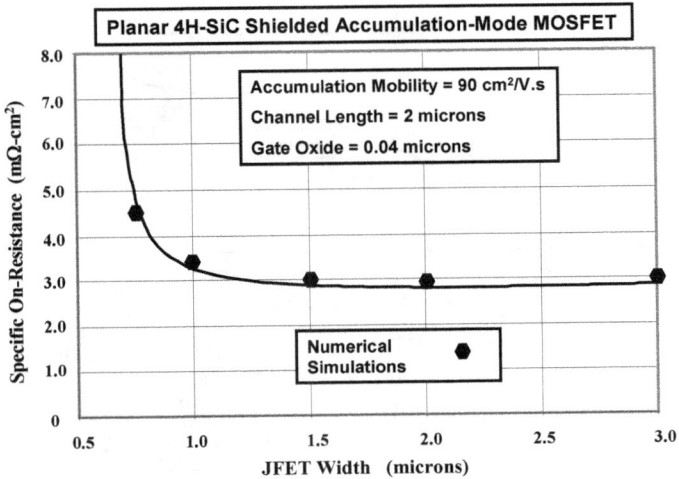

Fig. 10.30 Specific On-Resistances for the Planar 4H-SiC Shielded
Accumulation-Mode MOSFETs.

The results of the specific on-resistance obtained from the simulations performed on planar 4H-SiC shielded accumulation-mode MOSFET structures with various JFET widths is shown in Fig. 10.30 together with the values obtained using the analytical model. These results are based upon using a channel mobility of 90 cm^2/Vs with a gate oxide thickness of 0.04 microns. A good agreement between the analytical model and the simulations provides further validation of the ability to predict specific on-resistance using the analytical model.

The optimization of the JFET width requires taking into consideration both the on-resistance and the electric field developed at the gate oxide interface. The relative changes in these parameters can be observed in Fig. 10.31 for the case of a device with channel length of 2 microns. A gate oxide thickness of 0.04 microns was used here with a channel mobility of 90 cm^2/V-s. It can be seen that the electric field in the semiconductor below the gate oxide can be reduced to less than 1 MV/cm when the JFET width is reduced to 1 micron with a degradation of the specific on-resistance by about 10 percent.

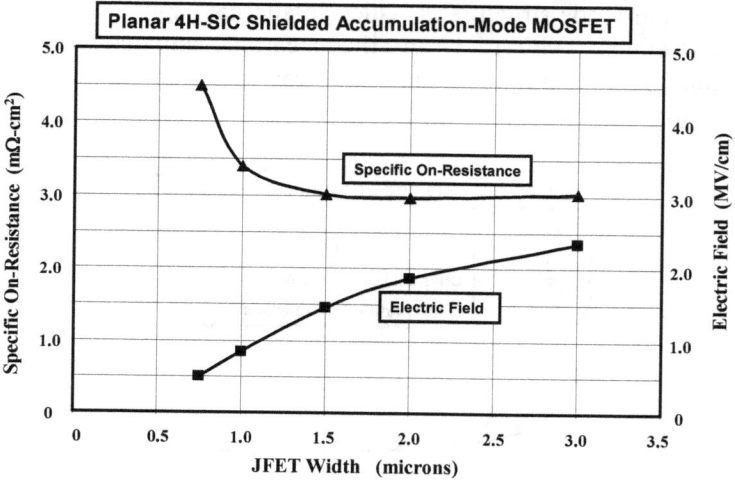

Fig. 10.31 Optimization of JFET Width in the Planar 4H-SiC Shielded
Accumulation-Mode MOSFETs.

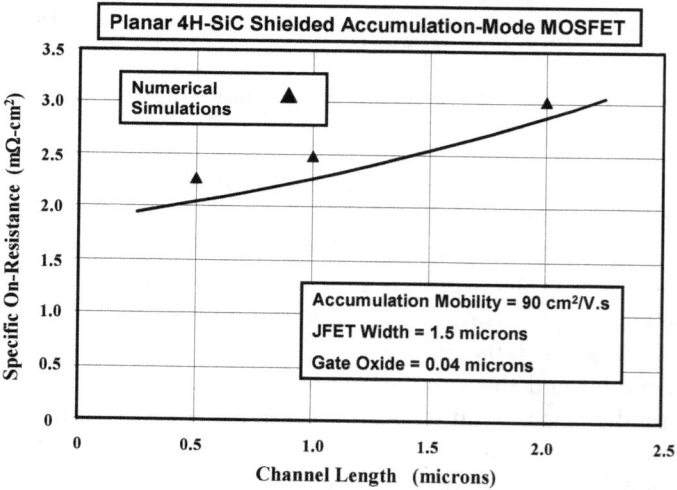

Fig. 10.32 Specific On-Resistances for the Planar 4H-SiC Shielded
Accumulation-Mode MOSFETs.

As discussed in the earlier section, the planar 4H-SiC shielded accumulation-mode MOSFET structure was found to be capable of blocking high drain bias voltages even when the channel length is reduced to 0.5 microns. This creates the opportunity to reduce the specific on-resistance by adjusting the channel length. The results of the specific on-resistance obtained from the simulations performed on planar 4H-SiC shielded accumulation-mode MOSFET structures with various channel lengths is shown in Fig. 10.32 together with the values obtained using the analytical model. These results are based upon using a channel mobility of 90 cm^2/Vs with a gate oxide thickness of 0.04 microns. A good agreement between the analytical model and the simulations provides further validation of the ability to predict specific on-resistance using the analytical model. With a channel length of 0.5 microns, the specific on-resistance for the planar 4H-SiC shielded accumulation-mode MOSFET structure is reduced to 2.3 mΩ-cm2, which is an extremely low value for a MOSFET capable of supporting 3000 volts.

10.3.3 Planar Shielded Accumulation-Mode MOSFET Structure: Output Characteristics

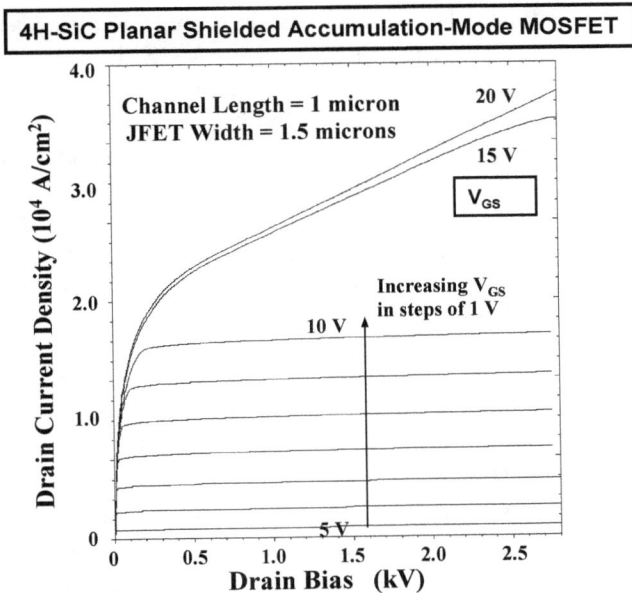

Fig. 10.33 Output Characteristics of a Planar 4H-SiC Shielded Accumulation-Mode MOSFET structure.

In a previous section, it was demonstrated that the shielding of the P-base region within the planar MOSFET structure suppresses the reach-through breakdown problem. The shielding effect can also be expected to improve the output characteristics of the accumulation-mode power MOSFET structure because it suppresses the lowering of the potential barrier formed in the channel. With adequate suppression of channel current flow over the potential barrier, the drain current is controlled by the gate bias induced charge in the accumulation layer. The output characteristics for the shielded 4H-SiC accumulation-mode MOSFET should therefore be similar to those observed for the inversion-mode device.

The output characteristics for the shielded 4H-SiC accumulation-mode MOSFET structure were obtained by using two-dimensional numerical simulations with sweeping of the drain voltage at various gate bias voltages. The results obtained for a structure with channel length of 1 micron, gate oxide thickness of 0.04 microns, and a JFET width of 1.5 microns are shown in Fig. 10.33. The device structure exhibits excellent drain current saturation for gate bias voltages upto 10 volts with a very high output resistance. The saturated drain current increases as the square of the gate bias voltage, which is typical for MOSFETs that operate with channel pinch-off. For the gate bias values of 15 and 20 volts, a pronounced quasi-saturation region is observed similar to that for the inversion-mode structure. However, when taking into consideration the very high drain current density obtained even for a gate bias of 10 volts, it is not necessary to operate the planar 4H-SiC shielded accumulation-mode MOSFET with gate bias voltages above 10 volts. For the same gate bias voltage, the accumulation-mode structure exhibits a higher drain current than the inversion-mode structure due to its lower threshold voltage and increased channel mobility.

In the previous section, it was demonstrated that the channel length of the planar 4H-SiC shielded accumulation-mode MOSFET structure could be reduced to 0.5 microns without encountering a blocking problem. However, it is not obvious that current flow through the N-base region will not occur under the application of a positive gate bias. Any current flow through the N-base region will degrade the output resistance of the device. However, the output characteristics for the accumulation-mode structure with channel length of 0.5 microns, shown in Fig. 10.34, were found to be quite similar to those of longer channel devices in terms of the output resistance. This confirms the ability to support high voltages using a depleted N-base region without short channel effects. A slight increase in the transconductance for this structure is apparent as expected due to the reduced channel length.

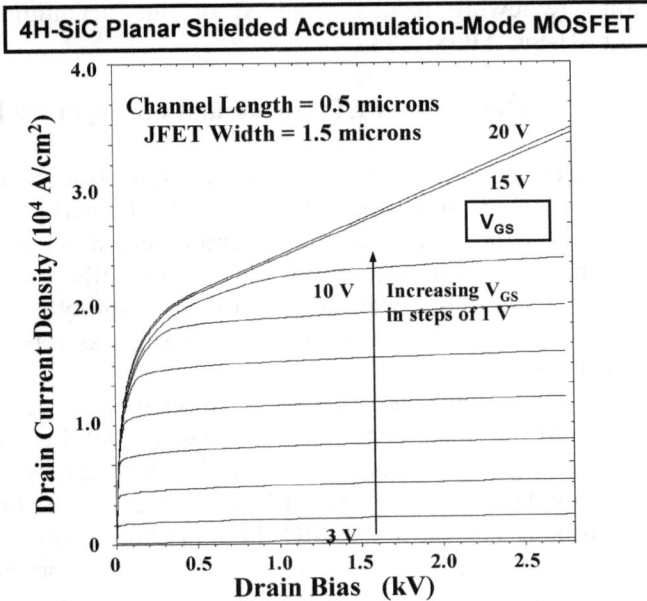

Fig. 10.34 Output Characteristics of a Planar 4H-SiC Shielded
Accumulation-Mode MOSFET structure.

10.4 Planar Shielded MOSFET Structures: Experimental Results

As pointed out in the previous chapter, the need to obtain a high quality interface between silicon carbide and the gate dielectric was identified as a challenging endeavor from the inception of interest in the development of unipolar transistors from this semiconductor material[7]. Initially, it was impossible to fabricate silicon carbide power MOSFETs with specific on-resistances below those reported for silicon devices until the interface was sufficiently improved to obtain adequate inversion layer mobility. Even in recent years, poor inversion layer mobility values have been reported due to trapping of charge at the oxide-semiconductor interface. To overcome this technological barrier and to obtain a lower threshold voltage, the accumulation-mode structure was proposed[1]. This structure was also designed to shield the gate oxide from high electric fields in the semiconductor drift region. Experimental results on the shielded inversion-mode and accumulation-mode structures are discussed in this section after reviewing results obtained on the accumulation layer mobility for 4H-SiC MOSFETs.

10.4.1 Silicon Carbide MOSFET: Accumulation Layer Mobility

Early investigations of the interface between P-type silicon carbide and thermally grown oxide indicated a high density of interface states and positive charge in the oxide. Significant improvement in the interface quality was achieved for 6H-SiC by anneal the thermally grown oxide in a NO ambient[8] leading to a peak inversion layer mobility of 70 cm^2/Vs. However, the inversion layer mobility for 4H-SiC was reported[9] to be less than 1 cm^2/Vs.

It has been established that the accumulation layer mobility is significantly greater than the inversion layer mobility in silicon MOSFET structures[10]. This encouraged the development of accumulation-mode MOSFETs in silicon carbide. The first planar accumulation-mode vertical power MOSFET structure, named the planar ACCUFET, was developed at PSRC using 6H-SiC as the semiconductor material[11]. The effective accumulation layer mobility extracted[12] from these structures was 120 cm^2/Vs for a structure fabricated using thermally grown gate oxide, which was considerably larger than the best inversion layer mobility (70 cm^2/Vs) reported[8] at that time.

More recently[13], the accumulation layer mobility has been measured in 4H-SiC MOSFETs using the Hall bar structure to differentiate between the effective mobility, which is limited by charge trapping, and the intrinsic mobility for free carriers in the channel. Note that the gate oxide process used for the fabrication of these MOSFETs was similar to that developed at PSRC[14] with the use of deposited oxides on 4H-SiC leading to a reported[15] record high inversion layer mobility of 165 cm^2/V-s. It was found that the accumulation mobility decreased from 350 cm^2/V-s in the weak accumulation regime of operation to 200 cm^2/V-s in the strong accumulation regime of operation. These results provided the justification for performing the simulations of the planar 4H-SiC shielded accumulation-mode MOSFET structures, as discussed in the previous section, using accumulation mobility values ranging upto 350 cm^2/V-s. However, the effective accumulation-layer mobility was found to be only 3-5 cm^2/V-s indicating substantial trapping of electrons at the gate oxide interface.

Better effective accumulation mobility values have been reported for 4H-SiC MOSFETs by using thermally grown oxide. In 1999, an effective accumulation-layer mobility of 20-30 cm^2/V-s was reported[16] by using nitrogen ion implantation to form the N-base region. In 2001, a significantly larger accumulation-layer effective mobility of 200 cm^2/V-s was reported[17] by using a stacked gate oxide process. The stacked gate oxide consisted of a thermally grown layer with thickness of 0.02 microns followed by a NSG-CVD deposited layer with a thickness of 0.03 microns. These results have established the ability to obtain relatively high accumulation-layer mobility in 4H-SiC indicating the possibility for fabrication of high performance vertical power MOSFETs from this material.

The accumulation-mode power MOSFET structure was proposed to circumvent the problem of low inversion-layer mobility observed in 4H-SiC in the early studies with this material. However, in recent years, considerable progress has been made to improve upon the inversion-layer mobility. Exceptionally high inversion-layer mobility was achieved at PSRC by using a deposited oxide that was subsequently annealed in a wet oxidation ambient[14,15]. A detailed study[18] of the process steps responsible for producing the improved interface resulting in the high channel mobility was also undertaken at PSRC. This work produced the first observation of phonon scattering limited inversion layer mobility (which decreased with increasing temperature), and demonstrated that a wet oxide anneal of the deposited oxide was the critical step for producing the high inversion layer mobility. The process proposed and demonstrated at PSRC was subsequently reproduced by the sponsors[19] as well by other research groups[20]. These results indicate that it is possible to achieve sufficiently high inversion layer mobility in 4H-SiC power MOSFETs to obtain low specific on-resistance in high voltage structures.

The low inversion layer mobility observed in 4H-SiC MOSFETs fabricated using thermally grown gate oxides has been traced to the trapping of the carriers at interface states. Hall effect measurements performed on lateral MOSFETs have demonstrated that the effective Hall mobility for carriers in the inversion layer is high when the trapping effect is taken into consideration[21]. Due to the trapping effect, as recently as 2001, the measured effective mobility in 4H-SiC MOSFETs has been reported[22] to be less than 10 cm^2/V-s when thermal oxidation is used to form the gate oxide. In these devices, the effective mobility was found to increase with temperature which is a signature of trap dominated current conduction in the channel. By using NO ambient anneals of the thermally grown gate oxide, an improvement in the effective channel mobility to 30-35 cm^2/V-s has been reported[23] in 2001.

In summary, inversion-layer effective mobility values in the range of 30-35 cm^2/V-s have been observed in 4H-SiC MOSFETs while significantly larger effective accumulative-layer mobility values of 200 cm^2/V-s have been observed in this material. These values indicate that it is possible to develop high voltage 4H-SiC power MOSFETs using either the inversion-mode or accumulation-mode architecture if the base is shielded from the high electric field in the drift region. The accumulation-mode structure is favored for designs with low channel density where the channel contribution becomes a significant fraction of the total on-resistance.

10.4.2 The Planar 4H-SiC Inversion-Mode MOSFET Structure: Experimental Results

As already mentioned earlier in the previous chapter, in recognition of the low diffusion coefficients for dopants in silicon carbide, it was proposed[24] that the P-base and N$^+$ source ion implantations for the planar silicon carbide power MOSFET be staggered by using photoresist masks rather that be defined by the gate edge as conventionally done for silicon devices. The idea to incorporate a sub-surface P-type layer to shield the channel was also proposed at PSRC and subsequently patented[1]. This idea was first successfully demonstrated[25] for 6H-SiC devices in 1997 under the DI-MOSFET moniker. The authors fabricated the devices using multiple energy boron implants to form a P-base region with a box profile at a depth of 1 micron. The energy and dose for the implants was chosen to produce a retrograde doping profile on one of the wafers. This creates a low doped P-base region with an under-lying highly doped P$^+$ region that shields it from the high electric fields developed in the drift region. However, the devices with retrograded doping profile had inferior breakdown when compared with the box profile indicating that the design did not provide adequate shielding.

The concept of shielding the P-base region has also been reported[26] using a triple ion-implantation process to fabricate 6H-SiC vertical power MOSFETs. The ion-implant steps were designed to produce a low P-base surface concentration of 7 x 10^{16} cm^{-3} with the doping increasing to 1 x 10^{18} cm^{-3} at a depth of 0.4 microns. This profile enabled achieving a threshold voltage of 3 volts with a thermally grown gate oxide thickness of 0.036 microns. A breakdown voltage of 1800 volts was obtained for this device by using a drift region doping concentration of 6.5 x 10^{15} cm^{-3} and thickness of 15 microns. A specific on-resistance of 82 mΩ-cm^2 was observed at a gate bias of 10 volts, which is an order of magnitude better than values that can be obtained with silicon technology. Good dynamic switching performance with ruggedness has been confirmed[27] for these MOSFETs.

Recently, a novel staggered implant process has been proposed to reduce the channel length and applied towards the fabrication of high voltage planar 4H-SiC MOSFETs[28]. In this process, oxidation of a sacrificial polysilicon layer is used to form a sidewall spacer that serves as a mask to stagger the P-base and N$^+$ source implants. The retrograde ion implant profile discussed earlier was also utilized for these devices to reduce the P-base doping concentration and suppress the reach-through problem. Planar 4H-SiC inversion-mode MOSFETs were successfully fabricated with breakdown voltage of 2 kV with a specific on-resistance of 27 mΩ-cm^2 at a gate bias of 20 volts.

10.4.3 The 4H-SiC Accumulation-Mode Planar MOSFET Structure: Experimental Results

The first planar shielded accumulation-mode MOSFET structure was fabricated[11] from 6H-SiC at PSRC in 1997. This device, named the planar ACCUFET, was conceived to circumvent the problems observed with obtaining high inversion layer mobility as well as to screen the channel and gate oxide from high electric fields developed in the drift region. The devices were fabricated using a drift region with doping concentration of 1 x 10^{16} cm^{-3} and thickness of 10 microns, corresponding to a breakdown voltage of 1500 volts. The sub-surface P$^+$ region used to shield the channel and gate oxide was formed by a single boron ion implant with energy of 380 keV and dose of 1 x 10^{14} cm^{-2}. This produced an N-base region with thickness of about 0.3 microns. The gate oxide with thickness of 0.0125 microns was formed using thermal oxidation followed by a re-oxidation anneal.

The un-terminated devices had a breakdown voltage of 350 volts. For the structure with cell pitch of 21 microns (channel length of 2.5 microns and JFET width of 4 microns), a specific on-resistance of 18 mΩ-cm^2 was measured at a gate bias of only 5 volts. This was possible because of the low threshold voltage ($\sim$ 1 volt) and high transconductance of the structure achieved with the small gate oxide thickness. The measured specific on-resistance was within 2.5 times the specific on-resistance of the drift region (8 mΩ-cm^2) despite the rather large cell pitch. In spite of the very thin gate oxide, no evidence of gate rupture was observed in the blocking state due to the shielding by the P^+ region. The measured specific on-resistance for these MOSFETs is about 30 times smaller than that for silicon MOSFETs with blocking voltage capability of 1500 volts.

A detailed analysis[29] of the operation and design of the 6H-SiC ACCUFET was published in 1999. In this paper, it was pointed out that a trade-off between reducing the electric field at the gate oxide and minimizing the specific on-resistance must be performed. It was also found that the specific on-resistance for the ACCUFET increases with increasing temperature. This had not been previously observed in silicon carbide power MOSFETs because the channel conductance improved rapidly with temperature due to trap limited effective inversion-layer mobility. In contrast, the extracted accumulation-layer mobility was found to remain independent of temperature resulting in a positive temperature coefficient for the specific on-resistance due to increase in resistance of the bulk components.

The ACCUFET structure has also been successfully fabricated using 4H-SiC with high blocking voltage capability[30]. These devices were labeled SIAFETs even though the operating principle for the devices was acknowledged by the authors to be identical to that of ACCUFET structures. The authors achieved a blocking voltage capability of 4580 volts with normally-off operation at zero gate bias by using a drift region with a doping concentration of 5×10^{14} cm^{-3} and thickness of 75 microns. These devices exhibited a specific on-resistance of 1200 mΩ-cm^2 in spite of using a gate bias of 40 volts because of a low accumulation-layer mobility of 0.5 cm^2/V-s. The specific on-resistance was found to reduce by a factor of 6 times by the application of a positive bias to the sub-surface P^+ region. Unfortunately, this entails a more complex package and gate drive circuit for the devices that departs from mainstream silicon technology. The performance of this structure, renamed SEMOSFET, has been extended[31] to a 5 kV blocking voltage capability with a specific on-resistance of 88 mΩ-cm^2 in the presence of a positive bias of 2 volts applied to the buried P^+ region with 20 volts applied to the MOS-gate electrode. Since the bias applied to the buried P^+ region was less than the junction potential, these is no minority carrier injection from the P^+ region. Consequently, these devices exhibit very fast switching speeds with turn-on and turn-off times of less than 50 nanoseconds while operating at on-state voltage drops slightly less than that for a 4.5 kV IGBT.

10.5 Summary

In this chapter, it has been established that the incorporation of a sub-surface P^+ region into the planar MOSFET structure enables shielding the P-base region from reach-through limited breakdown and preventing high electric fields from developing across the gate oxide during the blocking mode. Since relatively low inversion layer mobility has been reported for 4H-SiC MOSFETs, an accumulation-mode structure was proposed with an N-base region that is completely depleted by the built-in potential of the underlying P^+/N junction. This structure takes advantage of the much larger accumulation mobility observed in semiconductors.

    The operating principle of the planar shielded MOSFET structures has been reviewed in the chapter providing guidelines for the design of the structures. It has been demonstrated that the JFET width is a critical parameter that controls the electric field at the gate oxide interface as well as the specific on-resistance. Its optimization is important for obtaining high performance devices. In the case of the accumulation-mode structure, the appropriate combination of the doping concentration and thickness of the N-base region must be chosen to ensure that it is completely depleted by the built-in potential of the underlying P^+/N junction. With adequate shielding of the base region, it is found that short channel devices will support high blocking voltages limited only by the properties of the drift region. These devices have excellent safe-operating-area and fast switching speed. This technology has potential for use in systems operating at upto at least 5000 volts.

References

1 B. J. Baliga, "Silicon Carbide Semiconductor Devices having Buried Silicon Carbide Conduction Barrier Layers Therein", U. S. Patent 5,543,637, Issued August 6, 1996.

2 B. J. Baliga and D. A. Girdhar, "Paradigm Shift in Planar Power MOSFET Technology", Power Electronics Technology Magazine, pp. 24-32, November 2003.

3 S. T. Sheppard, M. R. Melloch, and J. A. Cooper, "Characteristics of Inversion-Channel and Buried-Channel MOS Devices in 6H-SiC", IEEE Transactions on Electron Devices, Vo. 41, pp. 1257-1264, 1994.

4 B. J. Baliga, "Power Semiconductor Devices", Chapter 7, pp. 357-362, PWS Publishing Company, 1996.

5 N. Thapar and B. J. Baliga, "Analytical Model for the Threshold Voltage of Accumulation Channel MOS-Gated Devices", Solid State Electronics, Vol. 42, pp. 1975-1979, 1998.

6 S. M. Sze, "Physics of Semiconductor Devices", Second Edition, pp. 74-79, John Wiley and Sons, 1981.

7 B. J. Baliga, "Impact of SiC on Power Devices", Proceedings of the 4th International Conference on Amorphous and Crystalline Silicon Carbide", pp. 305-313, 1991.

8 L. Lipkin and J. W. Palmour, "Improved Oxidation Procedures for Reduced SiO_2/SiC Defects", J. Electronic Materials, Vol. 25, pp. 909-915, 1996.

9 R. Schorner, et al, "Significantly improved performance of MOSFETs on Silicon Carbide using the 15R-SiC Polytype", IEEE Electron Device Letters, Vol. 20, pp. 241-244, 1999.

10 S. C. Sun and J. D. Plummer, "Electron Mobility in Inversion and Accumulation Layers on Thermally Oxidized Silicon Surfaces", IEEE Transactions on Electron Devices, Vol. 27, pp. 1497-1508, 1980.

11 P. M. Shenoy and B. J. Baliga, "The Planar 6H-SiC ACCUFET", IEEE Electron Device Letters, Vol. 18, pp. 589-591, 1997.

12 P. M. Shenoy and B. J. Baliga, "High Voltage Planar 6H-SiC ACCUFET", Silicon Carbide and Related Materials – 1997, Material Science Forum, Vol. 264-268, pp. 993-996, 1998.

13 K. Chatty, et al, "Accumulation-Layer Electron Mobility in n-Channel 4H-SiC MOSFETs", IEEE Electron Device Letters, Vol. 22, pp. 212-214, 2001.

14 S. Sridevan and B. J. Baliga, "Lateral N-channel Inversion Mode 4H-SiC MOSFETs", PSRC Technical Report TR-97-019, 1997.

[15] S. Sridevan and B. J. Baliga, "Lateral N-channel Inversion Mode 4H-SiC MOSFETs", IEEE Electron Device Letters, Vol. 19, pp. 228-230, 1998.

[16] K. Ueno and T. Oikawa, "Counter-Doped MOSFETs of 4H-SiC", IEEE Electron Device Letters, Vol. 20, pp. 624-626, 1999.

[17] S. Kaneko, et al, "4H-SiC ACCUFET with a Two-Layer Stacked Gate Oxide", Silicon Carbide and Related Materials – 2001, Material Science Forum, Vol. 389-393, pp. 1073-1076, 2002.

[18] S. Sridevan and B. J. Baliga, "Phonon Scattering Limited Mobility in SiC Inversion Layers", PSRC Technical Report TR-98-03, 1998.

[19] D. Alok, E. Arnold, and R. Egloff, "Process Dependence of Inversion Layer Mobility in 4H-SiC Devices", Silicon Carbide and Related Materials – 1999, Material Science Forum, Vol. 338-342, pp. 1077-1080, 2000.

[20] K. Chatty, et al, "Hall Measurements of Inversion and Accumulation-Mode 4H-SiC MOSFETs", Silicon Carbide and Related Materials – 2002, Material Science Forum, Vol. 389-393, pp. 1041-1044, 2002.

[21] N. S. Saks, S. S. Mani, and K. Agarwal, Applied Physics Letters, Vol. 76, pp. 2250-2251, 2000.

[22] S. Harada, et al, "Temperature Dependence of the Channel Mobility and Threshold Voltage in 4H and 6H-SiC MOSFETs", Material Society Research Symposium Proceedings, Vol. 640, pp. H5.37.1-H5.37.6, 2001.

[23] G. Y. Chung, et al, "Improved Inversion Channel Mobility for 4H-SiC MOSFETs following High Temperature Anneals in Nitric Oxide", IEEE Electron Device Letters, Vol. 22, pp. 176-178, 2001.

[24] B. J. Baliga and M. Bhatnagar, "Method of Fabricating Silicon Carbide Field Effect Transistor", U. S. Patent 5,322,802, Issued June 21, 1994.

[25] J. N. Shenoy, J. A. Cooper, and M. R. Melloch, "High Voltage Double-Implanted Power MOSFETs in 6H-SiC", IEEE Electron Device Letters, Vol. 18, pp. 93-95, 1997.

[26] D. Peters, et al, "An 1800V Triple Implanted Vertical 6H-SiC MOSFET", IEEE Transactions on Electron Devices, Vol. 46, pp. 542-545, 1999.

[27] R. Schorner, et al, "Rugged Power MOSFETs in 6H-SiC with Blocking Voltage Capability upto 1800V", Silicon Carbide and Related Materials – 1999, Material Science Forum, Vol. 338-342, pp. 1295-1298, 2000.

[28]   M. Matin, A. Saha, and J. A. Cooper, "Self-Aligned Short-Channel Vertical Power DMOSFETs in 4H-SiC", Silicon Carbide and Related Materials – 2003, Material Science Forum, Vol. 457-460, pp. 1393-1396, 2004.

[29]   P. M. Shenoy and B. J. Baliga, "Analysis and Optimization of the Planar 6H-SiC ACCUFET", Solid State Electronics, Vol. 43, pp. 213-220, 1999.

[30]   Y. Sugawara, et al, "4.5 kV Novel High Voltage High Performance SiC-FET", IEEE International Symposium on Power Semiconductor Devices and ICs, pp. 105-108, 2000.

[31]   Y. Sugawara, et al, "5.0 kV 4H-SiC SEMOSFET with Low RonS of 88 mOcm2", Silicon Carbide and Related Materials – 2001, Material Science Forum, Vol. 389-393, pp. 1199-1202, 2002.

Chapter 11

Trench-Gate Power MOSFETs

The silicon trench-gate power MOSFET was developed in the 1990s by borrowing the trench technology originally developed for DRAMs. Before the introduction of the trench-gate structure, it was found that the ability to reduce the specific on-resistance for silicon power MOSFETs was constrained by the poor channel density and the JFET region resistance[1]. Although the use of more advanced lithographic design rules enabled a steady reduction of the specific on-resistance in the 1970s and early 1980s, the improvements to the DMOSFET structure were saturating[2]. The trench-gate or UMOSFET structure enabled significant increase in the channel density and elimination of the JFET resistance contribution resulting in a major enhancement in low voltage (<50 volt) silicon power MOSFET performance. The first trench-gate devices[3,4] were fabricated in the mid-1980s and shown to have significantly lower specific on-resistance than the DMOSFET structure. However, problems with controlling the quality of the trench surface and oxide reliability problems needed to be solved before the introduction of commercial devices. Eventually, this technology overtook the planar DMOSFET technology in the 1990s and has now taken a dominant position in the industry for serving portable appliances, such as laptops, PDAs, etc.

In the silicon power DMOSFET structure, the channel and JFET resistances were found to become the dominant components when the breakdown voltage was reduced below 50 volts because of the low resistance of the drift region. Due to the much lower specific on-resistance for the drift region in silicon carbide, the channel and JFET contributions become dominant even when the breakdown voltage approaches 5000 volts, especially if the channel mobility is poor. This has motivated the development of the trench-gate architecture in silicon carbide. In addition, the P-base region for the trench-gate device could be fabricated using epitaxial growth which was at a more advanced state than the ability to use ion implantation to create P-type layers in silicon carbide.

This chapter reviews the basic principles of operation of the trench-gate MOSFET structure. The specific on-resistance for this structure is shown to be significantly lower than that for the DMOSFET structure. However, the gate oxide in the UMOSFET structure is exposed to the very high electric field developed in the silicon carbide drift region during the blocking-mode. This is a major limitation to adopting the basic UMOSFET structure from silicon to silicon carbide. Consequently, structures designed to reduce the electric field at the gate oxide by using a shielding region are essential to realization of practical silicon carbide MOSFET structures[5]. These structures are described and analyzed in the next chapter. In this chapter, the results of the analysis of the basic trench-gate silicon carbide MOSFET structure by using two-dimensional numerical simulations are provided followed by the description of experimental results on relevant structures to define the state of the development effort on these devices.

11.1 Trench-Gate Power MOSFET Structure

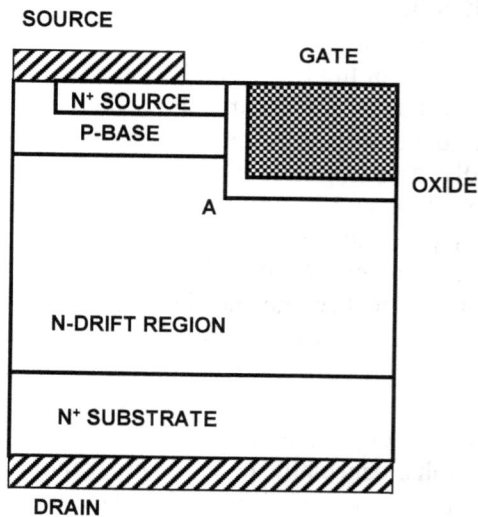

Fig. 11.1 The Trench-Gate Power MOSFET Structure.

The basic structure of the trench-gate power MOSFET is shown in Fig. 11.1. The structure can be fabricated by either the epitaxial growth of the P-base region over the drift region or by introducing the P-type dopants using ion-implantation. The first 4H-SiC UMOSFET structure was fabricated by the epitaxial growth of the P-base region due to problems with activation of P-type ion implanted dopants in silicon carbide[6]. However, this requires either removal of the P-type layer on the edges of the structure to form a mesa edge termination[6] or multiple trench-isolated guard rings[7]. The trench-gate structure was fabricated by using reactive-ion etching to form the U-shaped trenches. The gate oxide was then created by thermal oxidation followed by refilling the trench with polysilicon as done for silicon UMOSFET structures.

Note that the P-base region is short-circuited to the N^+ source region by the source metal. Although this is routinely done in silicon devices using a common ohmic contact metal for both the N^+ and P-type regions, this has been difficult to achieve in silicon carbide. It is usual to use different metals to form the ohmic contact to silicon carbide, typically with titanium for N-type regions and aluminum for P-type regions. In this case, it may be preferable to short circuit the source-base junction at a location orthogonal to the cross-section shown in the figure. This is also advantageous for reducing the cell pitch which results in a larger channel density.

11.1.1 Trench-Gate Power MOSFET Structure: Blocking Characteristics

When the trench-gate power MOSFET is operating in the forward blocking mode, the voltage is supported by a depletion region formed on both sides of the P-base/N-drift junction. The maximum blocking voltage can be determined by the electric field at this junction becoming equal to the critical electric field for breakdown if the parasitic $N^+/P/N$ bipolar transistor is completely suppressed. This suppression is accomplished by short-circuiting the N^+ source and P-base regions using the source metal as shown on the upper left hand side of the cross-section. However, a large leakage current can occur when the depletion region in the P-base region reaches-through to the N^+ source region. The doping concentration and thickness of the P-base region must be designed to prevent the reach-through phenomenon from limiting the breakdown voltage. The physics governing the reach-through process is identical to that already described in chapter 9 for the planar MOSFET structure.

Consequently, the design rules provided earlier with the aid of Fig. 9.3 can be applied to the trench-gate structure as well. This implies that the minimum channel length required for the silicon carbide trench-gate devices is much larger than for silicon resulting in a substantial increase in the on-resistance. The degradation of the on-resistance is compounded by the lower channel inversion layer mobility observed for silicon carbide.

The minimum thickness of the P-base region required to prevent reach-through breakdown decreases with increasing doping concentration as shown in Fig. 9.3. For 4H-SiC, it is necessary to increase the P-base doping concentration to above 2×10^{17} cm^{-3} to prevent reach-through with a 1 micron P-base thickness. This higher doping concentration increases the threshold voltage as already discussed in chapter 9.

The maximum blocking voltage capability of the trench-gate MOSFET structure is determined by the drift region doping concentration and thickness as already discussed in chapter 3. However, in the trench-gate MOSFET structure, the gate extends into the drift region exposing the gate oxide to the high electric field developed in the drift region under forward blocking conditions. For 4H-SiC, the electric field in the oxide reaches a value of 9×10^6 V/cm when the field in the semiconductor reaches its breakdown strength. This value not only exceeds the reliability limit but can cause rupture of the oxide leading to catastrophic breakdown. The problem is exacerbated by electric field enhancement at the corners of the trenches at location A in the figure. Novel structures[5] that shield the gate oxide from high electric field have been proposed and demonstrated to resolve this problem. These structures are discussed in the next chapter.

11.1.2 Trench-Gate Power MOSFET Structure: Forward Conduction

Current flow between the drain and source can be induced by creating an inversion layer channel on the surface of the P-base region. The current path is illustrated in Fig. 11.2 by the dotted area. The current flows from the source region into the drift region through the inversion layer channel formed on the vertical side-walls of the trench due to the applied gate bias. It then spreads into the N-drift region from the bottom of the trench at a 45 degree angle and becomes uniform through the rest of the structure. The total on-resistance for the trench-gate power MOSFET structure is determined by the resistance of these components in the current path:

$$R_{on,sp} = R_{CH} + R_D + R_{subs} \qquad [11.1]$$

where R_{CH} is the channel resistance, R_D is the resistance of the drift region after taking into account current spreading from the channel, and R_{subs} is the resistance of the N^+ substrate. These resistances can be analytically modeled by using the current flow pattern indicated by the shaded regions in Fig. 11.2.

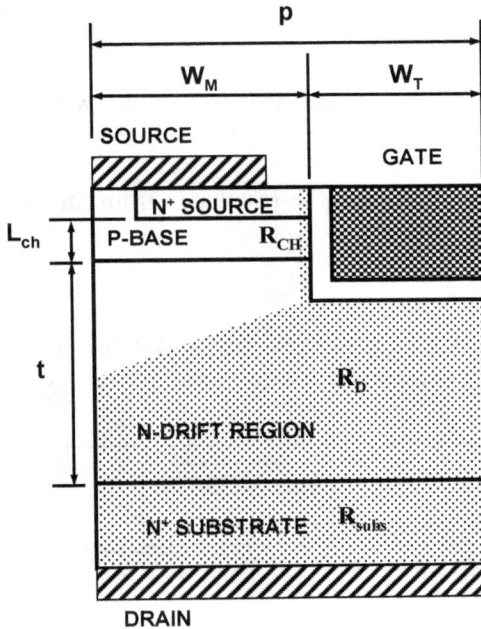

Fig. 11.2 Current Flow Path in the Trench-Gate 4H-SiC Power MOSFET.

The specific channel resistance is given by:

$$R_{CH} = \frac{(L_{CH} \cdot p)}{\mu_{inv} C_{ox}(V_G - V_T)} \qquad [11.2]$$

where L_{CH} is the channel length determined by the width of the P-base region as shown in Fig. 11.2, μ_{inv} is the mobility for electrons in the inversion layer channel, C_{ox} is the specific capacitance of the gate oxide,

V_G is the applied gate bias, and V_T is the threshold voltage. The specific capacitance can be obtained using:

$$C_{ox} = \frac{\varepsilon_{ox}}{t_{ox}}$$

[11.3]

where ε_{ox} is the dielectric constant for the gate oxide and t_{ox} is its thickness.

The drift region spreading resistance can be obtained by using:

$$R_D = \rho_D \cdot p \cdot \ln\left(\frac{p}{W_T}\right) + \rho_D \cdot (t - W_M)$$

[11.4]

where t is the thickness of the drift region below the P- base region and W_T, W_M are the widths of the trench and mesa regions, respectively, as shown in the figure.

The contribution to the resistance from the N^+ substrate is given by:

$$R_{subs} = \rho_{subs} \cdot t_{subs}$$

[11.5]

where ρ_{subs} and t_{subs} are the resistivity and thickness of the substrate, respectively. A typical value for this contribution is 4×10^{-4} Ω-cm^2.

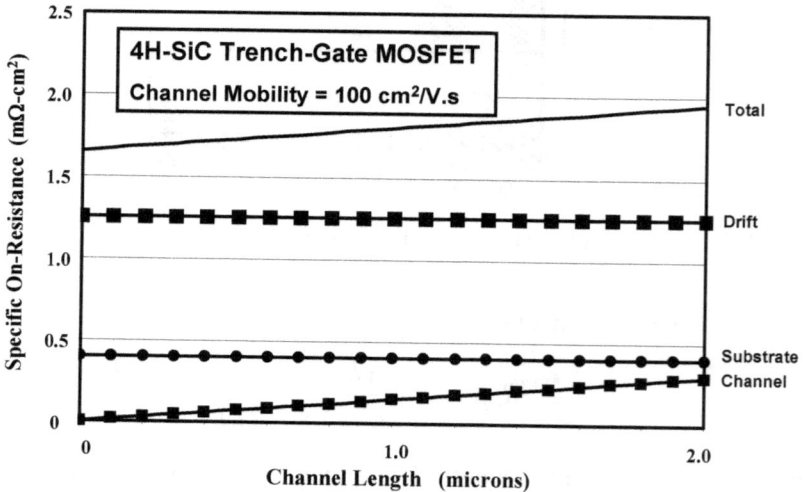

Fig. 11.3 On-Resistance Components for a 4H-SiC Trench-Gate MOSFET.

The specific on-resistances of 4H-SiC trench MOSFETs with a drift region doping concentration of 1×10^{16} cm^{-3} and thickness of 20 microns, capable of supporting 3000 volts, were modeled using the above analytical expressions. Since the trench-gate structure offers the opportunity to reduce the specific on-resistance with aggressive scaling, a gate oxide thickness of 0.05 microns was used. The mesa width (W_M) and trench width (W_T) for the structure shown in Fig. 11.2 were kept at 0.25 microns. These values can be achieved using 0.5 micron design rules. The effective gate drive voltage ($V_G - V_T$) was assumed to be 5 volts based upon a gate drive voltage of 10 volts and a threshold voltage of 5 volts. This threshold voltage is consistent with the model described in chapter 9 for the inversion-mode 4H-SiC MOSFET.

The various components of the on-resistance are shown in Fig. 11.3 when a channel inversion layer mobility of 100 cm^2/Vs was used. Inversion layer mobility values greater than this have been experimentally observed[8] in planar 4H-SiC MOSFET structures indicating the possibility to achieve them in trench gate devices as well. Note that only the channel resistance increases with increasing channel length. The other components are unaffected by the increase in channel length because the cell pitch remains unaltered. It can be seen that the drift region resistance is dominant here. At a channel length of 1 micron, a total specific on-resistance of 1.8 mΩ-cm^2 is obtained, which is within two times the ideal specific on-resistance of the drift region.

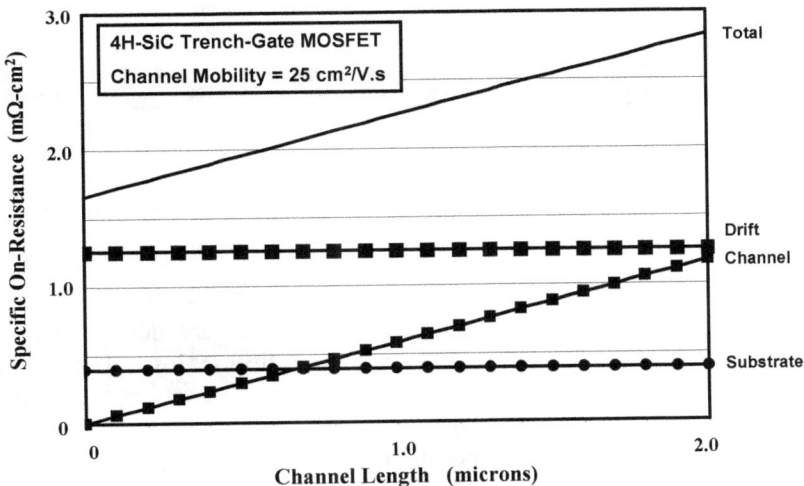

Fig. 11.4 On-Resistance Components for a 4H-SiC Trench-Gate MOSFET.

The above results are based upon using a channel mobility of 100 cm^2/Vs. Since most groups working on silicon carbide have reported much lower magnitudes for the inversion layer mobility in 4H-SiC structures[9], the impact of reducing the inversion layer mobility for the trench-gate MOSFET structure is depicted in Fig. 11.4 and 11.5 for the case of mobility values of 25 and 2.5 cm^2/Vs. With an inversion layer mobility of 25 cm^2/Vs, the channel contribution becomes comparable to the contribution from the drift region. This results in the specific on-resistance increasing to a value of 2.3 mΩ-cm^2 for a channel length of 1 micron. When the inversion layer mobility is reduced to 2.5 cm^2/Vs, the channel resistance becomes dominant as shown in Fig. 11.5, with an increase in the specific on-resistance to a value of 7.5 mΩ-cm^2 for a channel length of 1 micron.

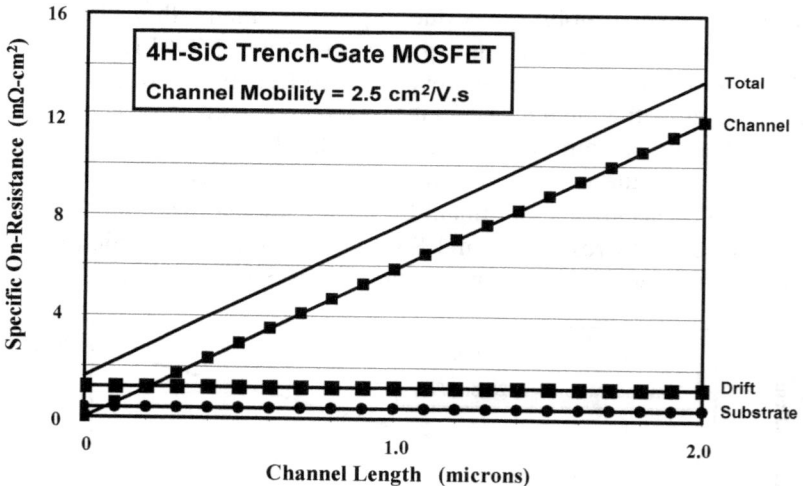

Fig. 11.5 On-Resistance Components for a 4H-SiC Trench-Gate MOSFET.

These results indicate that the channel resistance can become dominant even in the trench-gate MOSFET structure despite the high channel density if the inversion layer mobility is low. The relative contribution from the channel resistance in the trench-gate MOSFET structure depends upon the drift region resistance which is a function of the breakdown voltage of the device. In order to provide a better understanding of this, consider a 4H-SiC trench-gate MOSFET with a cell pitch of 0.5 microns (based upon 0.5 micron design rules), channel length of 1 micron and gate oxide thickness of 0.1 microns, while the properties of the underlying drift region are adjusted to obtain the desired breakdown voltage. Since the threshold voltage for this oxide thickness is about 15 volts, the gate drive voltage was assumed to be 20 volts in this analysis. The specific on-resistance for this trench-gate 4H-SiC MOSFET structure is plotted in Fig. 11.6 as a function of the breakdown voltage using a channel mobility of 25 cm²/V-s. In performing this modeling, it is important to recognize that the thickness of the drift region (parameter 't' in Fig. 11.2) can become smaller than the mesa width (W_M) at lower breakdown voltages (below 600 volts). Under these conditions, the current does not distribute at a 45 degree angle into the drift region from the bottom of the trench. Instead, the current flows from a cross-sectional width of (W_T) to a cross-section of (p = W_M + W_P). The drift region resistance for these cases can be modeled using:

$$R_D = \rho_D \cdot p \cdot \ln\left(\frac{p}{W_T}\right) \qquad [11.6]$$

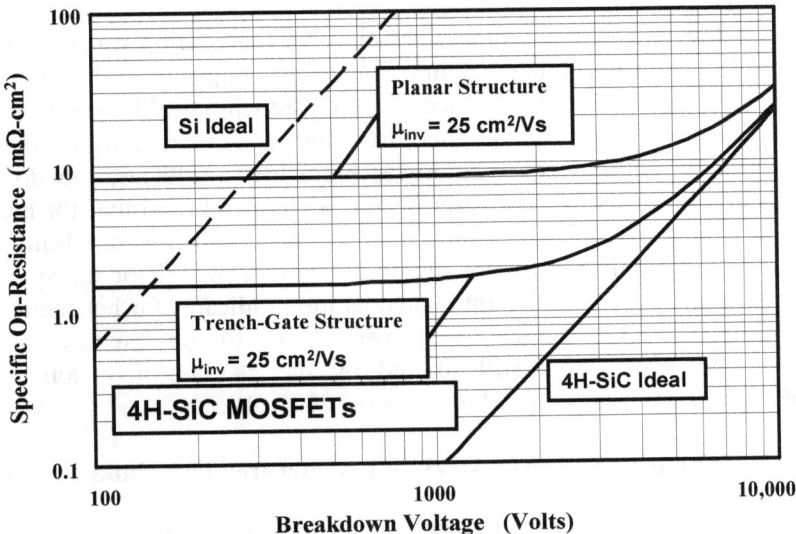

Fig. 11.6 Specific On-Resistance for 4H-SiC Trench-gate MOSFETs.

From Fig. 11.6, it can be concluded that the specific on-resistance of 4H-SiC trench-gate MOSFETs approaches the ideal specific on-resistance when the breakdown voltage exceeds 5000 volts even with a channel mobility of 25 cm^2/V-s. However, the channel resistance limits the performance of the trench-gate 4H-SiC MOSFET structure when the breakdown voltage falls below 1000 volts. For the case of a breakdown voltage of 1000 volts, the anticipated improvement in specific on-resistance over silicon power MOSFETs is then about 100 times (as opposed to the 2000x improvement in the specific on-resistance of the drift region). For a better perspective, the performance of the planar 4H-SiC inversion-mode MOSFET structure is also shown in the figure. The planar MOSFET analysis was based upon using the same channel length, gate oxide thickness, and channel mobility of 25 cm^2/V-s as the trench-gate structure. A cell-pitch of 3 microns was used for the planar structure based upon the same 0.5 micron design rules. It can be seen that the trench-gate structure offers a substantial (5-fold) reduction of the specific-on-resistance due to the higher channel density. These results highlight the need for development of aggressive processing techniques, like those routinely used for low voltage silicon power UMOSFET structures, to achieve a small cell pitch in the trench-gate 4H-SiC MOSFET structure.

The on-resistance model presented in this section was originally described[10] to highlight the importance of improving the channel inversion layer mobility for silicon carbide trench-gate MOSFETs. This model assumes that all the applied drain bias is supported within the N-drift region. For devices with lower breakdown voltages, the doping concentration in the N-drift layer becomes comparable to that for the P-base region. Consequently, a substantial fraction of the applied drain bias is supported within the P-base region as well. A model for the specific on-resistance that takes this into consideration indicates further reduction of the specific on-resistance[11]. However, the specific on-resistance of the trench-gate structure is still limited by the channel inversion layer mobility for devices designed to support 1000 volts.

11.1.3 Trench-Gate Power MOSFET Structure: Threshold Voltage

The threshold voltage of the trench-gate MOSFET structure is determined by the doping concentration of the P-base region along the sidewalls of the trench region. A minimum threshold voltage must be maintained at above 1 volt for most system applications to provide immunity against inadvertent turn-on due to voltage spikes arising from noise. At the same time, a high threshold voltage is not desirable because the voltage available for creating the charge in the channel inversion layer is determined by $(V_G - V_T)$ where V_G is the applied gate bias voltage and V_T is the threshold voltage. Most power electronic systems designed for high voltage operation (the most suitable application area for silicon carbide devices) provide a gate drive voltage of only upto 10 volts. Based upon this criterion, the threshold voltage should be kept below 3 volts in order to obtain a low channel resistance contribution.

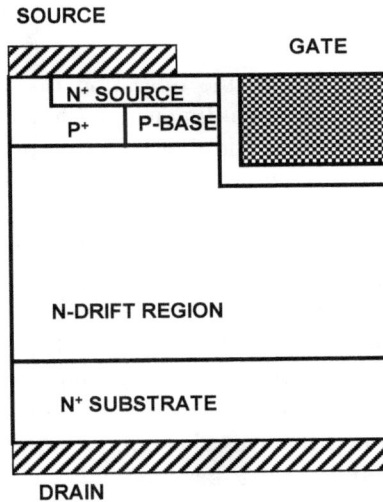

Fig. 11.7 Enhanced Trench-Gate Power MOSFET Structure.

The threshold voltage for the trench-gate inversion-mode MOSFET structure can be modeled using the same physics described in chapter 9 for the planar gate inversion-mode MOSFET structure. The threshold voltage for 4H-SiC devices can be obtained using the graphs provided in that chapter. Based upon that analysis, it is preferable to use a gate oxide thickness of 0.05 microns and a P-base doping concentration of 1×10^{17} cm^{-3} to obtain a threshold voltage of about 5 volts. However, the low P-base doping concentration can lead to a reach-through induced breakdown problem as discussed in chapter 9. A structural enhancement that can circumvent this problem is shown in Fig. 11.7. A P$^+$ region is incorporated into the trench-gate MOSFET structure under the source region to suppress the extension of the depletion region in this portion of the cell structure. A relatively thin, lightly doped P-base region is formed adjacent to the trench sidewalls to determine the threshold voltage. It has been found (see simulation section in this chapter) that the presence of the P^+ region also suppresses reach-through within the adjacent P-base region allowing the structure to support high drain voltages upto the breakdown voltage capability of the drift region.

11.2 Trench-Gate Power MOSFET Structure: Numerical Simulations

In order to gain insight into the operation of the trench-gate 4H-SiC power MOSFET structure, two-dimensional numerical simulations were performed with a drift region doping concentration of 1 x 10^{16} cm^{-3} and thickness of 20 microns corresponding to a parallel-plane breakdown voltage of 3000 volts. The baseline device had a gate oxide thickness of 0.05 microns. The P^+ and P-base regions had a depth of 1 micron. The N^+ source region had a depth of 0.1 microns leading to a channel length of 0.9 microns. The mesa and trench width were chosen to be 0.5 microns. The P^+ region occupied half of the mesa width leaving the other half for the lightly doped P-base region with a doping concentration of 9 x 10^{16} cm^{-3}. This is compatible with a fabrication design rule of 0.5 microns. A trench depth of 1.5 microns was chosen to ensure that the gate electrode extended past the depth of the P-base region.

11.2.1 Trench-Gate Power MOSFET Structure: Blocking Characteristics

The blocking capability of the trench-gate 4H-SiC power MOSFET structure was investigated by maintaining zero gate bias while increasing the drain voltage. It was found that the device could sustain a drain bias of upto 3000 volts, as shown in Fig. 11.8, without any reach-through current despite the low P-base doping concentration. A low leakage current is observed for gate bias voltages ranging from 0 to 5 volts with current increasing at a gate bias of 6 volts. This indicates that the threshold voltage is about 5 volts. Unlike the planar 4H-SiC inversion-mode structure described in chapter 9, there is no reach-through induced current in the trench-gate device because of the incorporation of the P^+ region.

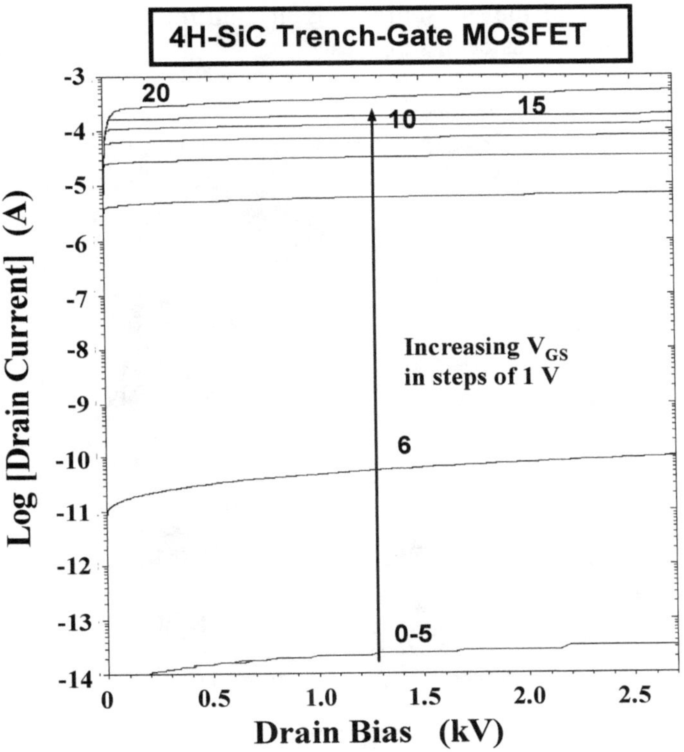

Fig. 11.8 Blocking Characteristics of the Trench-Gate 4H-SiC MOSFET Structure.

The potential distribution (solid-lines) within the upper portion of the trench-gate 4H-SiC MOSFET structure is shown in Fig. 11.9 at a drain bias of 3000 volts. It can be seen that there is a little electric field enhancement at the corner of the trench. Most notably, no potential lines are visible within the P-base region indicating that it is supporting less than 25 volts (the separation between the potential contours). This demonstrates that the presence of the P^+ region under the N^+ source region is suppressing the depletion of the more lightly doped P-base region. The current flow-lines observed under these voltage blocking conditions are also shown in the figure using dashed lines. It can be seen that the leakage current is collected at the contact to the P^+ region placed on the left-hand-side of the structure. Unlike in the case of the planar inversion-mode structure discussed in chapter 9 (see Fig. 9.21), there is no current flow into the N^+ source region confirming that the reach-through phenomenon has been suppressed.

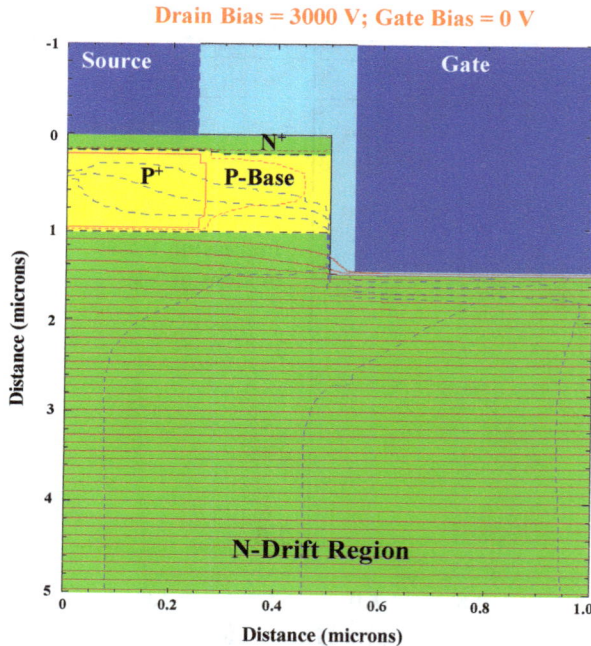

Fig. 11.9 Potential Distribution in the Trench-Gate 4H-SiC Power MOSFET.

A serious problem with the trench-gate silicon carbide MOSFET structure is associated with the high electric field generated at the gate oxide interface. In order for the channel to span the P-base region, it is necessary for the trench to penetrate through the P-base region into the N-drift region. This exposes the gate oxide located at the bottom of the trench to the high electric field developed in the silicon carbide drift region. The problem can be illustrated by examination of the electric field distribution within the trench-gate 4H-SiC power MOSFET structure under voltage blocking conditions. A three-dimensional view of the electric field distribution is shown in Fig. 11.10 for a drain bias of 1000 volts with zero bias applied to the gate electrode. The field distribution shown in this figure indicates that the electric field at the gate oxide has reached a magnitude of close to 1×10^7 V/cm. Consequently, the gate oxide will be prone to failure from rupture at a drain bias that is just one-third of the breakdown voltage capability of the drift region. Gate oxide reliability problems can arise at much lower drain voltages indicating a serious limitation to operating the trench-gate 4H-SiC MOSFET structure in the blocking mode. Fortunately, this problem has been overcome using innovative structures[5] to shield the gate oxide that will be described in the next chapter.

Drain Bias = 1000 Volts

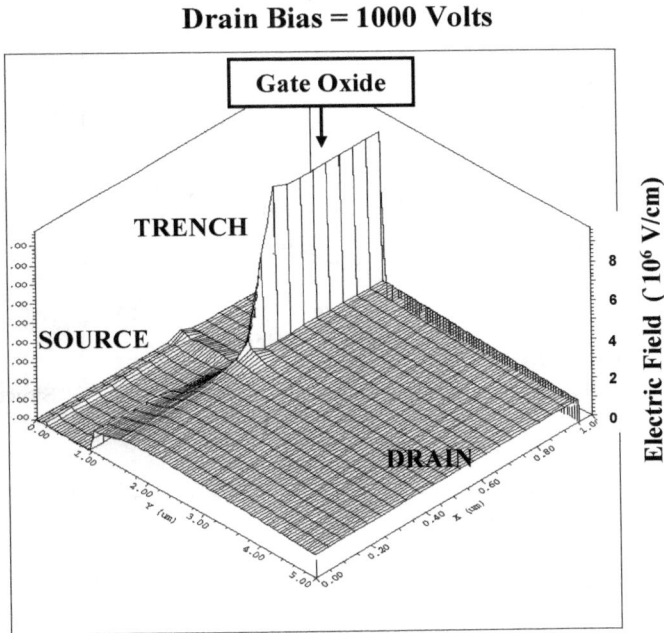

Fig. 11.10 Electric Field Distribution in the Trench-Gate 4H-SiC MOSFET Structure.

From Fig. 11.10, it can be seen that the largest electric field in the gate oxide in the trench-gate 4H-SiC power MOSFET structure occurs at the center of the trench region (x = 1 micron). The behavior of this electric field with increasing drain bias is shown in Fig. 11.11 for drain bias upto 1000 volts. It is obvious that the electric field in the oxide reaches its rupture strength at a drain bias of just 1000 volts. Meanwhile the maximum electric field in the bulk of the semiconductor occurs at the P^+/N junction at x = 0 microns. The development of this electric field is shown in Fig. 11.12, where it can be seen that the maximum electric field is far below the breakdown strength of 4H-SiC.

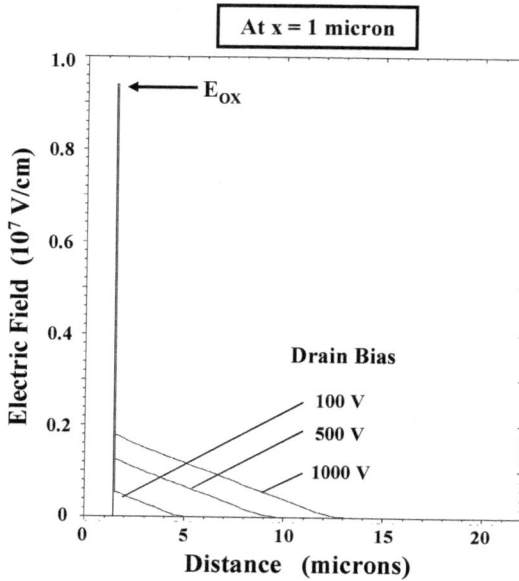

Fig. 11.11 Electric Field in the Trench-Gate 4H-SiC MOSFET Structure.

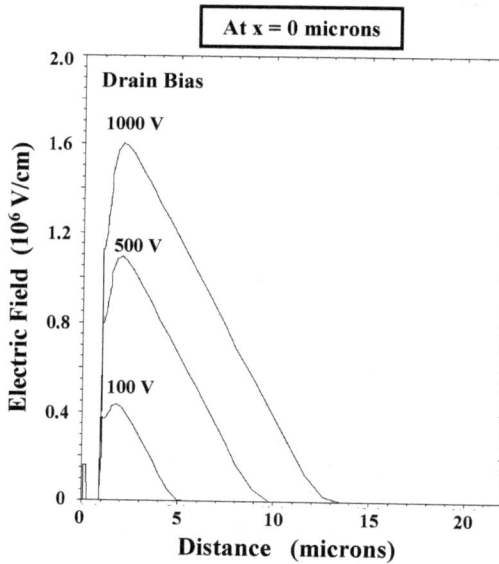

Fig. 11.12 Electric Field in the Trench-Gate 4H-SiC MOSFET Structure.

11.2.2 Trench-Gate Power MOSFET Structure: On-State Characteristics

The on-state voltage drop for the trench-gate 4H-SiC power MOSFET structure is determined by its specific on-resistance. The on-resistance for the structure was extracted by performing numerical simulations with a gate bias of 10 volts. The current distribution within the trench-gate 4H-SiC power MOSFET structure is shown in Fig. 11.13. It can be seen that the current flows through the channel and is then distributed from the bottom of the trench into the drift region. This is due to the formation of an accumulation region at the bottom of the trench because of the applied positive gate bias. The current is distributed into the drift region at a 45 degree angle and then becomes uniform from a depth of about 2 microns. This is consistent with the assumptions made in deriving the analytical model provided in the earlier section of this chapter.

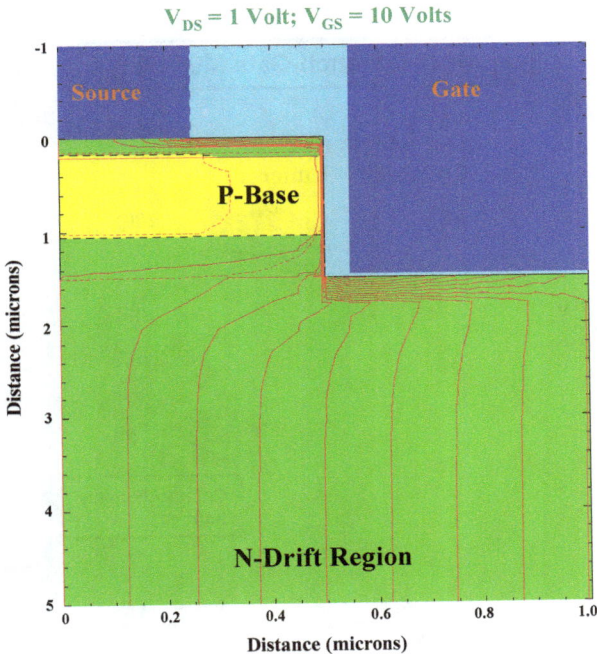

Fig. 11.13 On-State Current Distribution in the Trench-Gate 4H-SiC MOSFET.

The simulations of the specific on-resistance were done with different values for the channel inversion layer mobility. The specific on-resistance was obtained by using a drain bias of 1 volt and sweeping the gate bias to 20 volts to obtain the transfer characteristic. A drain bias of 1 volt was used for these simulations because the on-state voltage drop for these structures is typically in this range. The transfer characteristic obtained using this method is shown in Fig. 11.14 for the device with cell pitch of 1 micron. From the transfer characteristics, it can be seen that the threshold voltage that determines the on-resistance is approximately 5 volts (allowing operation of this device with a gate bias of 10 volts). This value is close to that calculated using the analytical model discussed in the previous chapter for the chosen P-base doping concentration of 9 x 10^{16} cm^{-3} and gate oxide thickness of 0.05 microns. The impact of changing the channel inversion layer mobility is also shown in this figure. As expected, there is a reduction in the drain current as the channel mobility is reduced.

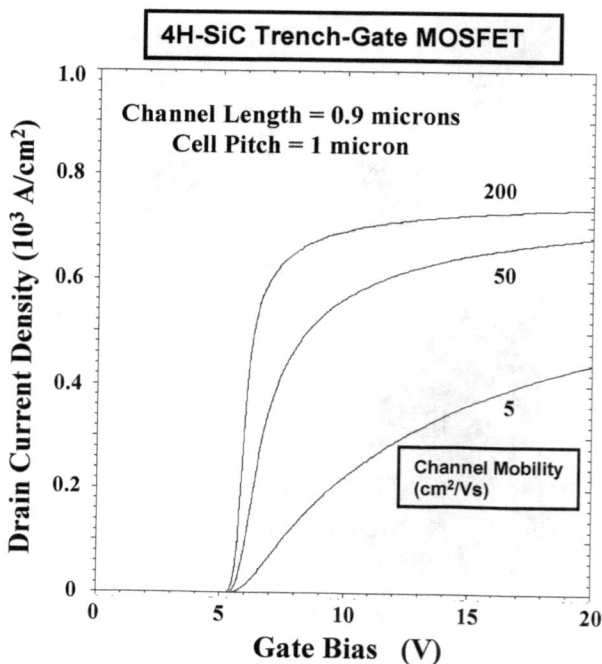

Fig. 11.14 Transfer Characteristics of the 4H-SiC Trench-Gate MOSFETs.

The specific on-resistances obtained from the on-state simulations are compared in Fig. 11.15 with the analytically calculated values obtained by using the model developed in the previous section. For the simulation cases, the channel mobility values were extracted from the *i-v* characteristics obtained for the lateral 4H-SiC MOSFET structure that was previously described in chapter 9 by using the same interface degradation factor as used during the simulation of the vertical 4H-SiC trench-gate MOSFET structure. It can be seen that the predictions of the analytical model are in excellent agreement with the values obtained from the simulations. These results provide confidence in the ability to use the analytical model during the design of the trench-gate 4H-SiC power MOSFET structure. From the figure, it can be concluded that a channel mobility of more than 50 cm^2/Vs is satisfactory for achieving a low specific on-resistance for a 3 kV trench-gate 4H-SiC power MOSFET structure with cell pitch of 1 micron.

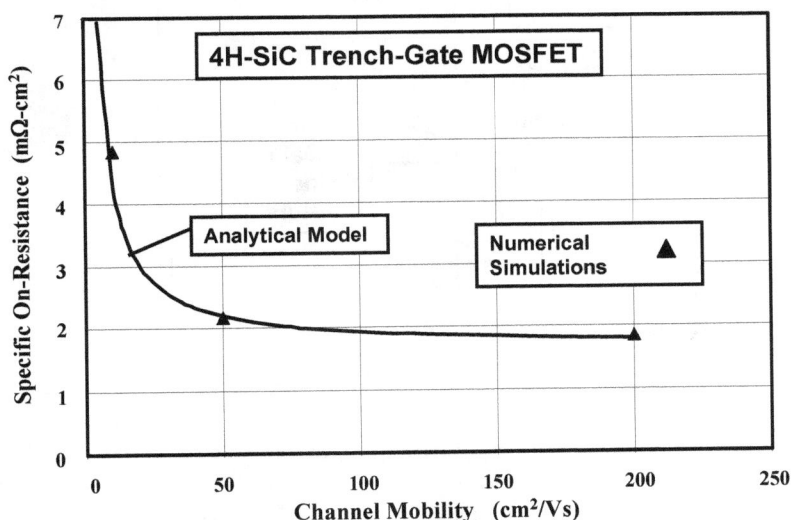

Fig. 11.15 Specific On-Resistance for the 4H-SiC Trench-Gate MOSFETs.

11.2.3 Trench-Gate Power MOSFET Structure: Output Characteristics

The output characteristics of the power MOSFET defines the region within which the device can operate during the switching transients in

applications. The current-voltage loci during the transient must be chosen
to remain within the boundary defined by the output characteristics. The
output characteristics for the 4H-SiC trench-gate power MOSFET
structure were determined by biasing the gate at various values and
sweeping the drain voltage upto 2800 volts. The characteristics obtained
for the structure with cell pitch of 1 micron are shown in Fig. 11.16. The
device exhibits excellent output characteristics upto a gate bias of 10
volts with high output resistance. However, it must be kept in mind that
the electric field in the gate oxide reaches its rupture strength at a drain
bias of 1000 volts. This is a serious limitation to the safe-operating-area
for the trench-gate 4H-SiC structure.

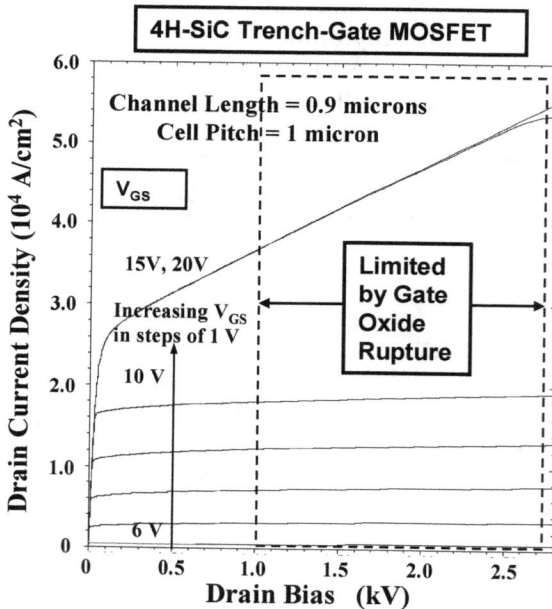

Fig. 11.16 Output Characteristics of a 4H-SiC Trench-Gate MOSFET.

11.3 Accumulation-Mode Trench-Gate Power MOSFET

With the recognition[10] of the difficulty for achieving high inversion layer
mobility in silicon carbide especially along the sidewalls of trenches, an
alternate device structure was proposed[12] based upon formation of a
channel using an accumulation layer. This device structure, named the

accumulation-mode trench-gate MOSFET, is illustrated in Fig. 11.17. In this structure, the P-base region of the conventional trench-gate inversion-mode MOSFET is replaced with a very lightly doped N-type base region. The width of the N-base region (i.e. the mesa width) is designed to be completely depleted by the work-function difference between the gate and the semiconductor. This establishes a potential barrier for transport of electrons from the source to the drain allowing normally-off operation. The device can be switched to the on-state by the application of a positive gate bias, which creates an accumulation layer channel on the trench sidewalls. This channel provides the path for current flow between the source and drain electrodes. As discussed in the previous chapter, a larger accumulation layer mobility was anticipated along the trench sidewalls enabling reduction of the specific on-resistance.

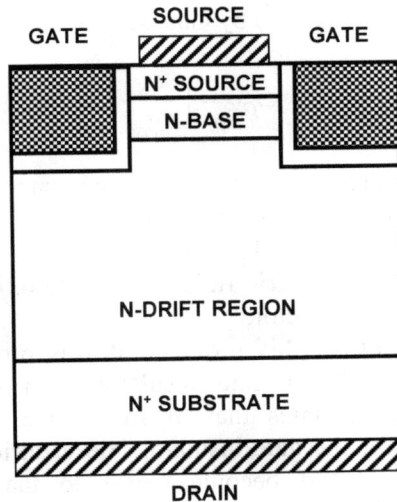

Fig. 11.17 Accumulation-Mode Trench-Gate Power MOSFET Structure.

One of the advantages of the accumulation-mode trench-gate MOSFET structure is the absence of P-N junctions in the path between the source and the drain. This eliminates the problems dealing with the parasitic P-N-P transistor within inversion-mode structures allowing reducing the mesa width to produce the desired normally-off behavior. The anti-parallel diode function can also be generated with the accumulation-mode structure by application of a positive gate bias to create the channel when the structure operates in the third quadrant. This approach is similar to that used in silicon synchronous rectifiers using low voltage MOSFETs.

11.4 Trench-Gate Power MOSFET: High Permittivity Gate Insulator

As already discussed in a previous section of this chapter, a high electric field is developed in the gate oxide at the bottom of the trench in the silicon carbide UMOSFET structure. The electric field in the gate oxide can exceed its rupture strength at relatively low drain bias voltages limiting the blocking voltage capability of the trench-gate structure. The electric field in the oxide (E_{OX}) is related to the electric field in the underlying semiconductor (E_S) by Gausses Law:

$$E_{OX} = \left(\frac{\varepsilon_S}{\varepsilon_{OX}} \right).E_S \qquad\qquad [11.7]$$

where ε_{OX} is the dielectric constant of the gate oxide and ε_S is the dielectric constant of the semiconductor. The relative dielectric constant (permittivity) for 4H-SiC is 9.7 while that for silicon dioxide is 3.85. Using these values, the electric field in the gate oxide is a factor of 2.5 times that in the semiconductor.

In recognition of this problem, it was proposed[13] that silicon dioxide be replaced by a gate insulator with high permittivity – preferably at least 10 times that of free space. With the increased dielectric constant for the gate insulator, the electric field developed in the gate dielectric would become closer to that developed in the semiconductor[14]. This implies that the maximum electric field in the gate dielectric can be reduced to 3×10^6 V/cm, which should be satisfactory for reliable operation. This approach can take advantage of the large effort being undertaken to develop high dielectric constant insulators for mainstream silicon DRAM chips.

A simple one-dimensional analysis of the metal-oxide-semiconductor structure provides interesting insight into the benefits of using high permittivity gate insulators in silicon carbide MOSFETs. Since the highest electric field occurs at the flat bottom of the trench region in the UMOSFET structure, a one dimensional analysis is a reasonable approximation for analytical purposes.

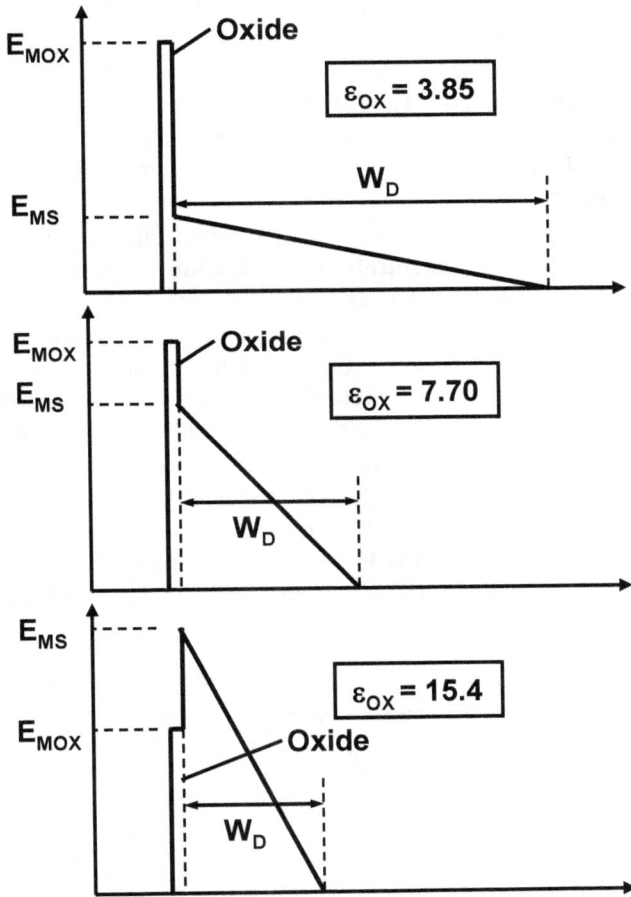

Fig. 11.18 Electric Field Distribution in the Gate Stack for the Trench-Gate Power MOSFET Structure.

The electric field profile in the MOS gate stack is illustrated in Fig. 11.18 for three cases of dielectric constants when it is supporting the same voltage (e.g. 3000 volts). In the first two cases with permittivity of 3.85 (corresponding to silicon dioxide) and 7.7 (corresponding to silicon nitride), it is assumed that the field distribution is constrained by reaching a maximum electric field of 3×10^6 V/cm in the gate oxide. This value was chosen based upon the maximum electric field in oxides for reliable operation. Under this constraint, the maximum electric field in the silicon carbide becomes much lower than its breakdown field strength (1.2×10^6 V/cm for the permittivity of 3.85 and 2.4×10^6 V/cm for the permittivity of 7.7). However, when the permittivity is increased to 15.4, the electric field distribution becomes constrained by the maximum allowable electric field in the silicon carbide (3.3×10^6 V/cm for 4H-SiC at a doping concentration of about 1×10^{16} cm^{-3}). Under this limitation, the electric field in the oxide is reduced to about 2×10^6 V/cm for the permittivity of 15.4.

The changes in the electric field distribution have a strong impact on the doping concentration and thickness of the drift region required to support the drain bias (e.g. 3000 volts) as indicated in the figure. For the lower dielectric constants, a much thicker drift region with lower doping concentration (reflected in the smaller slope of the electric field profile in the semiconductor) is required because of the reduced maximum electric field, when compared with the properties of the ideal drift region. For the high dielectric constant case, the electric field in the semiconductor becomes identical to that for the ideal one-dimensional parallel plane abrupt junction. Thus, the specific on-resistance is equal to that for the ideal drift region only for the case of the high permittivity gate dielectric and much larger for the other cases.

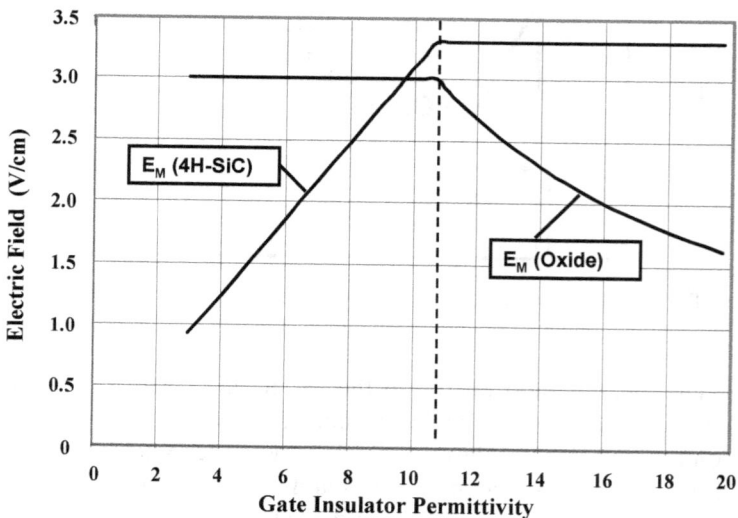

Fig. 11.19 Electric Field Distribution in the Gate Stack for the Trench-Gate Power MOSFET Structure.

In the above discussion, it was pointed out that the gate MOS stack operates in one of two regimes of operation – namely with the electric field constrained by a maximum allowable value in the gate insulator or by the electric field constrained by the maximum allowable value in the semiconductor. The results of the analytically calculated electric fields in the gate insulator and the semiconductor under these constraints are shown in Fig. 11.19. It can be seen that the transition occurs at a permittivity given by:

$$\varepsilon_{OX} = \left(\frac{E_{MS}(4H - SiC)}{E_{MOX}} \right).\varepsilon_S \qquad [11.8]$$

where E_{MS}(4H-SiC) is the critical electric field for breakdown in 4H-SiC and E_{MOX} is the maximum electric field in the dielectric for reliable operation. For the values for these parameters described above, the transition occurs at a permittivity of 10.7. This is the minimum permittivity needed to reduce the specific on-resistance of the drift region to that for the ideal drift region.

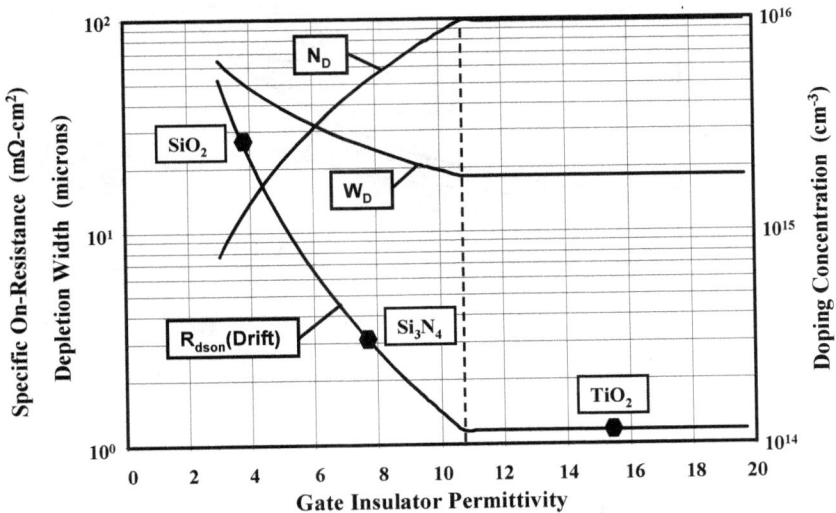

Fig. 11.20 Impact of Gate Insulator Permittivity for the Trench-Gate Power MOSFET Structure.

The permittivity of the gate insulator has a strong influence on the doping concentration and thickness of the drift region. This is shown in Fig. 11.20. The doping concentration of the drift region increases and its thickness decreases as the permittivity is increased upto a value of 10. This results in a drastic reduction of the specific on-resistance of the drift region. As examples, points are shown for selected insulators in the graph. In the case of the commonly used silicon dioxide gate insulator, the specific resistance of the drift region is degraded by a factor of 25 times. If silicon nitride is used as the gate insulator, the drift region resistance becomes only 3 times larger than the ideal case. With the use of a much higher dielectric constant of 15 with titanium dioxide as the insulator, the ideal drift region resistance can be obtained together with reduced electric field in the insulator.

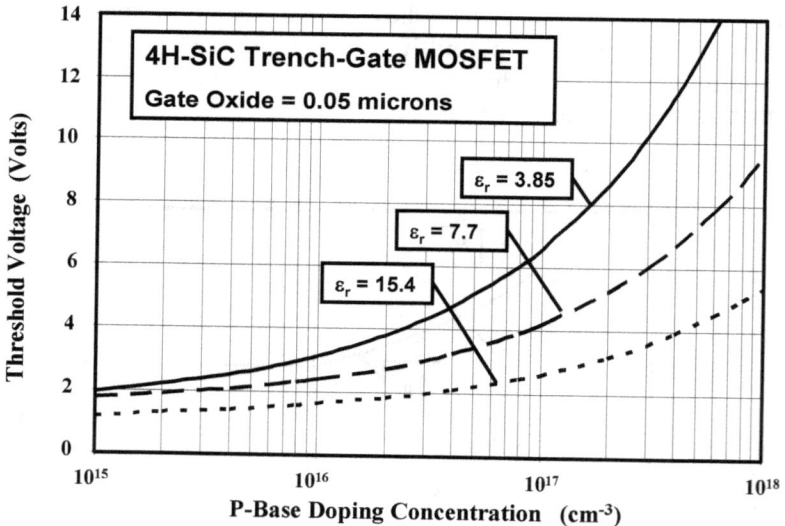

Fig. 11.21 Impact of Gate Insulator Permittivity on the Threshold Voltage of the Trench-Gate Power MOSFET Structure.

It is worth pointing out that a larger dielectric constant for the gate insulator is beneficial for achieving a lower threshold voltage with any given insulator thickness. This is shown in Fig. 11.21 for the case of a gate oxide thickness of 0.05 microns. The calculations of the threshold voltage were performed using the same assumptions regarding a work-function difference of 1 volt and the presence of a fixed oxide charge of

2 x 10^{11} cm^{-2} used in chapter 9 (see Fig. 9.11). At a P-base doping concentration of 3 x 10^{17} cm^{-3}, it can be seen that the threshold voltage is reduced from 10 volts for the silicon dioxide case (permittivity of 3.85) to 6 volts for the silicon nitride case (permittivity of 7.7), and even further to about 3 volts for a permittivity of 15.4. This is favorable for increasing the charge induced in the channel for an available gate bias of 10 volts in systems resulting in reducing the specific on-resistance.

The analysis described above highlights the benefits of using a gate insulator with high permittivity. It must be pointed out that the breakdown strength of insulators has been empirically found to reduce with increasing permittivity. Consequently, care must be taken to ensure that the gate insulator with high permittivity will operate in a reliable manner even when the electric field in it has been reduced. In addition, the band offset between the gate insulator and silicon carbide should be examined to make sure that the Fowler-Nordheim tunneling described in chapter 9 does not create instabilities.

11.5 Trench-Gate Power MOSFET: Experimental Results

The lack of significant diffusion of dopants in silicon carbide motivated the investigation of trench-gate structures before planar-gate structures. The need to obtain a high quality interface between silicon carbide and the gate dielectric was identified as a challenging endeavor from the inception of interest in the development of unipolar transistors from this semiconductor material[10,15]. In addition, the problem of high electric fields developed across the gate oxide, especially at the trench corners, was identified as a limitation to achieving high electric fields within the semiconductor[10]. These issues were indeed found to be a major limitation on the performance of the first trench-gate 4H-SiC power MOSFET structures.

The first reported[6] trench-gate power MOSFETs fabricated using 4H-SiC were limited to a blocking voltage of 260 volts due to rupture of the gate oxide. The specific on-resistance of 18 mΩ-cm^2, measured at a gate bias of 22 volts, was inferior to that of silicon power MOSFETs because of a large cell pitch and low channel mobility (~ 10 cm^2/V-s).

Subsequently, a trench-gate 4H-SiC power MOSFET structure with breakdown voltage of 1100 volts was reported[16] by using a drift region with thickness of 12 microns and doping concentration of 1 x 10^{15} cm^{-3}. The low doping concentration in the drift region was required to keep the maximum electric field in the silicon carbide to only 1 x 10^6 V/cm to prevent oxide rupture. This is consistent with the analysis in the previous section of this chapter. These devices exhibited a very high specific on-resistance (over 1 Ω-cm^2) at room temperature due to the poor channel mobility (~ 1.5 cm^2/V-s). The specific on-resistance reduced to about 180 mΩ-cm^2 (using data in Fig. 3 in the publication) due to an increase in the channel mobility. In these devices, a threshold voltage of 5 volts was obtained by using a low P-base doping concentration of 6.5 x 10^{16} cm^{-3} with a gate oxide thickness of 0.1 microns. However, this necessitated increasing the channel length to 4 microns to prevent reach-through problems. The large channel length, in conjunction with the poor channel mobility, was responsible for the poor (worse than silicon power MOSFETs) specific on-resistance of the devices.

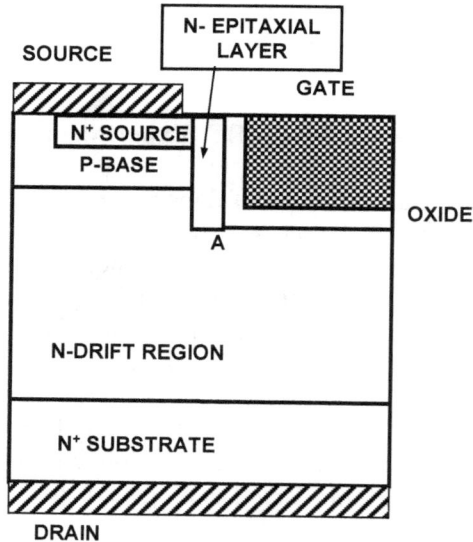

Fig. 11.22 Epitaxial-Channel Accumulation-Mode Trench-Gate Power MOSFET Structure.

In section 11.3, an accumulation-mode trench-gate MOSFET structure was described as an approach to reduce the specific on-resistance due to an increase in the channel mobility. An accumulation-mode structure that combines a trench-gate structure with a depleted N-base region located adjacent to a P-N junction (previously described in chapter 10 as the shielded accumulation-mode planar MOSFET structure) has been proposed[17] and experimentally demonstrated. This structure, shown in Fig. 11.22, contains an N-base region that is epitaxially grown on the trench sidewall surface. The authors observed an accumulation layer mobility of 108 cm^2/V-s for this structure resulting in a specific on-resistance of 11 $m\Omega$-cm^2 at a gate bias of 10 volts. The breakdown voltage of 500 volts for this structure is still limited by the development of high electric field in the gate oxide.

11.6 Summary

The trench-gate 4H-SiC vertical power MOSFET structure was the first approach explored by the silicon carbide community because the P-base region could be epitaxially grown rather than formed by ion implantation, which was a less mature technology. Although devices with breakdown voltages of upto 1100 volts have been fabricated, their performance was severely limited by the on-set of rupture of the gate oxide and the poor channel mobility. In order to overcome these problems, it is necessary to shield the gate oxide from the high electric field developed in the drift region. These device structures are discussed in the next chapter.

References

[1] B. J. Baliga, "Power Semiconductor Devices", Chapter 7, pp. 335-425, PWS Publishing Company, 1996.

[2] B. J. Baliga, "Evolution of MOS-Bipolar Power Semiconductor Technology", Proceeding of the IEEE, Vol. 74, pp. 409-418, 1988.

[3] D. Ueda, H. Takagi, and G. Kano, "A New Vertical Power MOSFET Structure with Extremely Reduced On-Resistance", IEEE Transactions on Electron Devices, Vol. 32, pp. 2-6, 1985.

[4] H-R. Chang, et al, "Ultra-Low Specific On-Resistance UMOSFET", IEEE International Electron Devices Meeting, Abstract 28.3, pp. 642-645, 1986.

[5] B. J. Baliga, "Silicon Carbide Switching Device with Rectifying Gate", U. S. Patent 5,396,085, Issued March 7, 1995.

[6] J. W. Palmour, et al, "4H-Silicon Carbide Power Switching Devices", Silicon Carbide and Related Materials – 1995, Institute of Physics Conference Series, Vol. 142, pp. 813-816, 1996.

[7] B. J. Baliga, "Silicon Carbide Power MOSFET with Floating Field Ring and Floating Field Plate", U. S. Patent 5,233,215, Issued August 3, 1993.

[8] S. Sridevan and B. J. Baliga, "Inversion Layer Mobility in SiC MOSFETs", Silicon Carbide and Related Materials – 1997, Material Science Forum, Vol. 264-268, pp. 997-1000, 1998.

[9] S-H Ryu, et al, "Design and Process Issues for Silicon Carbide Power DiMOSFETs", Material Research Society Symposium Proceeding, Vol. 640, pp. H4.5.1-H4.5.6, 2001.

[10] B. J. Baliga, "Critical Nature of Oxide/Interface Quality for SiC Power Devices", INFOS'95, Paper 1.1, June 1995. Published in Microelectronics Journal, Vol. 28, pp. 177-184, 1995.

[11] M. Bhatnagar, D. Alok, and B. J. Baliga, "SiC Power UMOSFET: Design, Analysis, and Technological Feasibility", Silicon Carbide and Related Materials – 1993, Institute of Physics Conference Series, Vol. 137, pp. 703-706, 1994.

[12] B. J. Baliga, "Silicon Carbide Field Effect Device", U. S. Patent 5,323,040, Issued June 21, 1994.

[13] S. Sridevan, P. K. McLarty, and B. J. Baliga, "Silicon Carbide Switching Devices having Near Ideal Breakdown Voltage Capability and Ultra-Low On-State Resistance", U. S. Patent 5,742,076, Issued April 21, 1998.

[14] S. Sridevan, P.K. McLarty, and B.J. Baliga, "Analysis of Gate Dielectrics for SiC Power UMOSFETs", IEEE International Symposium on Power Semiconductor Devices and ICs, pp. 153-156, 1997.

[15] B. J. Baliga, "Impact of SiC on Power Devices", Proceedings of the 4th International Conference on Amorphous and Crystalline Silicon Carbide", pp. 305-313, 1991.

[16] A. K. Agarwal, et al, "1.1 kV 4H-SiC Power UMOSFETs", IEEE Electron Device Letters, Vol. 18, pp. 586-588, 1997.

[17] K. Hara, "Vital Issues for SiC Power Devices", Silicon Carbide and Related Materials – 1997, Materials Science Forum, Vols. 264-268, pp. 901-906, 1998.

Chapter 12

Shielded Trench-Gate Power MOSFETs

In the previous chapter, it was demonstrated that the performance of the trench-gate silicon carbide power MOSFET structure is severely compromised by the development of a high electric field in the gate oxide during the blocking mode of operation. This problem occurs because the trench spans the P-base region exposing the gate oxide at the bottom of the trench to the high electric field in the silicon carbide drift region. The electric field in the gate oxide reaches its rupture strength well before the electric field in the semiconductor approaches its breakdown field strength. Consequently, in order to operate at any given blocking voltage, the drift region doping concentration has to be reduced and its thickness increased until the specific on-resistance becomes 25 times larger than that for the ideal drift region. This problem inhibited the performance of the first trench-gate 4H-SiC power MOSFETs.

In order to suppress the development of high electric fields in the gate oxide, a trench-gate power MOSFET structure has been proposed[1] with a shielding region incorporated at the bottom of the trench. This chapter reviews the basic principles of operation of this shielded trench-gate MOSFET structure. The impact of the JFET region, formed by the incorporation of the shielding region, on the specific on-resistance for this structure is analyzed here. It is demonstrated that the JFET region resistance can be reduced by enhancement of the doping concentration between the shielding regions while retaining a small cell pitch to obtain a high channel density.

The results of the analysis of the shielded trench-gate silicon carbide MOSFET structure by using two-dimensional numerical simulations are provided in this chapter. The analysis confirms the ability to reduce the electric field in the gate oxide. In addition, it demonstrates that the electric field at the P-base/N-drift junction is also reduced allowing a reduction of the channel length. This is beneficial for reducing the channel resistance contribution. The performance of shielded trench-gate power MOSFETs reported in the literature is summarized at the end of the chapter to define the state of the development effort on these devices.

12.1 Shielded Trench-Gate Power MOSFET Structure

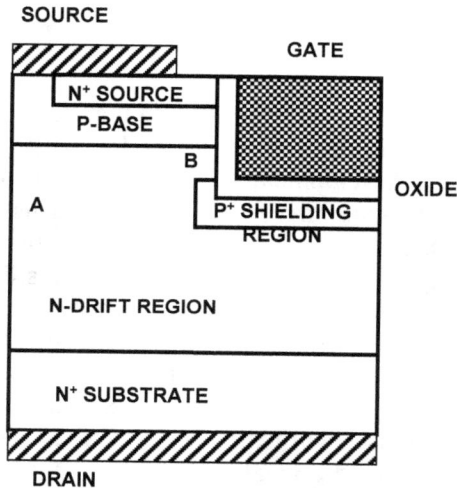

Fig. 12.1 The Shielded Trench-Gate Power MOSFET Structure.

The basic structure of the shielded trench-gate power MOSFET is shown in Fig. 12.1. The shielding region consists of a heavily doped P-type (P$^+$) region located at the bottom of the trench. The P$^+$ shielding region is connected to the source electrode at a location orthogonal to the device cross-section shown above. The structure then behaves like a monolithic version of the *Baliga-Pair* configuration discussed in chapter 8 with both the JFET and MOSFET formed in the same semiconductor, namely, silicon carbide. From this point of view, the portion of the N-drift region located just below the P-base region serves as both the drain of the MOSFET and the source of the JFET. For this reason, both the *Baliga-Pair* configuration and the monolithic option were patented together[1].

The shielded trench-gate power MOSFET structure can be fabricated by using the same process described earlier for the conventional trench-gate device structure in chapter 11 with the addition of an ion implantation step to form the P^+ shielding region. The ion implant used to form the P^+ shielding region must be performed after etching the trenches. It is worth pointing out that this implantation step should not dope the sidewalls of the trenches. This can be accomplished by doing the ion implant orthogonal to the wafer surface. Alternately, a conformal oxide can be deposited on the trench sidewalls and removed from the trench bottom using anisotropic reactive ion etching to selectively expose the trench bottom to the P-type dopant during the ion implantation. Another option to reduce the introduction of P-type doping on the trench sidewalls is by using a relatively low P-type doping concentration for the shielding region. The shielding is effective as long as the gate oxide is buffered from the high electric field in the N-drift region.

Although the P-type shielding region can be confined to just the bottom of the trench, it is preferable that it overlaps the trench corner as illustrated in Fig. 12.1. The overlap can be a natural outcome of the straggle in the ion implant or the removal of some of the masking oxide on the trench sidewall if its profile is slightly tapered.

12.1.1 Shielded Trench-Gate Power MOSFET Structure: Blocking Characteristics

The shielded trench-gate power MOSFET operates in the forward blocking mode when the gate electrode is shorted to the source by the external gate drive circuit. At low drain bias voltages, the voltage is supported by a depletion region formed on both sides of the P-base/N-drift junction. Consequently, the drain potential appears across the MOSFET located at the top of the structure. This produces a positive potential at location 'A' in Fig. 12.1, which reverse biases the junction between the P^+ shielding region and the N-drift region because the P^+ shielding region is held at zero volts. The depletion region that extends from the P^+/N junction pinches off the JFET region producing a potential barrier at location 'A'. The potential barrier tends to isolate the P-base region from any additional bias applied to the drain electrode. Consequently, a high electric field can develop in the N-drift region below the P^+ shielding region while the electric field at the P-base region remains low. This has the beneficial effects of mitigating the reach-through of the depletion region within the P-base region and in keeping the electric field in the gate oxide low at location 'B' where it is exposed to the N-drift region.

The maximum blocking voltage of the shielded trench-gate power MOSFET is determined by the properties of the drift region. This allows reduction of the drift region resistance close to that of the ideal case. In addition, the reduction of the electric field in the vicinity of the P-base region allows reduction of the channel length as well as the gate oxide thickness. This is beneficial for further reduction of the device on-state resistance.

It is worth pointing out that the P^+ shielding region must be adequately short-circuited to the source terminal in order for the shielding to be fully effective. The location of the P^+ region at the bottom of the trench implies that contact to it must be provided at selected locations orthogonal to the cross-section of the device shown in Fig. 12.1. Since the sheet resistance of the ion implanted P^+ region can be quite high, it is important to provide the contact to the P^+ region frequently in the orthogonal direction during chip design. This must be accomplished without significant loss of channel density if low specific on-resistance is to be realized.

12.1.2 Shielded Trench-Gate Power MOSFET Structure: Forward Conduction

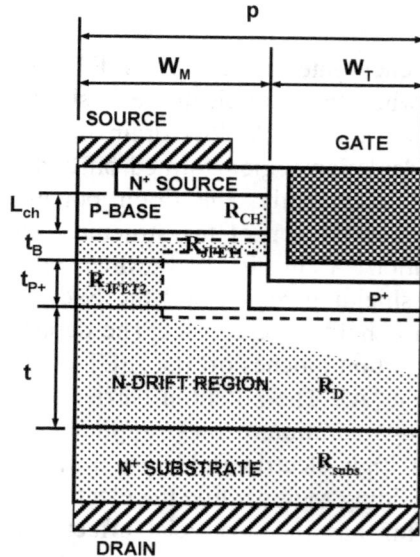

Fig. 12.2 Current Flow Path in the Shielded Trench-Gate 4H-SiC Power MOSFET Structure.

In the shielded trench-gate MOSFET structure, current flow between the drain and source can be induced by creating an inversion layer channel on the surface of the P-base region. The current flows from the source region into the drift region through the inversion layer channel formed on the vertical side-walls of the trench due to the applied gate bias. It must then flow from point 'B' in the cross-section through the first JFET region which constricts the current into location 'A' shown in the cross-section. The current then spreads into the N-drift region at a 45 degree angle and becomes uniform through the rest of the structure.

The current path is illustrated in Fig. 12.2 by the shaded area together with the zero-bias depletion boundaries of the junctions indicated by the dashed lines. It can be seen that the addition of the P$^+$ shielding region introduces *two* JFET regions into the basic trench-gate MOSFET structure. The first one, labeled R$_{JFET1}$ in the figure, is formed between the P-base region and the P$^+$ shielding region with the current constricted by their zero-bias depletion boundaries. The spacing between these regions (labeled t$_B$ in the figure) must be chosen to prevent it from becoming completely depleted. For a typical N-drift region doping concentration of 1 x 10^{16} cm^{-3}, corresponding to a breakdown voltage of 3000 volts, the zero-base depletion width is about 0.6 microns. In this case, the spacing (t$_B$) would the have to be about 1.5 microns to ensure the existence of an un-depleted path for the transport of electrons. This spacing can be reduced if the doping concentration in the JFET region is selectively increased when compared with the N-drift region.

The second JFET region, labeled R$_{JFET2}$ in the figure, is formed between the P$^+$ shielding regions. Its resistance is determined by the thickness of the P$^+$ shielding region (labeled t$_{P+}$ in the figure), which can be assumed to be twice the junction depth of the P$^+$ shielding region. Since the cross-section for current flow through this region is constricted by the zero-bias depletion width of the P$^+$/N junction, it is again advantageous to increase the doping concentration in the JFET region to avoid having to enlarge the mesa width. A smaller mesa width allows maintaining a smaller cell pitch which reduces the specific on-resistance due to a larger channel density.

The total on-resistance for the shielded trench-gate power MOSFET structure is determined by the resistance of all the components in the current path:

$$R_{on,sp} = R_{CH} + R_{JFET1} + R_{JFET2} + R_D + R_{subs} \qquad [12.1]$$

where R_{CH} is the channel resistance, R_{JFET1} and R_{JFET2} are the resistances of the two JFET regions, R_D is the resistance of the drift region after taking into account current spreading from the channel, and R_{subs} is the resistance of the N^+ substrate. These resistances can be analytically modeled by using the current flow pattern indicated by the shaded regions in Fig. 12.2.

The specific channel resistance is given by:

$$R_{CH} = \frac{(L_{CH} \cdot p)}{\mu_{inv} C_{ox} (V_G - V_T)}$$ [12.2]

where L_{CH} is the channel length determined by the width of the P-base region as shown in Fig. 12.2, μ_{inv} is the mobility for electrons in the inversion layer channel, C_{ox} is the specific capacitance of the gate oxide, V_G is the applied gate bias, and V_T is the threshold voltage. The specific capacitance can be obtained using:

$$C_{ox} = \frac{\varepsilon_{ox}}{t_{ox}}$$ [12.3]

where ε_{ox} is the dielectric constant for the gate oxide and t_{ox} is its thickness.

The specific resistance of the first JFET region can be calculated using:

$$R_{JFET1} = \rho_{JFET} \cdot p \cdot \left(\frac{x_{P+} + W_P}{t_B - 2W_P} \right)$$ [12.4]

where ρ_{JFET} is the resistivity of the JFET region, x_{P+} is the junction depth of the P^+ shielding region, and W_P is the zero-bias depletion width *in the JFET region*. The resistivity and zero-bias depletion width used in this equation must be computed using the enhanced doping concentration of the JFET region.

The specific resistance of the second JFET region can be calculated using:

$$R_{JFET2} = \rho_{JFET} \cdot p \cdot \left(\frac{t_{P+} + 2W_P}{W_M - x_{P+} - W_P} \right)$$ [12.5]

The drift region spreading resistance can be obtained by using:

$$R_D = \rho_D . p . \ln\left(\frac{p}{W_M - x_{P+} - W_P}\right) + \rho_D . \left(t - W_T - x_{P+} - W_P\right) \quad [\,12.6\,]$$

where t is the thickness of the drift region below the P$^+$ shielding region and W_T, W_M are the widths of the trench and mesa regions, respectively, as shown in the figure.

The contribution to the resistance from the N$^+$ substrate is given by:

$$R_{subs} = \rho_{subs} . t_{subs} \qquad\qquad [\,12.7\,]$$

where ρ_{subs} and t_{subs} are the resistivity and thickness of the substrate, respectively. A typical value for this contribution is 4×10^{-4} Ω-cm^2.

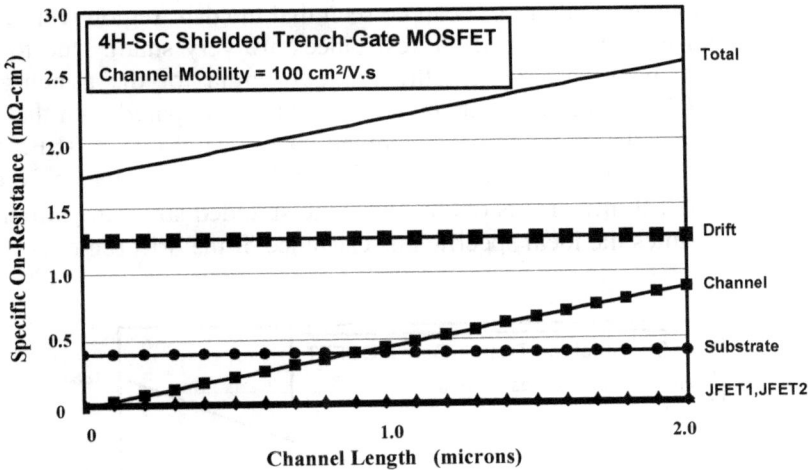

Fig. 12.3 On-Resistance Components for a 4H-SiC Shielded Trench-Gate MOSFET.

The specific on-resistances of 4H-SiC shielded trench MOSFET structures with a drift region doping concentration of 1×10^{16} cm^{-3} and thickness of 20 microns, capable of supporting 3000 volts, were modeled using the above analytical expressions. The same gate oxide thickness of 0.05 microns was used as in the case of the trench-gate (unshielded) MOSFET structure discussed in the previous chapter to allow comparison of the structures. For this reason, the trench width (W_T) for the structure was kept at 0.25 microns. However, the mesa width (W_M)

for the shielded trench-gate structure had to be enlarged to 1.25 microns to allow for the presence of the P^+ region and the resulting depletion layer formed in the JFET region. These values are also consistent with 0.5 micron design rules. The effective gate drive voltage ($V_G - V_T$) was assumed to be 5 volts based upon a gate voltage of 10 volts and a threshold voltage of 5 volts. This threshold voltage is consistent with the model described in chapter 9 for the inversion-mode 4H-SiC MOSFET. The JFET region doping concentration was enhanced to 5 x 10^{16} cm^{-3} to reduce its resistance contribution.

The various components of the on-resistance are shown in Fig. 12.3 when a channel inversion layer mobility of 100 cm^2/Vs was used. As in the case of the trench-gate MOSFET structure, the channel resistance increases with increasing channel length. The other components are unaffected by the increase in channel length because the cell pitch remains unaltered. It can be seen that the drift region resistance is dominant here while the JFET resistances are very small. Due to the shielding of the P-base region, the channel length for the trench-gate shielded MOSFET structure can be reduced when compared with that for the unshielded conventional trench-gate MOSFET structure. Consequently, with a channel length of 0.4 microns, a total specific on-resistance of 1.9 mΩ-cm^2 is obtained for the shielded structure, which is within two times the ideal specific on-resistance of the drift region.

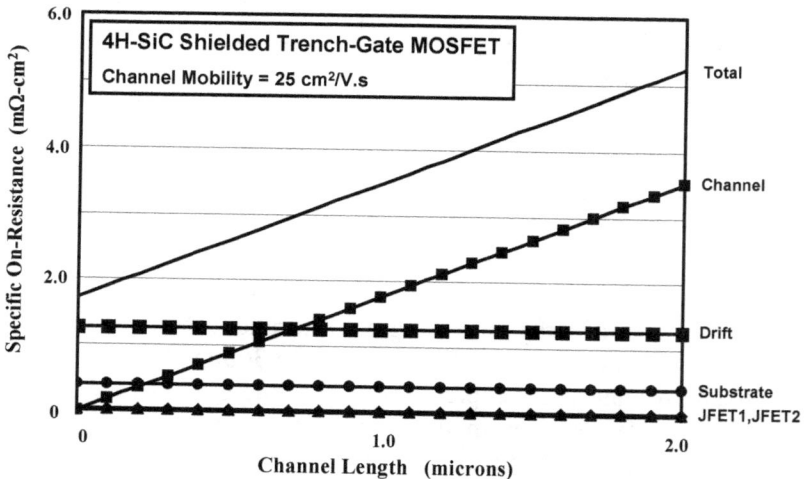

Fig. 12.4 On-Resistance Components for a 4H-SiC Shielded Trench-Gate MOSFET.

The above results are based upon using a channel mobility of 100 cm^2/Vs. The impact of reducing the inversion layer mobility for the shielded trench-gate MOSFET structure is depicted in Fig. 12.4 and 12.5 for the case of mobility values of 25 and 2.5 cm^2/Vs. With an inversion layer mobility of 25 cm^2/Vs, the channel contribution becomes comparable to the contribution from the drift region. This results in the specific on-resistance increasing to a value of 2.4 mΩ-cm^2 for a channel length of 0.4 microns. When the inversion layer mobility is reduced to 2.5 cm^2/Vs, the channel resistance becomes dominant as shown in Fig. 12.5, with an increase in the specific on-resistance to a value of 8.8 mΩ-cm^2 for a channel length of 0.4 microns.

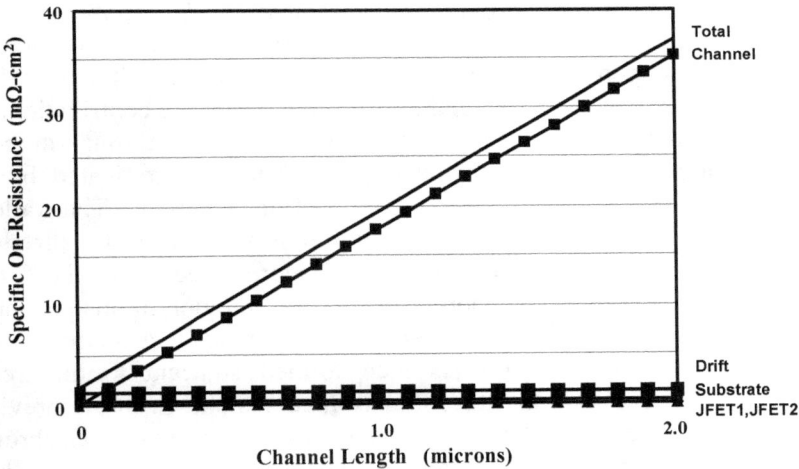

Fig. 12.5 On-Resistance Components for a 4H-SiC Shielded Trench-Gate MOSFET.

In all the inversion layer mobility cases, the specific on-resistances of the shielded trench-gate 4H-SiC MOSFET structure with a channel length of 0.4 microns are found to be very close to those for the conventional trench-gate structure with a channel length of 1 micron. This demonstrates the ability to obtain low specific on-resistance with the shielded trench-gate structure while resolving the problems of P-base reach-through and high electric field in the gate oxide observed for the conventional trench-gate structure. The behavior of the shielded structure with respect to other blocking voltages is similar to that provided for the conventional-trench gate structure in the previous chapter.

12.1.3 Shielded Trench-Gate Power MOSFET Structure: Threshold Voltage

The threshold voltage of the trench-gate MOSFET structure is determined by the doping concentration of the P-base region along the sidewalls of the trench region. A minimum threshold voltage must be maintained at above 1 volt for most system applications to provide immunity against inadvertent turn-on due to voltage spikes arising from noise. At the same time, a high threshold voltage is not desirable because the voltage available for creating the charge in the channel inversion layer is determined by $(V_G - V_T)$ where V_G is the applied gate bias voltage and V_T is the threshold voltage. Most power electronic systems designed for high voltage operation (the most suitable application area for silicon carbide devices) provide a gate drive voltage of only upto 10 volts. Based upon this criterion, the threshold voltage should be kept below 3 volts in order to obtain a low channel resistance contribution.

As discussed in the previous sections, the reach-through in the P-base region of the shielded trench-gate structure is mitigated by the reduced potential under the P-base/N-drift junction. This allows decreasing the P-base doping concentration to reduce the threshold voltage. In addition, a P^+ region can be incorporated under the source region to suppress the extension of the depletion region in this portion of the cell structure, as previously described in chapter 11. A relatively thin, lightly doped P-base region is then formed adjacent to the trench sidewalls to determine the threshold voltage. As shown in the previous chapter, the presence of the P^+ region also suppresses reach-through within the adjacent P-base region allowing the structure to support high drain voltages upto the breakdown voltage capability of the drift region.

12.2 Shielded Trench-Gate Power MOSFET Structure: Numerical Simulations

In order to gain insight into the operation of the shielded trench-gate 4H-SiC power MOSFET structure, two-dimensional numerical simulations were performed with a drift region doping concentration of 1×10^{16} cm^{-3} and thickness of 20 microns corresponding to a parallel-plane breakdown voltage of 3000 volts. The baseline device had a gate oxide thickness of 0.05 microns. In order to take advantage of the reduced channel length in the shielded structure, the depth of the P-base region was reduced to 0.5

microns. The N^+ source region had a depth of 0.1 microns leading to a channel length of 0.4 microns. A trench width (W_T) of 0.55 microns was chosen with a mesa width (W_M) of 1.25 microns, resulting in a cell pitch of 1.8 microns. The P^+ region located at the bottom of the 1.5 micron deep trench had a junction depth of 0.2 microns, resulting in a t_{P+} parameter of 0.4 microns. The doping concentration of the JFET region was increased to 5×10^{16} cm^{-3} to reduce the zero-bias depletion widths of the junctions. The use of enhanced doping concentration in the JFET region was pointed out in reference 1 (see column 8 in the patent).

12.2.1 Shielded Trench-Gate Power MOSFET Structure: Blocking Characteristics

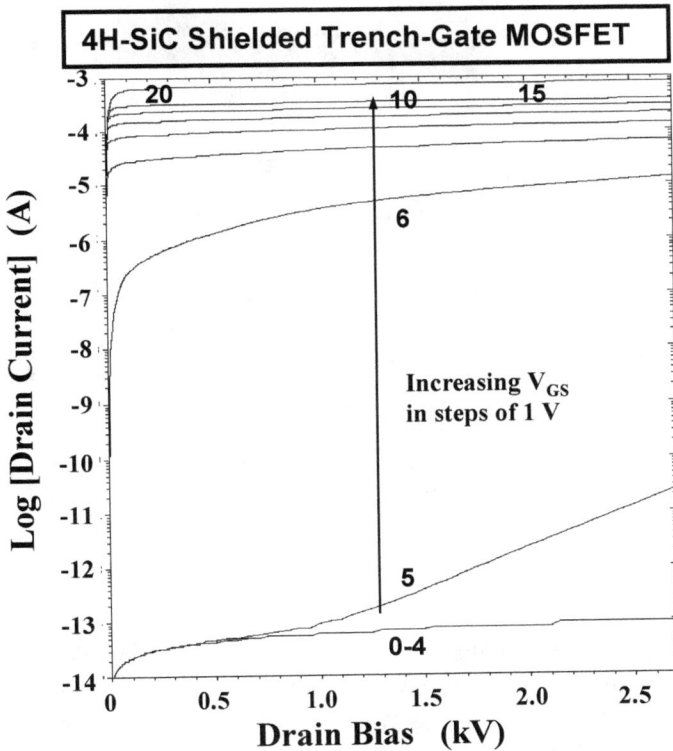

Fig. 12.6 Blocking Characteristics of the Shielded Trench-gate 4H-SiC MOSFET Structure.

The blocking capability of the shielded trench-gate 4H-SiC power MOSFET structure was investigated by maintaining zero gate bias while increasing the drain voltage. It was found that the device could sustain a drain bias of upto 3000 volts, as shown in Fig. 12.6, without any reach-through current despite the narrow width of the P-base region. A low leakage current is observed for gate bias voltages ranging from 0 to 5 volts with current increasing at a gate bias of 6 volts. This indicates that the threshold voltage is about 5 volts.

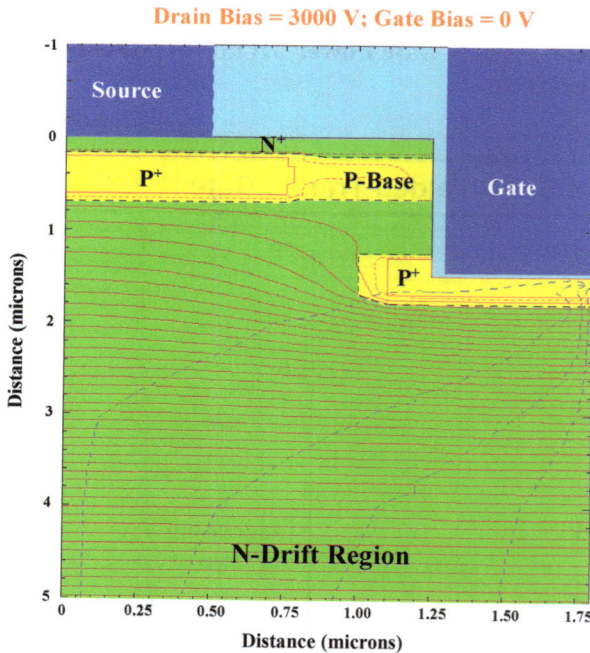

Fig. 12.7 Potential Distribution in the Shielded Trench-Gate 4H-SiC Power MOSFET Structure.

The potential distribution (solid-lines) within the upper portion of the shielded trench-gate 4H-SiC MOSFET structure is shown in Fig. 12.7 at a drain bias of 3000 volts with the gate held at zero volts. It can be seen that there is an electric field enhancement at the corner of the P^+ shielding region. No potential lines are visible within the P-base region indicating that it is supporting less than 25 volts (the separation between the potential contours). This demonstrates that the presence of the P^+

region under the N^+ source region together with the formation of the JFET region suppresses the depletion of the relatively thin P-base region without the occurrence of reach-through. In addition, the potential contours are widely separated in the vicinity of the gate oxide indicating a low electric field within the gate insulator.

The current flow-lines observed under these voltage blocking conditions are also shown in the figure using dashed lines. It can be seen that the leakage current is collected at the contact to the P^+ shielding region placed on the right-hand-side of the structure. No current flow is observed into the N^+ source region confirming that the reach-through phenomenon has been suppressed. The contact made to the P^+ shielding region in the orthogonal direction to the cross-section must be such that any voltage drop in the P^+ shielding region due to the leakage current does not produce a significant voltage drop. This may not be a critical problem for 4H-SiC structures because of the low leakage current levels. However, the high sheet resistance of the P^+ shielding region could be an issue during the switching transient when a large displacement current density may be generated at high dV/dt rates.

Drain Bias = 3000 V

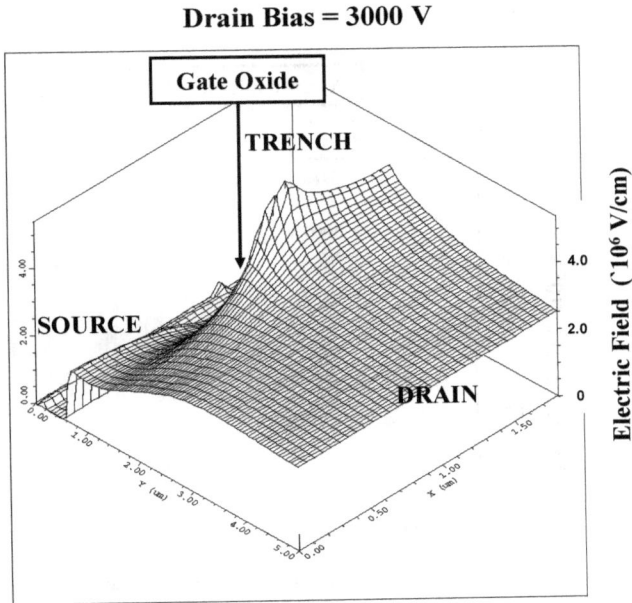

Fig. 12.8 Electric Field Distribution in the Shielded Trench-Gate 4H-SiC MOSFET Structure.

In the previous chapter, it was demonstrated that a serious problem with the conventional trench-gate silicon carbide MOSFET structure is associated with the high electric field generated at the gate oxide interface. In order for the channel to span the P-base region, it is necessary for the trench to penetrate through the P-base region into the N-drift region. This exposes the gate oxide located at the bottom of the trench to the high electric field developed in the silicon carbide drift region. This problem is overcome in the shielded trench-gate structure. A three-dimensional view of the electric field distribution for the shielded trench-gate 4H-SiC MOSFET structure is shown in Fig. 12.8 for a drain bias of 3000 volts with zero bias applied to the gate electrode. The maximum electric field is generated at the junction between the P^+ shielding region and the N-drift region. The field distribution shown in this figure indicates that the electric field at the junction is close to the breakdown field strength for 4H-SiC (about 3×10^6 V/cm). Concurrently, the electric field at the P-base region is greatly reduced to about 1×10^6 V/cm and it is even smaller in the vicinity of the gate oxide. These results demonstrate the ability to fully utilize the breakdown field strength of the semiconductor without problems of rupture or reliability for the gate oxide.

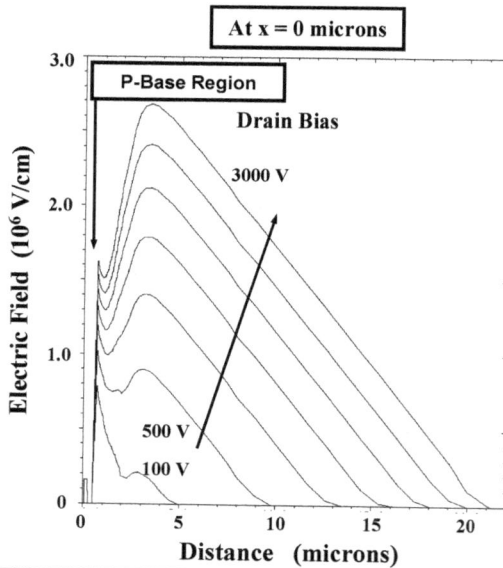

Fig. 12.9 Electric Field in the Shielded Trench-Gate 4H-SiC MOSFET Structure.

From Fig. 12.8, it can be seen that the electric field at the P-base region is significantly reduced when compared with the maximum electric field generated in the semiconductor. The behavior of the electric field at x = 0 microns in the cell structure with increasing drain bias is shown in Fig. 12.9 for drain bias upto 3000 volts. It can be observed that the formation of a potential barrier due to the presence of the JFET region suppresses the electric field at the P-base region to about one-half of the electric field generated in the drift region. This is beneficial for preventing reach-through even with a narrow width for the P-base region.

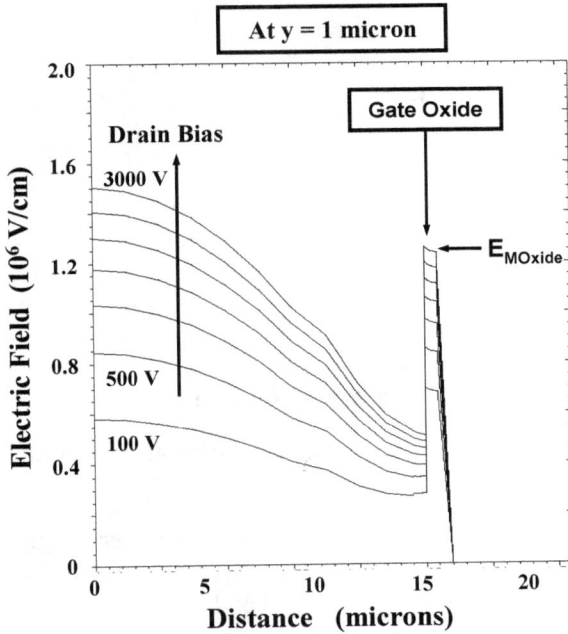

Fig. 12.10 Electric Field in the Shielded Trench-Gate 4H-SiC MOSFET Structure.

The reduction of the electric field in the gate oxide by the shielding provided by the JFET region can be observed in the electric field profile shown in Fig. 12.10 for various drain bias voltages. This profile was taken at a depth of 1 micron from the surface at a location below the P-base region. It can be seen that the electric field in the oxide is only 1.3×10^6 V/cm - far below its rupture strength - even when the drain bias reaches 3000 volts. The effective screening of the gate oxide in the shielded trench-gate structure from the high electric fields in the drift region provide wide latitude in choice of the gate oxide thickness and fabrication processes.

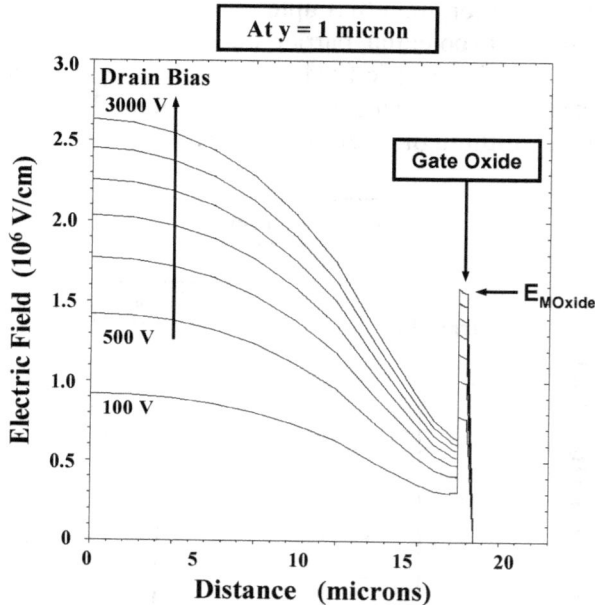

Fig. 12.11 Electric Field in the Shielded Trench-Gate 4H-SiC MOSFET Structure.

The impact of the JFET width can be observed in Fig. 12.11 where the electric field below the P-base region is shown for the case of mesa width of 2.25 microns. Although the shielding by the JFET region has been diminished, the electric field in the gate oxide remains below 2 x 10^6 V/cm when the drain bias is increased to 3000 volts. This is sufficiently low for reliable operation of the 4H-SiC shielded trench-gate MOSFET structure. However, the electric field below the P-base region increases from 1.5 x 10^6 V/cm for the mesa width of 1.25 microns to 2.6 x 10^6 V/cm. This aggravates the base reach-through problem. It was found that the specific on-resistance obtained with the larger mesa width was very close that for the mesa width of 1.25 microns. Consequently, it is preferable to reduce the mesa width to 1.25 microns to simultaneously increase the channel density and reduce the electric fields in the vicinity of the P-base region and the gate oxide.

12.2.2 Shielded Trench-Gate Power MOSFET Structure: On-State Characteristics

V_{DS} = 1 Volt; V_{GS} = 10 Volts

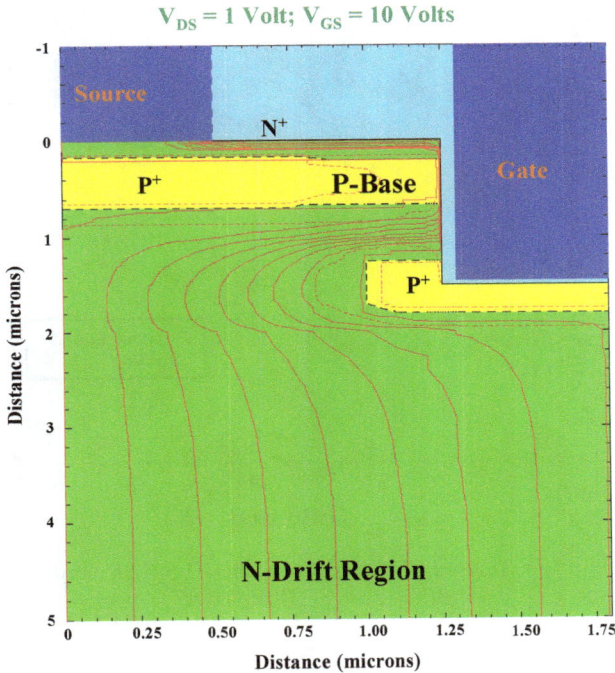

Fig. 12.12 On-State Current Distribution in the Shielded Trench-Gate
4H-SiC MOSFET Structure.

The on-state voltage drop for the shielded trench-gate 4H-SiC power MOSFET structure is determined by its specific on-resistance. The on-resistance for the structure was extracted by performing numerical simulations with a gate bias of 10 volts. The current distribution within the shielded trench-gate 4H-SiC power MOSFET structure is shown in Fig. 12.12. It can be seen that the current flows through the channel and is then constricted by the two JFET regions. This current flow pattern is consistent with the shaded region in Fig. 12.2 used to model the on-resistance. The current flow from the second JFET region is distributed into the drift region at a 45 degree angle and then becomes uniform from a depth of about 3 microns.

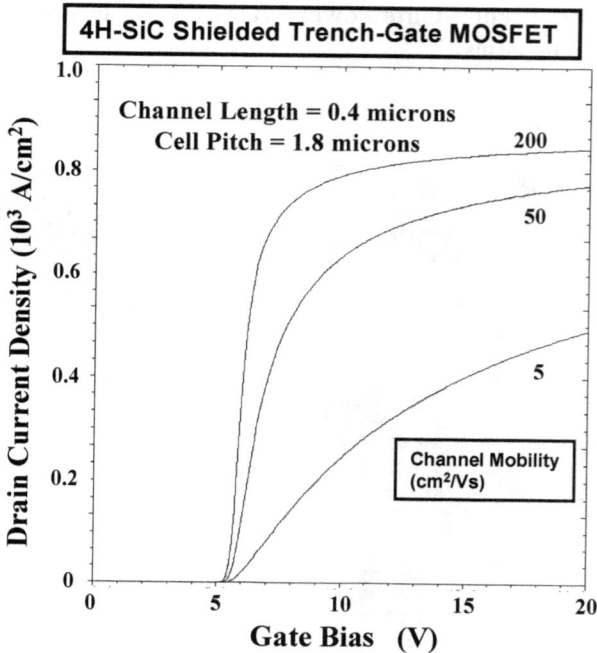

Fig. 12.13 Transfer Characteristics of the 4H-SiC Shielded Trench-Gate MOSFET Structure.

The simulations of the specific on-resistance for the shielded trench-gate 4H-SiC MOSFET structure were performed with different values for the channel inversion layer mobility. The specific on-resistance was obtained by using a drain bias of 1 volt and sweeping the gate bias to 20 volts to obtain the transfer characteristic. A drain bias of 1 volt was used for these simulations because the on-state voltage drop for these structures is typically in this range. The transfer characteristics obtained by using this method are shown in Fig. 12.13 for the device with cell pitch of 1.8 microns. From the transfer characteristics, it can be seen that the threshold voltage that determines the on-resistance is approximately 5 volts (allowing operation of this device with a gate bias of 10 volts). The impact of changing the channel inversion layer mobility is also shown in this figure. As expected, there is a reduction in the drain current as the channel mobility is reduced.

The specific on-resistances obtained from the on-state simulations are compared in Fig. 12.14 with the analytically calculated values obtained by using the model developed in the previous section. For the simulation cases, the channel mobility values were extracted from the *i-v* characteristics obtained for the lateral 4H-SiC MOSFET structure that was previously described in chapter 9 by using the same interface degradation factor as used during the simulation of the vertical 4H-SiC shielded trench-gate MOSFET structure. It can be seen that the predictions of the analytical model are in excellent agreement with the values obtained from the simulations. These results provide confidence in the ability to use the analytical model during the design of the shielded trench-gate 4H-SiC power MOSFET structure. From the figure, it can be concluded that a channel mobility of more than $50 \text{ cm}^2/\text{Vs}$ is satisfactory for achieving a low specific on-resistance for a 3 kV shielded trench-gate 4H-SiC power MOSFET structure with cell pitch of 1.8 microns.

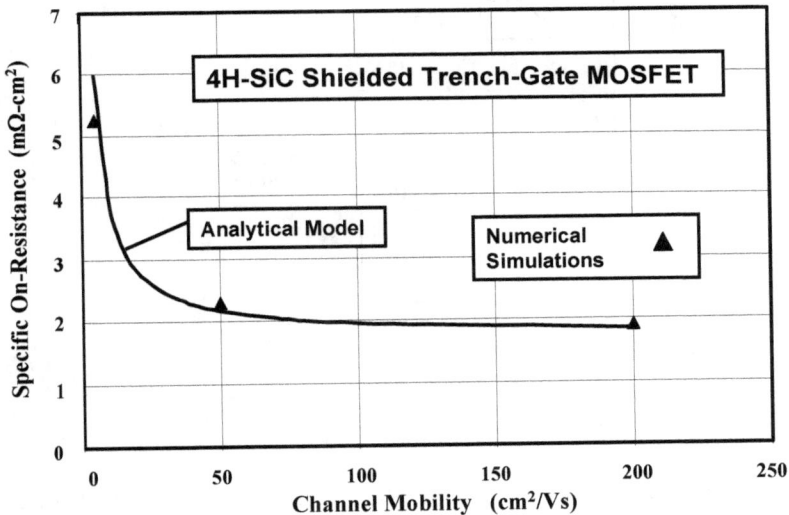

Fig. 12.14 Specific On-Resistance for the 4H-SiC Shielded Trench-Gate MOSFETs.

12.2.3 Shielded Trench-Gate Power MOSFET Structure: Output Characteristics

The output characteristics of the power MOSFET defines the region within which the device can operate during the switching transients in applications. The current-voltage loci during the transient must be chosen to remain within the boundary defined by the output characteristics. The output characteristics for the 4H-SiC shielded trench-gate power MOSFET structure were determined by biasing the gate at various values and sweeping the drain voltage upto 2800 volts. The characteristics obtained for the structure with cell pitch of 1.8 microns are shown in Fig. 12.15. The device exhibits excellent output characteristics upto a gate bias of 10 volts with high output resistance. Since the gate oxide is shielded by the JFET region in this structure, it is possible to fully utilize the excellent safe-operating-area shown in the figure.

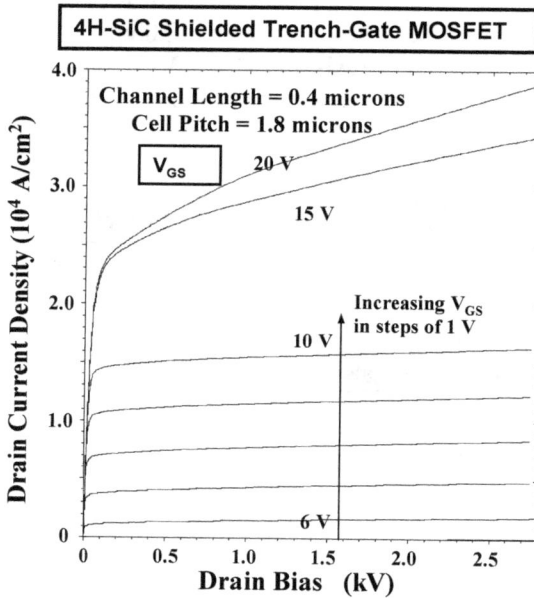

Fig. 12.15 Output Characteristics of a 4H-SiC Shielded Trench-Gate MOSFET.

12.3 Shielded Trench-Gate Power MOSFET: Experimental Results

Inversion-mode trench-gate 4H-SiC power MOSFETs with a P^+ shielding region incorporated at the bottom of the trench were first reported[2] in 2002. The structures were fabricated using 50 micron thick N-type drift regions with doping concentration of 8.5×10^{14} cm^{-3}. An N-type layer with doping concentration of 2×10^{17} cm^{-3} was grown on the drift region to provide the enhanced doping in the JFET region. The P-base region was formed by growth of a 1 micron thick P-type layer with doping concentration of 2×10^{17} cm^{-3} followed by the growth of an N$^+$ source region with doping concentration of 1×10^{19} cm^{-3}. The P$^+$ shielding regions were formed (after etching the trenches) with a junction depth of 0.8 microns by using aluminum ion implantation with a dose of 4×10^{13} cm^{-2}. The authors used two sacrificial thermal oxidation steps to remove any surface residue remaining from the implant anneal step. This may have also removed any P-type doping on the trench sidewalls. A relatively thick gate oxide of 0.275 microns was prepared by the thermal oxidation of a deposited layer of polysilicon into the trench to avoid the non-uniformity of thermally grown oxide. The thick gate oxide resulted in a high threshold voltage of 40 volts. The devices exhibited a breakdown voltage of 3000 volts by the use of a JTE edge termination. A specific on-resistance of 120 mΩ-cm^2 was observed at a gate bias of 100 volts. The relatively high value is due to the poor channel density in the cell design, the large gate oxide thickness, and the low inversion layer channel mobility (2 cm^2/V-s) observed by the authors. However, these results demonstrated the ability to support high voltage in the drift region without encountering gate oxide rupture in a trench-gate device. The authors subsequently reported[3] the fabrication of devices using 115 micron thick N-type drift regions with doping concentration of 7.5×10^{14} cm^{-3}. These devices exhibited a specific on-resistance of 228 mΩ-cm^2 at a gate bias of 40 volts. The epitaxial layer was stated to be capable of supporting 14 kV although the actual measured breakdown voltage was only 5 kV.

In the previous chapter, an accumulation-mode trench-gate MOSFET structure was described with an epitaxially grown N-base region on the trench sidewalls (see Fig. 11.22) as an approach to reduce the specific on-resistance due to an increase in the channel mobility. This concept has been supplemented with shielding provided by the P-type region implanted at the bottom of the trenches to prevent gate oxide rupture[4]. These devices were fabricated using 10 micron thick N-type drift regions with doping concentration of 2.5×10^{15} cm^{-3} to obtain a breakdown voltage of 1400 volts. The rest of the process was similar to that described in the previous paragraph. A gate oxide thickness of 0.13 microns was formed by thermal oxidation of a polysilicon layer deposited in the trenches. A specific on-resistance of 16 mΩ-cm^2 was observed at a gate bias of 40 volts. This improved specific on-resistance was correlated with an accumulation layer mobility of 9-30 cm^2/V-s by the authors. Accumulation-mode shielded trench-gate MOSFETs fabricated from 4H-SiC were also reported in reference [2] with breakdown voltage of 3360 volts. These devices had a specific on-resistance of 199 mΩ-cm^2 at a gate bias of 100 volts, which was worse than that for the inversion-mode structures. In addition, the leakage current for the accumulation-mode structure was reported to be about 100 times worse than for the inversion-mode structure. These results indicate the need for further optimization of the epitaxially grown N-base region.

12.4 Summary

The shielded trench-gate power MOSFET structure was proposed for silicon carbide in order to shield the P-base region and the gate oxide from the high electric field generated in the drift region during the blocking mode. The shielding is provided by the addition of a P-type region located at the bottom of the trenches. This P-type shielding region must be connected to the source electrode. Its presence creates two JFET regions in the trench gate structure, which can increase the on-resistance unless the doping concentration in the vicinity of the trenches is enhanced. With the enhanced doping concentration, it has been found that the full blocking voltage capability of the drift region in 4H-SiC can be utilized with low electric field at the P-base region and the gate oxide. This provides a very attractive power MOSFET structure that can be made from 4H-SiC with specific on-resistance approaching the ideal specific on-resistance of the drift region if a channel mobility of over 50 cm^2/Vs is achieved.

References

[1]   B. J. Baliga, "Silicon Carbide Switching Device with Rectifying Gate", U. S. Patent 5,396,085, Issued March 7, 1995.

[2]   Y. Li, J. A. Cooper, and M. A. Capano, "High Voltage (3kV) UMOSFETs in 4H-SiC", IEEE Transactions on Electron Devices, Vol. 49, pp. 972-975, 2002.

[3]   Y. Sui, T. Tsuji, and J. A. Cooper, "On-State Characteristics of SiC Power UMOSFETs on 115 micron Drift Layers", IEEE Electron Device Letters, Vol. 26, pp. 255-257, 2005.

[4]   J. Tan, J. A. Cooper, and M. R. Melloch, "High Voltage Accumulation-Layer UMOSFETs in 4H-SiC", IEEE Electron Device Letters, Vol. 19, pp. 487-489, 1998.

Chapter 13

Charge Coupled Structures

Throughout the previous sections of this book, the benefits of achieving a low specific on-resistance for the drift region in silicon carbide devices, associated with the high critical electric field for breakdown, has been emphasized. It has been demonstrated that the high critical electric field in the 4H-SiC drift region allows increasing its doping concentration by about 100 times and reducing its thickness by about 10 times when compared with a silicon device with the same breakdown voltage. This results in an improvement in the specific on-resistance by a factor of 2000 times. It has been shown that unipolar device structures can be constructed from 4H-SiC with excellent on-state voltage drop if the blocking voltage capability is below 3000 volts.

While awaiting sufficient maturity of the silicon carbide technology, advances have continued to be made in silicon devices. One very effective approach taken in the 1980's was the merger of MOS and bipolar physics[1] to create a new generation of devices, including the Insulated Gate Bipolar Transistor (IGBT) which has been widely accepted for medium and high power system applications. This approach has had a revolutionary impact on the efficiency of high voltage systems. Concurrently, the concept of charge coupling was introduced in the 1990s to improve upon the specific on-resistance of the drift region in unipolar silicon devices. The charge coupling idea has been based upon either using alternating layers of P-type and N-type regions[2] co-located within the drift region or by the use of trenches with embedded electrodes[3] introduced into the drift region. The ability to increase the doping concentration of the drift region, by utilizing the charge coupling to suppress impact ionization in the drift region, has enabled decreasing its specific on-resistance by an order of magnitude.

Due to the low specific on-resistance of the drift region for the drift region in 4H-SiC, it is unnecessary to introduce the complexity of the charge coupling concept when the breakdown voltage is below 3000 volts. However, when the breakdown voltage exceeds 3000 volts, the charge coupling approach becomes worth exploring even for silicon carbide because it enables preserving the unipolar mode of operation. However, the charge coupling concepts proposed for silicon devices do not translate readily to silicon carbide. For the silicon devices with alternating P-type and N-type columns in the drift region, multiple N-type epitaxial layers are grown over patterned ion implanted P-type regions to create alternating stripes[4] of N and P-type material. The P-type regions merge due to the diffusion of the dopant (Boron) during the epitaxial growth process. This is not feasible in silicon carbide due to the very low diffusion rates for dopants. In the case of charge coupling using trench based MOS electrodes, the high electric field in silicon carbide produces an unacceptably larger electric field in the oxide precluding reliable operation. This was demonstrated in chapter 6 in the section on the TMBS rectifier.

In this chapter, a novel method to create the charge coupling is described based upon forming P-type regions within trenches[5] etched into the silicon carbide N-type drift region. It is demonstrated that this idea can be applied to create high performance Schottky rectifiers and high voltage JFET/MOSFET structures. The results of two-dimensional numerical simulations are provided in this chapter to quantify the benefits of the charge coupling concept. These ideas have not yet been experimentally demonstrated.

13.1 Charge Coupled Silicon Carbide Structures

The proposed trench-based charge coupled structure is shown in Fig. 13.1 for a Schottky rectifier. Here, a P-type charge coupling region has been introduced into the N-type drift region by locating it within a trench etched at regular intervals. The etching of trenches, followed by refill, has been demonstrated in silicon using both vapor-phase epitaxial growth[6] and liquid-phase epitaxial growth[7]. Although an equivalent trench refill process has not yet been reported for silicon carbide, the growth of epitaxial layers on trench surfaces has been utilized for the fabrication of high voltage accumulation-mode power MOSFETs[8,9]. This process can be extended to the refill of trenches with P-type regions containing precisely controlled doping concentration. The device fabrication is then completed by the formation of the Schottky barrier contact to the N-drift region and an Ohmic contact to the P-type region.

A heavily doped P-type region can be formed at the top of the refill either during epitaxial growth or with ion implantation of aluminum to assist in obtaining a good ohmic contact. Although stripes of metal are indicated in the figure with a small spacing, this is not required in practice because the Schottky and ohmic contacts are both operating at the same potential. Consequently, only one of the metal layers must be patterned as small stripes while the other metal layer can overlap it and cover the entire cell surface to make a large metal pad for external connections during packaging.

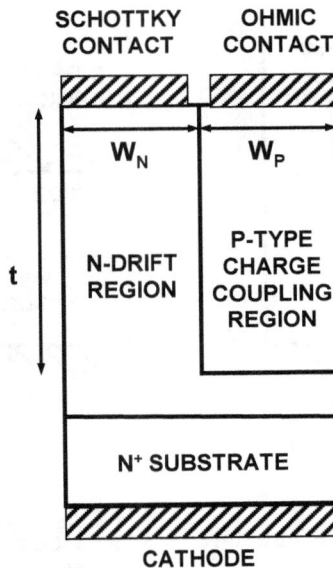

Fig. 13.1 Basic Charge Coupled Silicon Carbide Structure.

The above basic charge coupled structure will be analyzed in detail in this chapter. However, several alternate embodiments have also been proposed[5]. These structures are illustrated in Fig. 13.2. In the structure on the left-hand-side, the charge coupling region is formed using semi-insulating material. This can be composed of a high resistivity material such as SIPOS (semi-insulating oxygen doped polysilicon). SIPOS layers have been developed[10] for fabrication of high voltage lateral silicon devices with resistivity values ranging from 1×10^{10} to 1×10^{18} ohm-cm. Alternately, the semi-insulating regions could be fabricated using vanadium doped silicon carbide which is commonly used to prepare high resistivity substrates[11]. A unique arc-shaped doping profile[5] has been found to provide the best electric field distribution in this structure.

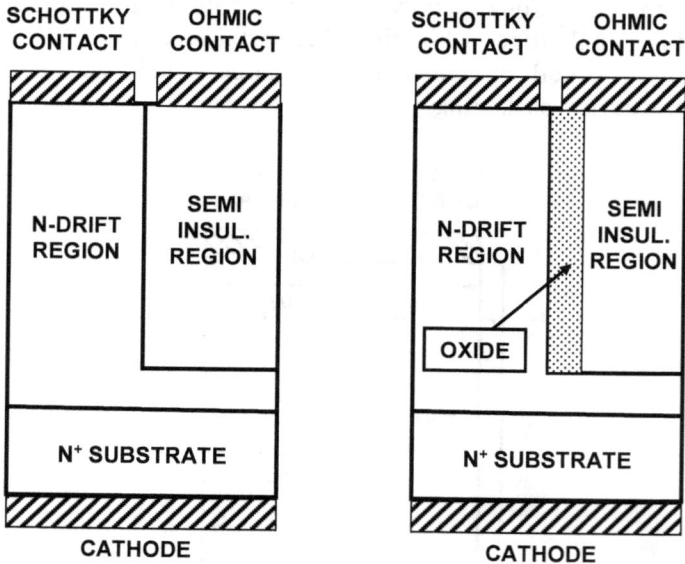

Fig. 13.2 Alternate Charge Coupled Silicon Carbide Structures.

Another alternate structure is shown on the right-hand-side of Fig 13.2. Here, an oxide layer is preserved on the sidewalls of the trenches to isolate the SIPOS from the N-type drift region. Note that contact is made to the SIPOS layer at both the top and bottom allowing for a small leakage current to flow through it in order to homogenize the potential distribution. A linearly graded doping profile was found to be the most suitable in this case for enhancing the breakdown voltage[5]. Recently, the use of a semi-insulating layer in the trenches, with oxide isolation, has also been utilized for the fabrication of 80 volt silicon devices[12]. This structure was fabricated using conformal coating of the trench with oxide, and its removal from the trench bottom by using a spacer etch process. The trench refill was performed using SIPOS with a resistivity of 10^8 ohm-cm. A breakdown voltage above the parallel-plane value was observed when the charge in the N-drift region was optimized.

13.2 Charge Coupled Silicon Carbide Structure: Ideal Specific On-Resistance

In the conventional structures that were discussed in previous chapters, the depletion region extends in one-dimension from a junction or Schottky contact during the blocking mode. The voltage blocking capability in the charge coupled structure is enhanced by the extension of depletion layers in two-dimensions. This effect is created by the formation of a horizontal Schottky contact on the top surface in Fig. 13.1 which promotes the extension of a depletion region along the vertical or y-direction. Concurrently, the presence of the vertical P-N junction created by the alternate N and P-type regions promotes the extension of a depletion region along the horizontal or x-direction. These depletion regions conspire to produce a two-dimensional charge coupling in the N-drift region which alters the electric field profile.

The optimization of the charge coupled structure requires proper choice of the doping concentration and thickness of the N and P-type regions. It has been found that the highest breakdown voltage occurs when the charge in these regions is given by:

$$Q_{optimum} = q.N_D.W_N = \varepsilon_S.E_C \qquad [13.1]$$

where q is the charge of an electron (1.6 x 10^{-19} Coulombs), N_D is the doping concentration of the N-Type drift region, W_N is the width of the N-type drift region as shown in Fig. 13.1, ε_S is the dielectric constant of the semiconductor, and E_C is the critical electric field for breakdown in the semiconductor. For silicon, the optimum charge is found to be 3.11 x 10^{-7} Coulombs/cm^2 based upon a critical electric field of 3 x 10^5 V/cm. The optimum charge is often represented as a dopant density per unit area, in which case it takes a value of about 2 x 10^{12} per cm^2 for silicon. For 4H-SiC, the critical electric field for breakdown has a much larger value of 3 x 10^6 V/cm. The optimum charge calculated using this value for the critical electric field is 2.58 x 10^{-6} Coulombs/cm^2 with a corresponding dopant density of about 1.6 x 10^{13} per cm^2 for 4H-SiC. The larger doping density for 4H-SiC reduces the resistance of the drift region when compared with silicon even for the charge coupled devices. A slightly lower value for the doping concentration in the drift region may be warranted as discussed later in this section of the chapter.

The specific on-resistance for the drift region in the charge coupled structures is given by:

$$R_{D,sp} = \rho_D \,.t.\left(\frac{p}{W_N}\right) \qquad [13.2]$$

where ρ_D is the resistivity of the N-type drift region, t is the trench depth and p is the cell pitch. Here, the uniform electric field is assumed to be produced only along the trench where the charge coupling occurs and the resistance of the remaining portion of the N-drift region is neglected. Using the relationship between the resistivity and the doping concentration, this equation can be written as:

$$R_{D,sp} = \frac{t.p}{q.\mu_N .N_D .W_N} \qquad [13.3]$$

Combining this expression with Eq. [13.1]:

$$R_{D,sp} = \frac{t.p}{\mu_N .Q_{optimum}} \qquad [13.4]$$

If the electric field along the trench, at the on-set of breakdown in the charge coupled device structure, is assumed to be uniform at a value equal to the critical electric field of the semiconductor:

$$t = \frac{BV}{E_C} \qquad [13.5]$$

Using this expression, as well as the second part of Eq. [13.1], in Eq. [13.4] yields:

$$R_{D,sp} = \frac{BV.p}{\mu_N .\varepsilon_S .E_C^2} \qquad [13.6]$$

This is a fundamental expression for the *ideal specific on-resistance of vertical charge coupled devices*. In contrast, the equation for the specific on-resistance for the drift region in the conventional drift region (derived earlier in chapter 1) is:

$$R_{D,sp} = \frac{4.BV^2}{\mu_N .\varepsilon_S .E_C^3} \qquad [13.7]$$

By comparison of these expressions, it can be observed that the specific on-resistance for the charge coupled devices increases linearly with the breakdown voltage unlike the more rapid quadratic rate for the conventional drift region. In addition, it is worth pointing out that the specific on-resistance for the drift region in the charge coupled structure can be reduced by decreasing the pitch. This occurs because the doping concentration in the drift region increases when the pitch is reduced in order to maintain the same optimum charge. The larger doping concentration reduces the resistivity and hence the specific on-resistance.

However, the analysis of the specific on-resistance for the drift region in charge coupled device structures must be tempered by several considerations. Firstly, it must be recognized that the mobility will become smaller when the doping concentration becomes larger. Secondly, the critical electric field for breakdown becomes smaller for the charge coupled structures because the high electric field in the drift region extends over a larger distance producing enhanced impact ionization. If a critical electric field for breakdown in the drift region for charge coupled structures is reduced to 2×10^6 V/cm, an optimum charge of 1.72×10^{-6} Coulombs/cm^2, with a corresponding dopant density of about 1.07×10^{13} per cm^2, is more appropriate for 4H-SiC.

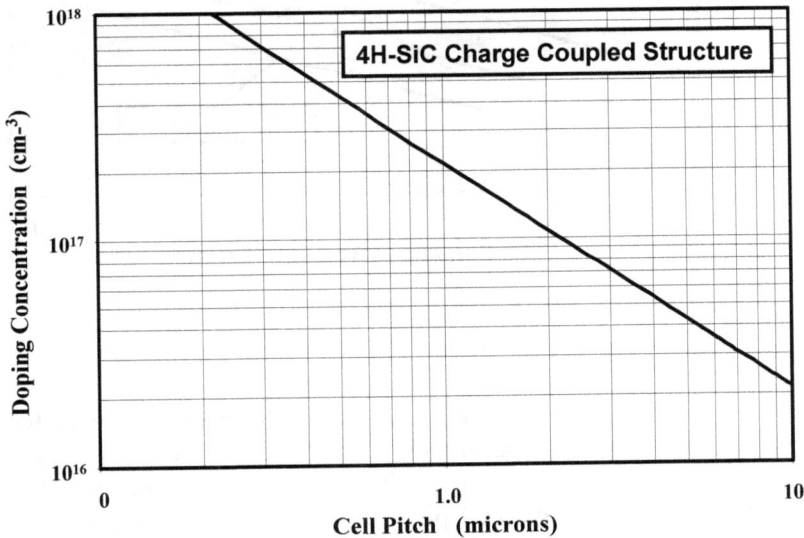

Fig. 13.3 Doping Concentration in the N-drift Region for the Charge Coupled Structure.

In designing the drift region for charge coupled structures, it is important to recognize that, unlike in the conventional one-dimensional case, the doping concentration of the drift region is dictated by the cell pitch and not the breakdown voltage. The breakdown voltage in the charge coupled structure is determined solely by the depth of the trench used to provide the charge coupling effect. The doping concentration for the N-type drift region is provided in Fig. 13.3 for the case of equal widths for the N-type and P-type charge coupling regions. For a typical cell pitch of 1 micron, the doping concentration in the N-type drift region is about 2×10^{17} cm^{-3}.

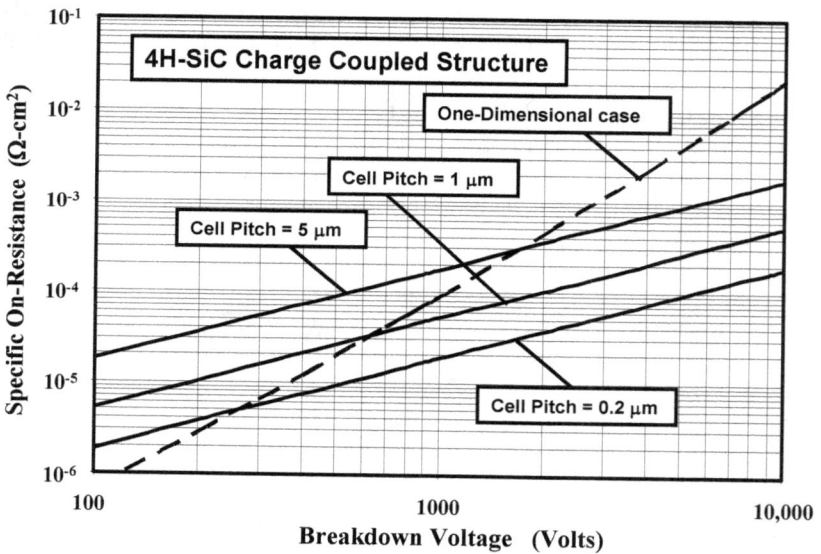

Fig. 13.4 Ideal Specific On-Resistance for the Charge Coupled Structure.

(Solid Lines: Charge Coupled Structures; Dashed Line: One Dimensional Case)

It is interesting to compare the ideal specific on-resistance for the drift region in the charge coupled structures to that for the one-dimensional parallel-plane case. This comparison is done in Fig. 13.4 using three values for the cell pitch in the case of the charge coupled structures. The doping concentration in the N-drift region increases when the cell pitch is reduced from 5 microns to 0.2 microns, as already shown in Fig. 13.3. The reduction of the mobility with increasing doping concentration was included during the calculation of the specific on-resistance in Fig. 13.4. There is a cross-over in the specific on-resistance for the two types of structures. For the cell pitch of 1 micron, the cross-over occurs at a breakdown voltage of about 600 volts. The cross-over moves to a breakdown voltage of about 1600 volts when the cell pitch is increased to 5 microns. If a small cell pitch of 0.2 microns is used, the cross-over occurs at a breakdown voltage of about 250 volts. Consequently, the charge coupled structure is more attractive for reducing the specific on-resistance when the cell pitch is smaller.

As a particular example, consider the case of devices designed to support 3000 volts. In the case of the conventional structure with a one-dimensional junction, the specific on-resistance of the drift region is found to be 1.5 mΩ-cm^2 if a critical electric field for breakdown of 3 x 10^6 V/cm is used. In contrast, the specific on-resistance for the drift region of the charge coupled structure with a cell pitch of 1 micron is found to be only 0.14 mΩ-cm^2 if a critical electric field for breakdown of 2 x 10^6 V/cm is used. In this calculation, a bulk mobility of 570 cm^2/V-s was used corresponding to a doping concentration of 2 x 10^{17} cm^{-3} in the N-type portion of the drift region. In this example, the drift region for the charge coupled structure would have a thickness of 15 microns when compared with 20 microns needed in the conventional structure.

13.3 Charge Coupled Schottky Rectifier Structure

The basic structure of the charge coupled Schottky rectifier was already depicted in Fig. 13.1. In the blocking-mode of operation, a depletion region extends in the vertical direction from the Schottky contact as well as in the horizontal direction from the P-N junctions for positive bias applied to the cathode. These depletion regions produce two-dimensional charge coupling resulting a uniform electric field in the y-direction. Both the N-type and P-type regions are simultaneously depleted by the voltage applied to the cathode. The breakdown voltage of the structure is decided by the depth of the trench (dimension 't' in the figure). The trench must not penetrate into the N$^+$ substrate to avoid the development of a high electric field at the bottom of the trench.

On-state current flow occurs when a negative bias is applied to the cathode. In this case, electrons are transported across the Schottky barrier contact and through the N-type drift region. No current conduction occurs through the P-type region unless the voltage exceeds 3 volts. Unipolar operation of the structure is favored by the low specific on-resistance of the drift region which keeps the on-state voltage drop below 2 volts even for structures designed to support very high voltages.

13.3.1 Charge Coupled Schottky Rectifier Structure: Blocking Characteristics

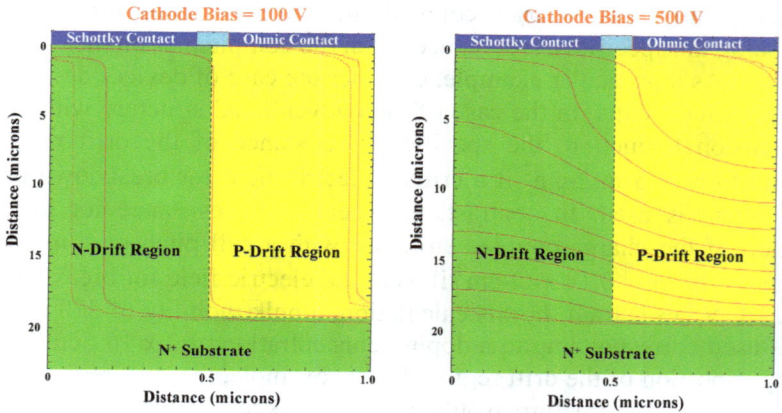

Fig. 13.5 Potential Distribution in the Charge Coupled 4H-SiC Schottky Rectifier Structure.

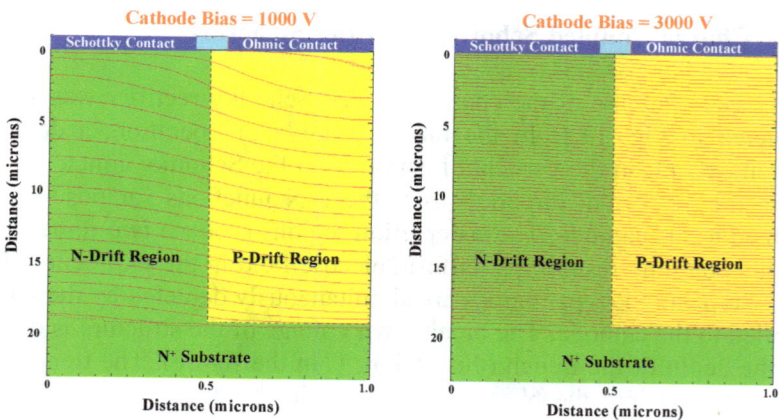

Fig. 13.6 Potential Distribution in the Charge Coupled 4H-SiC Schottky Rectifier Structure.

In order to provide insight into the operation of the charge coupled Schottky rectifier structure, two-dimensional numerical simulations were performed using a cell pitch of 1 micron, a trench depth of 18 microns, and a total epitaxial layer thickness of 20 microns. A work function of 4.5 eV was used for the Schottky contact. A uniform doping concentration was assumed for both the N-type and P-type portions of the drift region.

The potential distribution for the case of the charge coupled structure with doping concentration of 3×10^{17} cm^{-3} in both the N-type and P-type portions of the drift region is shown in Fig. 13.5 for cathode bias of 100 and 500 volts. At a bias of 100 volts, the N and P-type drift regions are not yet completely depleted by the horizontal extension of the depletion layers. At a bias of 500 volts, the drift region is completely depleted but the potential lines are crowded at both the top and bottom of the structure. The potential distributions for the cases of cathode bias voltages of 1000 and 4000 volts are shown in Fig. 13.6. At a bias of 1000 volts, the potential lines begin to get uniform with a very uniform spacing observed between the potential contours at a bias of 4000 volts. This indicates a fairly constant electric field within the drift region, which improves the breakdown voltage capability of the structure.

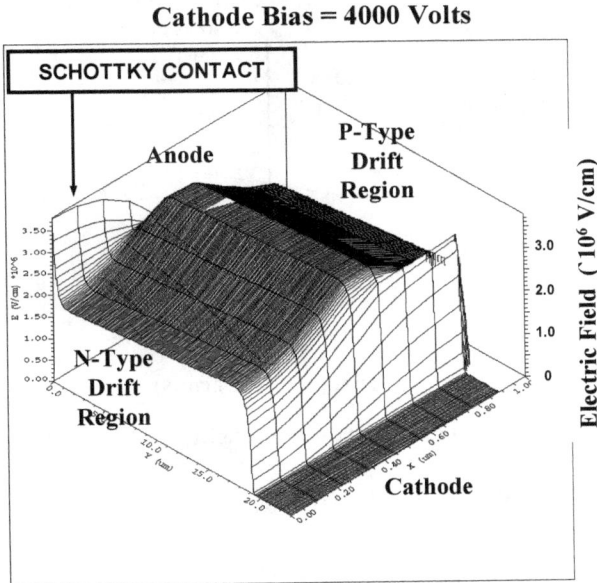

Fig. 13.7 Electric Field Distribution in the Charge Coupled Schottky Rectifier Structure.

A three-dimensional view of the electric field distribution for the charge coupled Schottky rectifier structure is shown in Fig. 13.7 for a bias of 4000 volts applied to the cathode electrode. A uniform electric field is observed along the N-type drift region indicating optimum two-dimensional charge coupling. However, a high electric field is also observed at the Schottky contact located above the N-drift region. The development of the electric field at the Schottky contact can be seen more clearly in a one-dimensional plot taken at x = 0 microns, as shown in Fig. 13.8. A very high electric field is developed at the Schottky contact beginning at the lower cathode voltages and becoming close to the breakdown field strength of 4H-SiC at higher voltages. This high electric field will exacerbate the barrier lowering and tunneling phenomena resulting in a high leakage current for the structure.

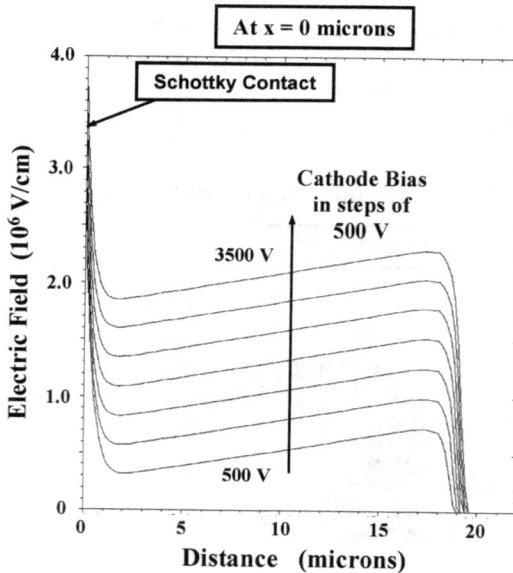

Fig. 13.8 Electric Field in the Charge Coupled Schottky Rectifier Structure.

A high electric field is also developed at the abrupt P-N junction located at the bottom of the trench in the charge coupled Schottky rectifier structure. This can be observed in Fig. 13.9, where the electric field is plotted at x = 1 micron. Although the electric field is again uniform through most of the P-type drift region, a sharp peak is observed at the P-N junction at the bottom of the trench. The breakdown in the charge coupled structure can therefore be initiated by impact ionization within either the N-type or P-type drift regions.

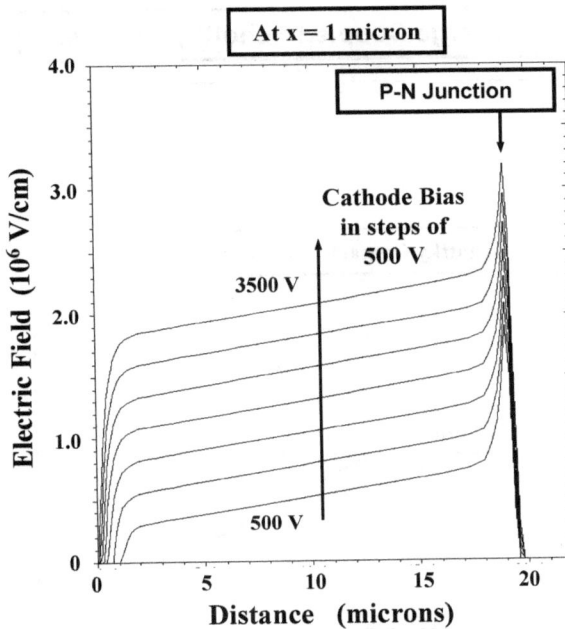

Fig. 13.9 Electric Field in the Charge Coupled Schottky Rectifier Structure.

The reverse blocking characteristic of the charge coupled Schottky rectifier structure is shown in Fig. 13.10. The current flowing through the Schottky contact is shown by the solid line while that through the ohmic contact to the P-type region is shown by the dashed line. At low cathode bias voltages, the current through the Schottky contact is dominant as expected due to thermionic emission over the barrier. However, when the cathode voltage exceeds 2500 volts, the current through the P-type region becomes visible indicating the on-set of impact ionization induced current. When the cathode bias exceeds 4000 volts, the current through the P-type region becomes larger than through the Schottky contact. The ultimate breakdown voltage therefore occurs by the impact ionization within the P-type region. It is worth pointing out that the barrier lowering and tunneling effects were not included in these simulations. When these effects are included, the leakage current through the Schottky contact can increase by many orders of magnitude as discussed in chapter 5. The impact ionization may shift to the N-drift region under these conditions.

4H-SiC Charge Coupled Schottky Rectifier

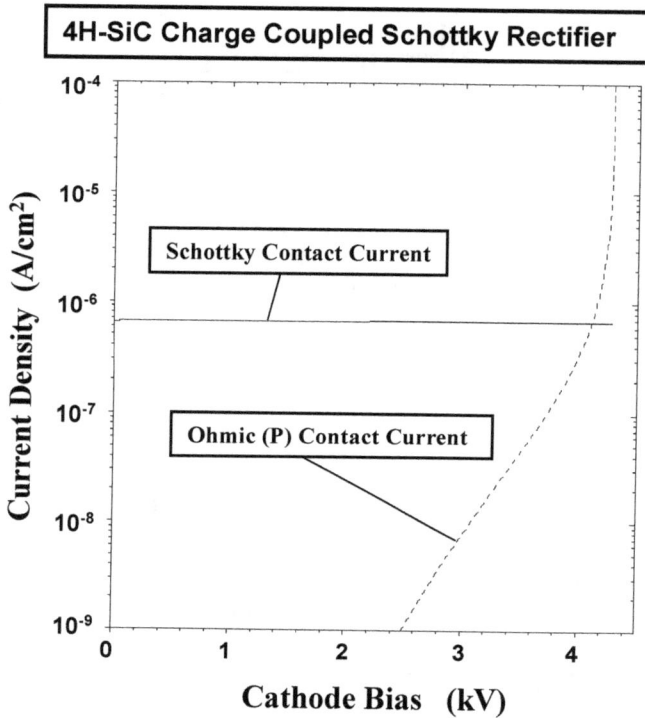

Fig. 13.10 Reverse Blocking Characteristics of the Charge Coupled Schottky Rectifier Structure.

It is worth emphasizing that the breakdown voltage for the charge coupled Schottky rectifier is 4300 volts despite the very high doping concentration of 3×10^{17} cm^{-3} in the N-type drift region. The charge coupling physics allows the blocking voltage to increase from 230 volts for the parallel-plane one-dimensional case to 4300 volts, an increase by a factor of nearly 20 times. This remarkable improvement in breakdown voltage capability with high drift layer doping concentration is very useful for development of devices that must operate at above 3000 volts. However, it is necessary to reduce the electric field at the Schottky contact to suppress the barrier lowering and tunneling phenomena. A structural enhancement that enables this is discussed in the section 13.4 of this chapter.

13.3.2 Charge Coupled Schottky Rectifier Structure: On-State Characteristics

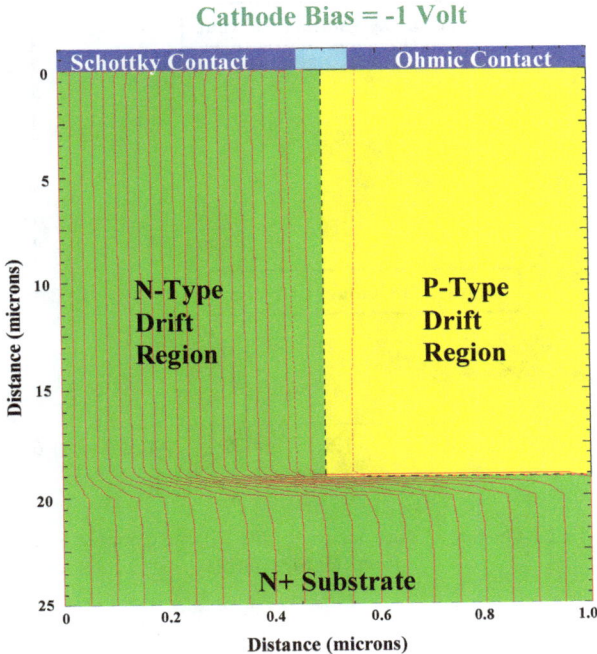

Fig. 13.11 On-State Current Distribution in the Charge Coupled Schottky Rectifier Structure.

The on-state characteristic of the charge coupled Schottky rectifier structure was obtained by application of a negative bias to the cathode electrode. Both the Schottky and Ohmic contact were held at zero potential and the current through them was monitored. The current distribution within the structure is shown in Fig. 13.11 at a cathode bias of 1 volt. It can be observed that all the current is constrained to the Schottky contact and the N-type drift region. This confirms that current flow occurs via unipolar conduction with no injection across the P-N junction. Once the current passes through the first 18 microns (the trench depth), it is rapidly distributed and becomes uniform in the N^+ substrate.

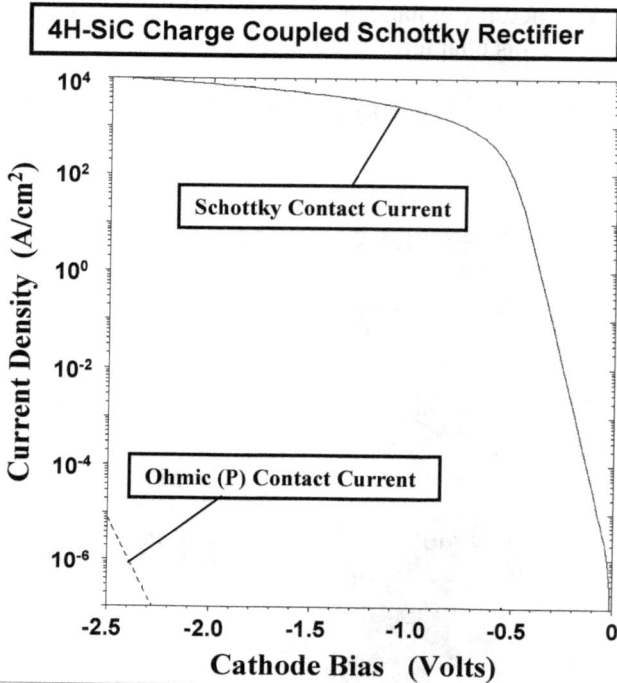

Fig. 13.12 On-State Characteristic of the Charge Coupled Schottky
Rectifier Structure.

The on-state characteristic for the charge coupled Schottky
rectifier is shown in Fig. 13.12. The current through the Schottky contact
is shown by the solid line while the current through the ohmic contact is
shown by the dashed line. It can be seen that the current through the
ohmic contact is negligible until the forward bias voltage exceeds 2.5
volts. At a forward current density of 100 A/cm^2, the on-state voltage
drop is only 0.5 volts. This is extremely low for a Schottky rectifier
capable of blocking over 4000 volts. Even at a current density of 1000
A/cm^2, the on-state voltage drop is less than 1 volt indicating that the P-
N junction will not become active under forward bias operating
conditions. It can also be concluded that the work function of the
Schottky contact can be increased from 4.5 eV (corresponding to a
barrier height of 0.8 eV) to 5.0 eV to reduce the leakage current while
retaining an on-state voltage drop of about 1 volt.

13.4 Enhanced Charge Coupled Schottky Rectifier Structure

In the previous section, it was found that the electric field at the Schottky contact in the basic charge coupled Schottky rectifier structure is very high during the blocking mode of operation. The high electric field can produce a severe increase in the leakage current, as discussed in chapter 5, due to barrier lowering and tunneling phenomena. For addressing this problem, an enhanced charge coupled Schottky rectifier structure was proposed[5]. This structure, shown in Fig. 13.13, contains a lightly doped N-type region located adjacent to the Schottky contact and a highly doped P-type region located adjacent to the ohmic contact. The lightly doped N-type region can be formed during the initial epitaxial growth of the N-type drift layer. The highly doped P-type region can be formed either during the epitaxial trench refill or by ion implantation of P-type dopant (aluminum) selectively into the trench opening after the refill.

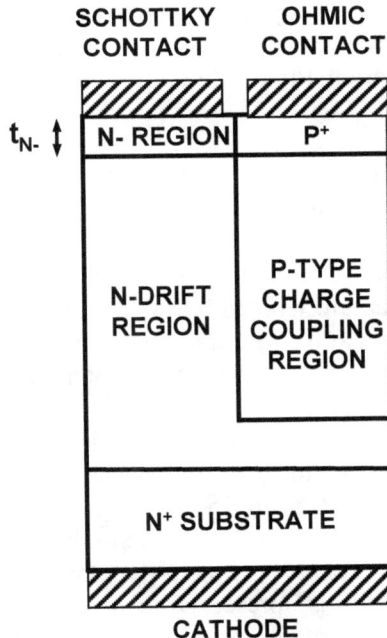

Fig. 13.13 Enhanced Charge Coupled Silicon Carbide Schottky
Rectifier Structure.

In the enhanced charge coupled Schottky rectifier structure, the lightly doped N- region is designed to be depleted at low cathode reverse bias voltages. Its depletion is promoted by the presence of the more highly doped P$^+$ region located above the P-type drift region in the trench. A potential barrier is formed within the N- region which suppresses the build up of the electric field at the Schottky contact. The presence of the lightly doped N- region adds to the series resistance of the N-type drift region. This can increase the on-state voltage drop of the Schottky rectifier. The increase in the series resistance depends upon the thickness (t_{N-}) and doping concentration of the N- region. Consequently, a trade-off must be made between increasing the on-state voltage drop and reducing the electric field at the Schottky contact.

In order to illustrate the improvements obtained with the addition of the N- region into the charge coupled Schottky rectifier structure, the results of two-dimensional numerical simulations are provided below. These simulations were done with same cell structure, namely a 1 micron cell pitch and 18 micron trench depth, as described in the previous section. The N-type and P-type regions were also doped at the same concentration of 3×10^{17} cm^{-3} to facilitate comparison of the structures. The same work function for the Schottky contact of 4.5 eV was also used.

13.4.1 Enhanced Charge Coupled Schottky Rectifier Structure: Blocking Characteristics

The potential distribution for the case of an enhanced charge coupled Schottky rectifier structure with an N- region doping concentration of 5×10^{16} cm^{-3} and thickness at 0.5 microns is shown in Fig. 13.14 at a cathode bias of 4000 volts. In comparison with the potential distribution previous shown on the right hand side of Fig. 13.6 for the basic charge coupled Schottky rectifier structure, a larger spacing between the potential contours is noticeable in the vicinity of the Schottky contact for the enhanced charge coupled Schottky rectifier structure. This implies a reduction of the electric field at the Schottky contact. The uniform potential distribution in the rest of the N-type drift region indicates that a uniform electric field is preserved demonstrating that the two-dimensional charge coupling phenomenon is still applicable. The improved electric field distribution near the Schottky contact can be clearly observed in the three-dimensional view of the field shown in Fig. 13.15 at a cathode bias of 4000 volts.

Cathode Bias = 4000 Volts

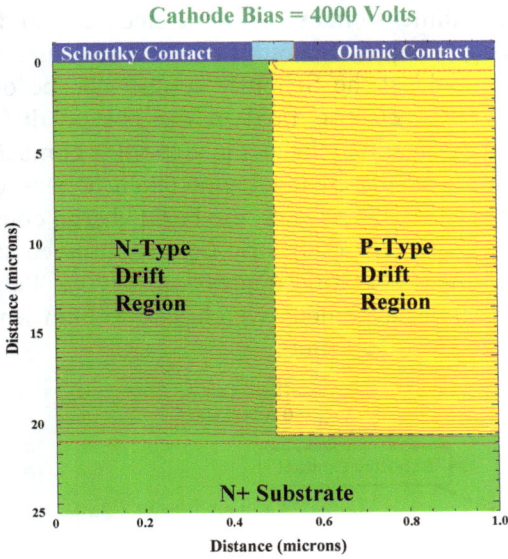

Fig. 13.14 Potential Distribution in the Enhanced Charge Coupled 4H-
SiC Schottky Rectifier Structure.

Cathode Bias = 4000 Volts

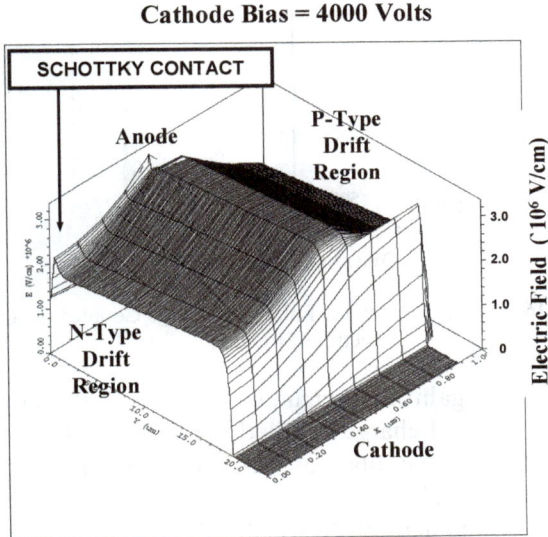

Fig. 13.15 Electric Field Distribution in the Enhanced Charge Coupled
Schottky Rectifier Structure.

In the three-dimensional view of the electric field distribution for the enhanced charge coupled Schottky rectifier structure shown in Fig. 13.15, the electric field at the Schottky contact can be observed to be much smaller than the electric field in the N-type drift region. The development of the electric field at the Schottky contact can be seen more clearly in the one-dimensional plot taken at x = 0 microns, as shown in Fig. 13.16. In contrast with the basic charge coupled Schottky rectifier structure, a much lower electric field is observed at the Schottky contact. This is important for suppression of the Schottky barrier lowering and tunneling phenomena which lead to high leakage currents in silicon carbide Schottky rectifiers.

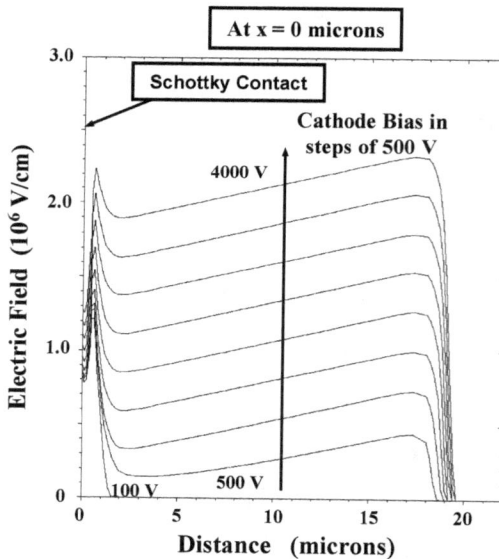

Fig. 13.16 Electric Field in the Enhanced Charge Coupled Schottky Rectifier Structure.

In order to gain a better understanding of the design of the N-region in the enhanced charge coupled Schottky rectifier structure, two-dimensional numerical simulations were also performed with a lower N-doping concentration of 1×10^{16} cm^{-3} while keeping its thickness at 0.5 microns. The development of the electric field at the Schottky contact is shown in Fig. 13.17 at x = 0 microns for this case. The electric field at the Schottky contact has been reduced by the decrease in the N- region doping because it becomes depleted at a smaller cathode bias.

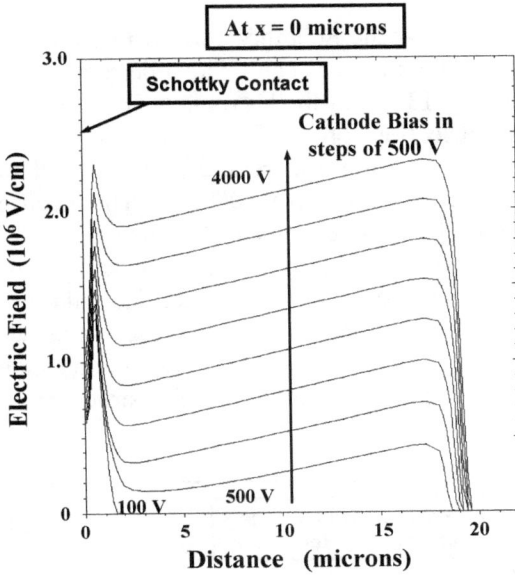

Fig. 13.17 Electric Field in the Enhanced Charge Coupled Schottky
Rectifier Structure.

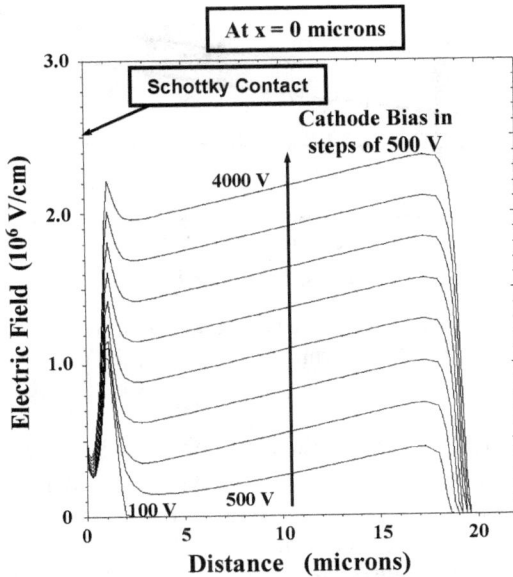

Fig. 13.18 Electric Field in the Enhanced Charge Coupled Schottky
Rectifier Structure.

The electric field at the Schottky contact in the enhanced charge coupled Schottky rectifier structure can be also be reduced by increasing the thickness (t_{N-} in Fig. 13.13) of the N- region. This is illustrated in Fig. 13.18 where the electric field profile is shown at x = 0 microns for a structure with N- region thickness increased from 0.5 to 1 micron while keeping its doping concentration at 5×10^{16} cm^{-3}. The electric field at the Schottky contact for this case becomes less than 5×10^{5} V/cm at a cathode bias of 4000 volts. It can also be seen that the width of low electric field zone below the Schottky contact has enlarged due to the increased thickness of the N- region. These improvements will essentially eliminate the Schottky barrier lowering and tunneling effects in the enhanced charge coupled Schottky rectifier structure.

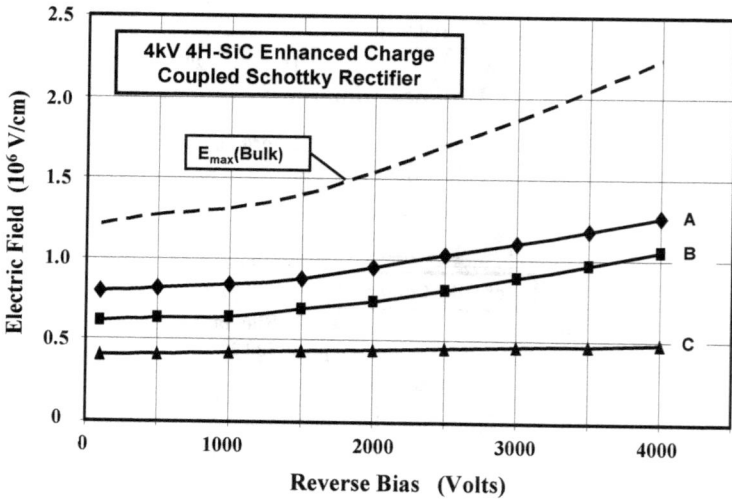

Fig. 13.19 Electric Field in the Enhanced Charge Coupled Schottky Rectifier Structure. (Case A: t_{N-} = 0.5 μm, N_D = 5×10^{16} cm^{-3}; Case B: t_{N-} = 0.5 μm, N_D = 1×10^{16} cm^{-3}; Case C: t_{N-} = 1 μm, N_D = 5×10^{16} cm^{-3})

For comparison of these designs for the N- region within the enhanced charge coupled Schottky rectifier structure, it is instructive to plot the development of the electric field at the Schottky contact as a function of the applied cathode voltage. This plot, provided in Fig. 13.19, clearly demonstrates the reduction of the electric field at the Schottky contact when compared with the maximum electric field observed in the bulk just below the N- region. It can be seen that the thicker N- region provides the greatest reduction of the electric field at the Schottky contact.

It should be pointed out that the addition of the N- region resulted in a slight reduction of the breakdown voltage from 4300 volts to 4200 volts. This change is associated with the loss of about 1 micron from the length of the drift region over which the electric field is uniform. An increase in the trench depth by 1 micron would of course bring the breakdown voltage back to that for the basic structure. The influence of the addition of the lightly doped N- region to the on-state characteristic of the enhanced charge coupled Schottky rectifier structure is discussed in the next section.

13.4.2 Enhanced Charge Coupled Schottky Rectifier Structure: On-State Characteristics

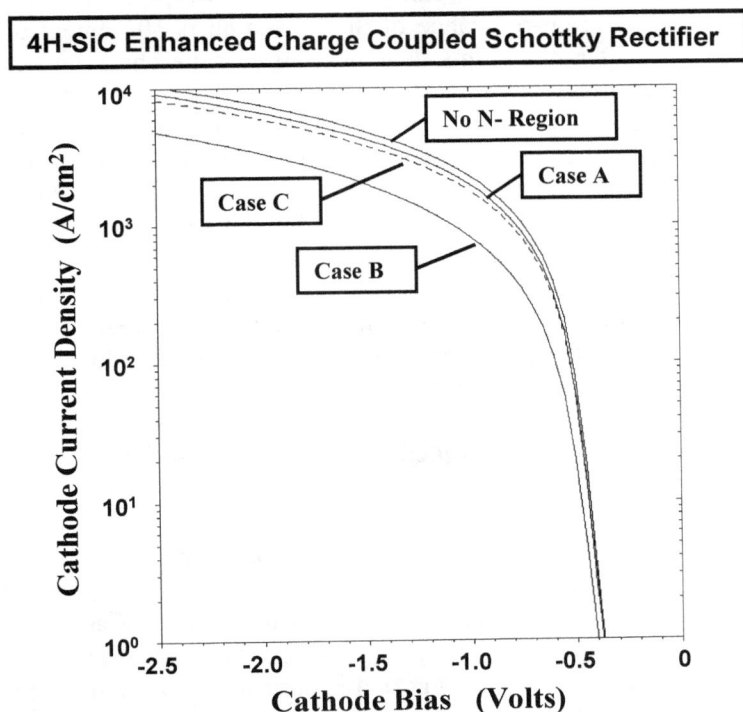

Fig. 13.20 On-State Characteristics of the Enhanced Charge Coupled Schottky Rectifier Structures. (Case A: $t_{N-} = 0.5$ μm, $N_D = 5 \times 10^{16}$ cm^{-3}; Case B: $t_{N-} = 0.5$ μm, $N_D = 1 \times 10^{16}$ cm^{-3}; Case C: $t_{N-} = 1$ μm, $N_D = 5 \times 10^{16}$ cm^{-3})

The on-state characteristic of the enhanced charge coupled Schottky rectifier structure was obtained by application of a negative bias to the cathode electrode. Both the Schottky and Ohmic contact were held at zero potential and the current through them was monitored. The on-state characteristics are shown in Fig. 13.20 for the various cases of N- region doping concentration and thickness together with the characteristics of the basic charge coupled structure for comparison. At a forward current density of 100 A/cm^2, the on-state voltage drop is close to 0.5 volts for all the structures indicating the impact of the additional resistance from the N- region is quite small. This can be confirmed with a simple calculation of the specific resistance for the N- region. The calculated specific resistance for the N- region is found to be 1.6 x 10^{-5} Ω-cm^2, 6.4 x 10^{-5} Ω-cm^2, and 3.1 x 10^{-5} Ω-cm^2 for case A, B and C, respectively. The values for case A and C are small when compared with the specific on-resistance of 1.3 x 10^{-4} Ω-cm^2 calculated for the N-type drift region. In case B, the specific on-resistance of the N- region is 50 percent of the value for the drift region. This degrades the on-state characteristics at high current densities.

When the three cases for the N- region design are compared with each other, it can be concluded that the case C with the thicker N- region produces the most reduction of the electric field at the Schottky contact while having a minimal effect on the on-state voltage drop. Consequently, this is the recommended design for the N- region in the enhanced charge coupled Schottky rectifier structure. With this design, the Schottky barrier lowering and tunneling induced increase in leakage current is completely mitigated while obtaining a very low on-state voltage drop of 0.5 volts for a 4 kV Schottky rectifier.

13.4.3 Enhanced Charge Coupled Schottky Rectifier Structure: Charge Optimization

In the previous sections, the simulations were performed using a doping concentration of 3 x 10^{17} cm^{-3} for both the N-type and P-type drift regions. This doping concentration was chosen based upon an optimum charge balance within the drift region. In the case of silicon charge coupled structures, it has been reported[13] that small deviations from the optimum charge can produce a drastic reduction in the breakdown voltage. This creates a challenging environment for the fabrication of these types of device structures.

In order to evaluate the impact of charge within the N-type and P-type regions upon the breakdown voltage, two-dimensional numerical simulations were performed with various values for the doping concentration in the drift region while maintaining the same cell pitch. In these simulations, the doping concentration in the N-type region was always kept the same as in the P-type region. All the structures considered here were the enhanced charge coupled Schottky rectifiers with a N- region having a doping concentration of 5×10^{16} cm^{-3} and thickness of 1 micron because this provided the best performance as demonstrated in the previous section.

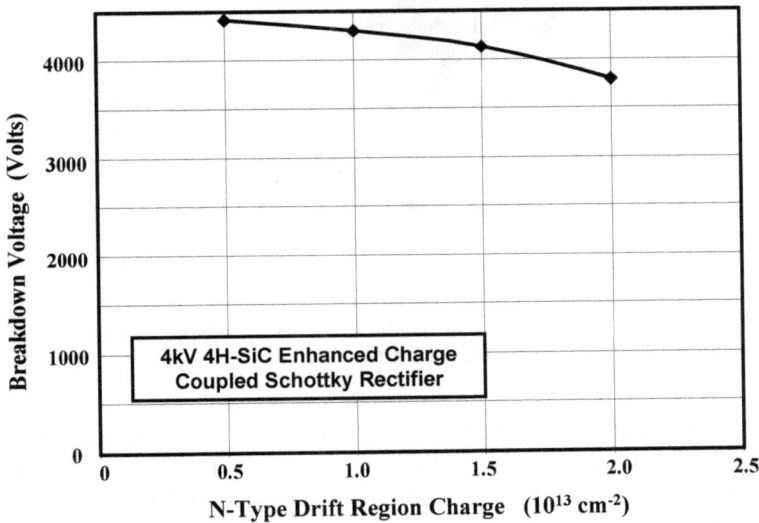

Fig. 13.21 Breakdown Voltage of the Enhanced Charge Coupled Schottky Rectifier Structures.

The breakdown voltages obtained from the simulations using various doping concentrations for the N-type drift region are shown in Fig. 13.21. It can be seen that there is considerable latitude in the choice of the doping concentration as long as the doping in the N-type and P-type regions are equal in magnitude. However, when the doping concentration exceeds 3.5×10^{17} cm^{-3}, corresponding to a charge of 1.75×10^{13} cm^{-2}, the breakdown voltage falls below 4000 volts. These results are favorable from the point of view of device fabrication tolerances. A larger charge in the N-type drift region is of course preferred to obtain a smaller specific on-resistance for the drift region. These results are valid for all device structures that utilize the charge coupling phenomenon with trench based P-type regions.

Cathode Bias = 3500 Volts

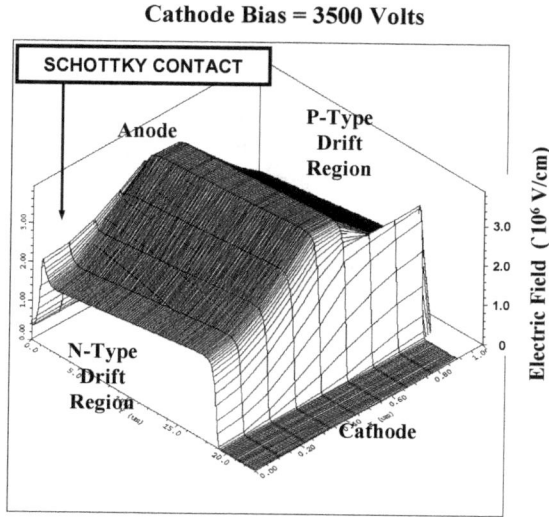

Fig. 13.22 Electric Field Distribution in the Enhanced Charge Coupled Schottky Rectifier Structure.

In order to understand the reasons for the reduction of the breakdown voltage with higher doping concentration in the drift region for the enhanced charge coupled Schottky rectifier structure, it is instructive to look at the electric field distribution within the device. A three dimensional view of the electric field distribution for the structure with doping concentration of 4×10^{17} cm^{-3} is provided in Fig. 13.22. In comparison with the field distribution shown in Fig. 13.15 for a lower doping concentration, it can be seen that the electric field at the junction between the N and P-type drift regions has become enhanced. This is related to the reduced rate of lateral depletion of the drift regions with increased doping concentration. The impact of the doping concentration on the development of the electric field within the N and P-type drift regions can be observed most clearly by examination of the electric field at a fixed depth from the surface. This electric field profile is shown in Fig. 13.23 and Fig. 13.24 at position y = 5 microns for the case of structures with doping concentrations of 1×10^{17} cm^{-3} and 4×10^{17} cm^{-3}.

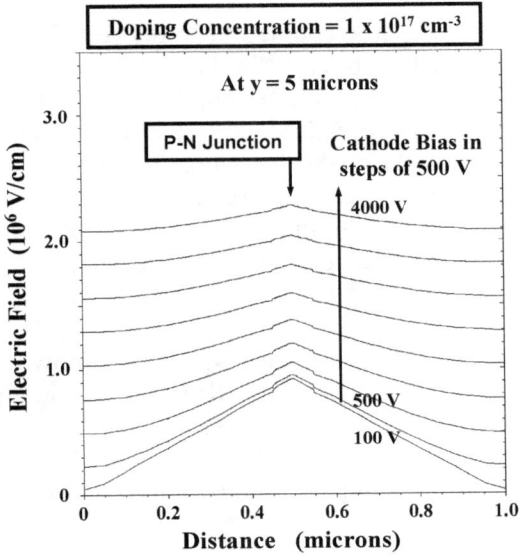

Fig. 13.23 Electric Field in the Enhanced Charge Coupled Schottky
Rectifier Structure.

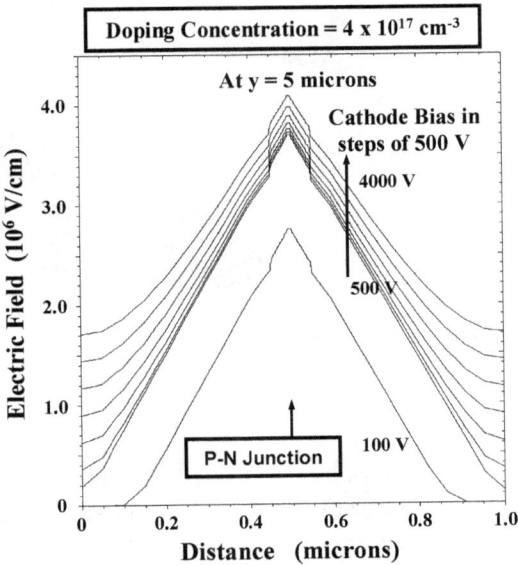

Fig. 13.24 Electric Field in the Enhanced Charge Coupled Schottky
Rectifier Structure.

In the enhanced charge coupled Schottky rectifier structure with doping concentration of 1 x 10^{17} cm^{-3} in the drift regions, it can be seen in Fig. 13.23 that the drift regions are completely depleted at a cathode bias of 100 volts. The electric field profile is quite flat leading to a relatively low electric field at the junction. In contrast to this, in the enhanced charge coupled Schottky rectifier structure with doping concentration of 4 x 10^{17} cm^{-3} in the drift regions, it can be seen in Fig. 13.24 that the drift regions are not completely depleted at a cathode bias of 100 volts. The electric field profile is triangular in shape resulting a relatively high electric field at the junction. This shift in the electric field profile is responsible to the degradation of the breakdown voltage with increasing doping concentration of the drift regions in the enhanced charge coupled Schottky rectifier structures.

13.4.4 Enhanced Charge Coupled Schottky Rectifier Structure: Charge Imbalance

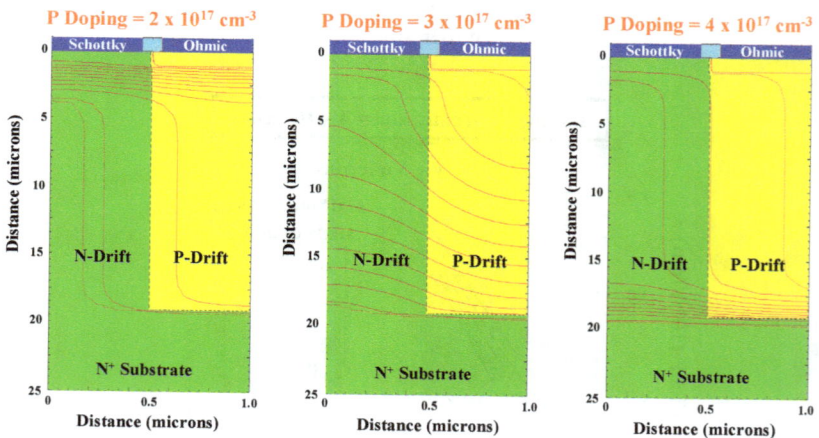

Fig. 13.25 Potential Distribution in the Charge Coupled 4H-SiC Schottky Rectifier Structure.

In the previous section, it was demonstrated that there is considerable latitude in the doping concentration for the N and P-type drift regions in the enhanced charge coupled Schottky rectifiers structure as long as both regions have the same magnitude. In practice, it can be expected that the doping concentration of the N and P-type regions will not be exactly matched. Even a 10 percent imbalance in the charge has been reported[13] to have a significant (> 30 percent) impact on the breakdown voltage for silicon devices.

In order to determine the sensitivity of the breakdown voltage on the charge imbalance within the enhanced charge coupled Schottky rectifier structure, two dimensional numerical simulations were performed with various doping concentrations for the P-type portion of the drift region while maintaining the doping concentration of the N-type portion at 3×10^{17} cm^{-3}. The rest of the structural parameters were held constant: cell pitch of 1 micron; trench depth of 18 microns; N- region doping concentration of 5×10^{16} cm^{-3} with thickness of 1 micron; work function for Schottky contact of 4.5 eV.

The impact of a charge imbalance between the N-type and P-type drift regions within the enhanced charge coupled Schottky rectifier structure can be observed in the potential contours shown in Fig. 13.25 at a cathode bias of 500 volts for the case of P-type drift region doping concentration of 2×10^{17} cm^{-3}, 3×10^{17} cm^{-3}, and 4×10^{17} cm^{-3}. In contrast with the perfectly balanced case shown in the middle, the potential lines become crowded at the top of the structure when the doping concentration of the P-type drift region is reduced while they become crowded at the bottom of the structure when the doping concentration of the P-type drift region is increased. The resulting electric field enhancement produces a decrease in the breakdown voltage.

The localized enhancement of the electric field due to the charge imbalance in the enhanced charge coupled Schottky rectifier structure can be observed in the three dimensional electric field plots shown in Fig. 13.26 and Fig. 13.27 at a cathode bias of 500 volts for the case of P-type drift region doping concentrations of 2×10^{17} cm^{-3} and 4×10^{17} cm^{-3}, respectively. In both cases, the electric field at the Schottky contact is suppressed by the inclusion of the lightly doped N- region in the N-type drift region. When the doping concentration of the P-type drift region is made less than the doping concentration of the N-type region, a high electric field is clearly observed towards the top of the structure. The triangular electric field profile in this portion is reminiscent of the field distribution for the one-dimensional parallel-plane junction. Even at a low cathode bias of 500 volts, the maximum electric field at this location has reached a magnitude of 2.5×10^6 V/cm, which approaches the critical electric field for breakdown in the one dimensional case. A similar behavior is observed for the case of charge imbalance created by increasing the doping concentration of the P-type drift region. However, the peak of the electric field now occurs at the bottom of the trench.

Cathode Bias = 500 Volts

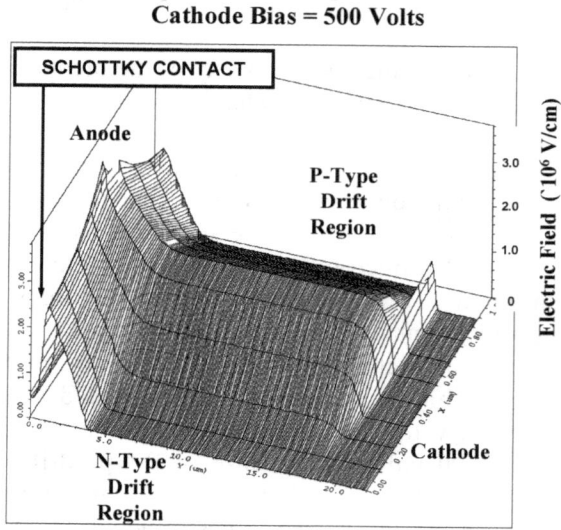

Fig. 13.26 Electric Field Distribution in the Enhanced Charge Coupled
Schottky Rectifier Structure.

Cathode Bias = 500 Volts

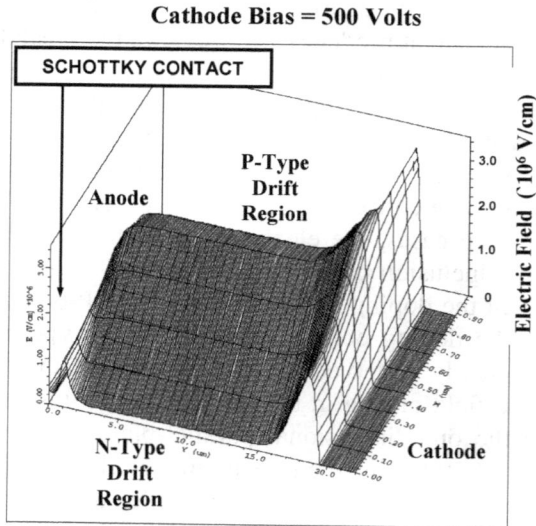

Fig. 13.27 Electric Field Distribution in the Enhanced Charge Coupled
Schottky Rectifier Structure.

The change in the breakdown voltage for the enhanced charge couple Schottky rectifier structure is shown in Fig. 13.28 as a function of the charge (and hence the doping concentration) in the P-type drift region. The charge in the P-type drift region has been varied on either side of the optimum value of 1.5×10^{13} cm^{-2}. It can be seen that the breakdown voltage is reduced in half by deviation of the charge in the P-type region by as little as 10 percent. Consequently, the silicon carbide charge coupled structure with trench based P-type regions is just as sensitive to charge imbalance as reported for silicon structures. The same degree of precision in controlling the doping concentration of the epitaxial layers will have to be achieved in silicon carbide to make the manufacturing of these structures feasible.

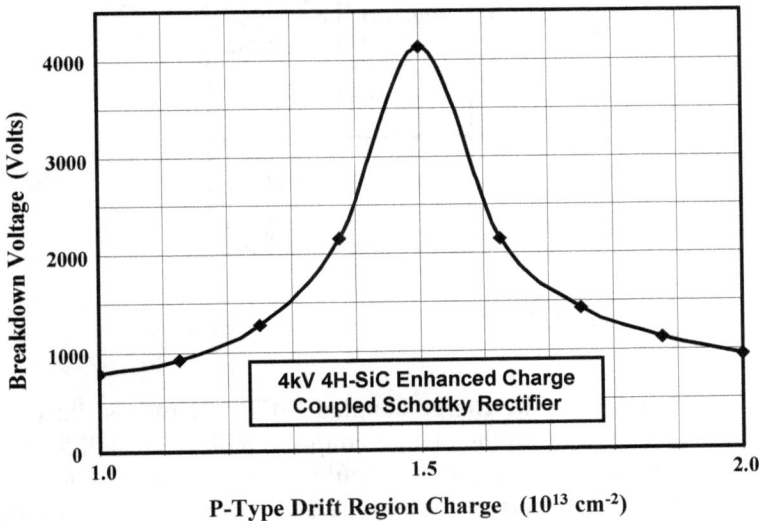

Fig. 13.28 Breakdown Voltage of the Enhanced Charge Coupled Schottky Rectifier Structures.

The discussion of the electric field distribution and breakdown voltage for the enhanced charge coupled Schottky rectifier structure is equally applicable to any other devices that utilize charge coupling based upon a P-type drift region located in a trench adjacent to the N-type drift region. The performance of high voltage JFETs and MOSFETs with the same charge coupling structure is provided in the following sections. Charge optimization and imbalance will not be considered separately for these structures because their behavior is similar to that provided above for the Schottky rectifier structure.

13.5 Charge Coupled JFET Structure

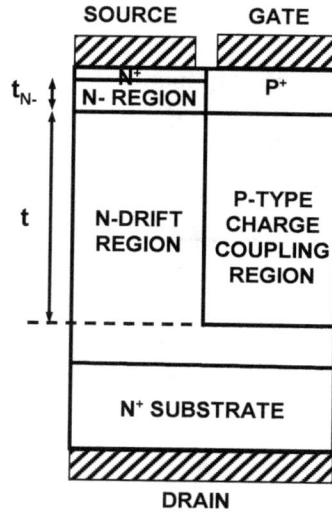

Fig. 13.29 The Charge Coupled JFET Structure.

The basic structure of the charge coupled JFET structure is illustrated in Fig. 13.29. In addition to the charge coupling regions, which are similar to those already shown for the Schottky rectifier, the JFET structure contains separate contacts to the N-type and P-type regions to form the source and gate regions, respectively. This requires definition of a highly interdigitated metal geometry in order to obtain good ohmic contacts for the source and gate. A more lightly doped N- region is incorporated in the JFET structure to improve upon the blocking gain[5].

The JFET structure discussed in this section is designed for normally-on operation because it can then be utilized in the Baliga-Pair circuit configuration (see chapter 8). In order to achieve this, the doping concentration of the N- region must be sufficiently high so that it does not get completely depleted by the built-in potential of the P^+/N-junction. The zero-bias depletion width for 4H-SiC was provided in Fig. 2.5. A doping concentration of more than 1×10^{16} cm^{-3} must be used to prevent complete depletion of a 0.5 micron wide N- region. The on-state current can then flow through the un-depleted portion of the N- region and via the N-type drift region into the N$^+$ substrate. A low specific on-resistance is achieved in the charge coupled JFET structure due to the very high doping concentration of the N-type drift region.

In order to block current flow from the drain to the source, a potential barrier must be established in the N- region to prevent the transport of electrons, as in the case of the trench gate JFET structure discussed in chapter 7. The potential barrier can be formed by the application of a negative bias to the gate electrode. The magnitude of the potential barrier will depend upon the doping concentration and thickness of the N- region as well as the applied gate bias. When the drain bias increases, the potential barrier can be reduced promoting the injection of electrons and inducing drain current flow. This produces an exponential increase in the drain current with increasing drain bias in a manner similar to that discussed in chapter 7 for the trench gate JFET structures.

In order to fully utilize the breakdown voltage capability of the charge coupled drift region, the potential barrier must be sufficiently large to prevent the injection of electrons over the barrier when the drain bias approaches the breakdown voltage of the charge coupled drift region. In principle this can be achieved by application of larger negative gate bias voltages. However, the maximum gate bias voltage is limited by the on-set of breakdown between the gate and source. For the structure shown in Fig. 13.29, the gate-source breakdown will first occur at the P$^+$/N$^+$ junction due to the high doping concentration on both sides. It was found, by using two dimensional numerical simulations, that the gate-source breakdown voltage was greater than 20 volts when the doping concentration of the regions was chosen as 1 x 10^{19} cm^{-3}. This is sufficiently high to obtain a blocking voltage limited by the breakdown voltage of the charge coupled drift regions.

In order to illustrate the operating principles of the charge coupled JFET structure, two-dimensional numerical simulations were performed with a cell pitch of 1 micron. The doping concentration of the N-type and P-type drift regions was 3 x 10^{17} cm^{-3} based upon the optimization discussed for the enhanced charge coupled Schottky rectifier structure. The N$^+$ source region had a depth of 0.1 microns. A trench depth of 18 microns was used, with a thickness of 1 micron for the P$^+$ and N- regions. These parameters are suitable for obtaining a breakdown voltage of 4000 volts for the charge coupled drift region. As discussed in the previous section for the enhanced charge coupled Schottky rectifier, the specific on-resistance for the N-type drift region is only 0.13 mΩ-cm^2. This extremely low value can be taken advantage of only if the resistance contributed by the N- region can be made comparable while achieving the desired blocking voltage capability.

13.5.1 Charge Coupled JFET Structure: Blocking Characteristics

In the charge coupled JFET structure, the ability to block current flow between the drain and source is produced by the formation of a potential barrier for the transport of electrons. The potential barrier can be formed by the application of a negative gate bias to initially deplete the N- region and to then enhance the magnitude of the barrier. The development of the potential barrier is shown in Fig. 13.30 for the case of an N- region with doping concentration of 2×10^{16} cm^{-3} and thickness of 1 micron. A potential barrier of over 8 volts is created by the application of a gate bias of -20 volts. As expected, the maxima for the barriers are formed at the center of the N- region at a depth of 0.7 microns from the surface.

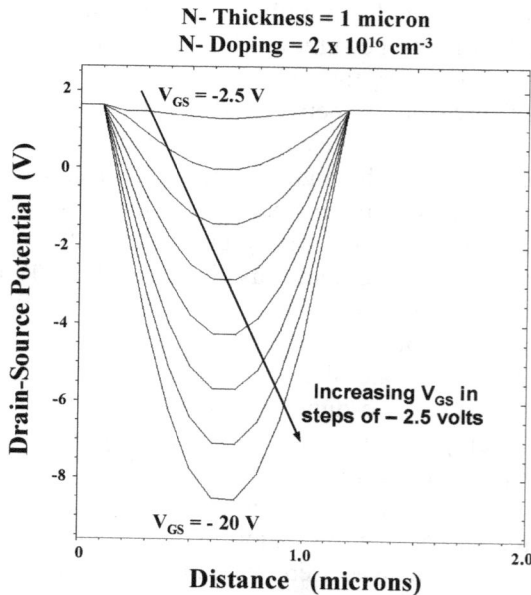

Fig. 13.30 Potential Barrier in a Charge Coupled JFET Structure.

The on-set of drain-source current flow in the JFET structure begins to occur when the potential barrier in the channel is reduced due to the applied drain bias. For the structure described above, the change in the potential barrier with increasing drain bias is shown in Fig. 13.31 for the case of a gate bias of -15 volts. It can be seen that the barrier is greatly diminished when the drain bias exceeds 2000 volts. This implies that electron injection over the potential barrier will begin to produce drain current flow when the drain bias is increased to over 2000 volts.

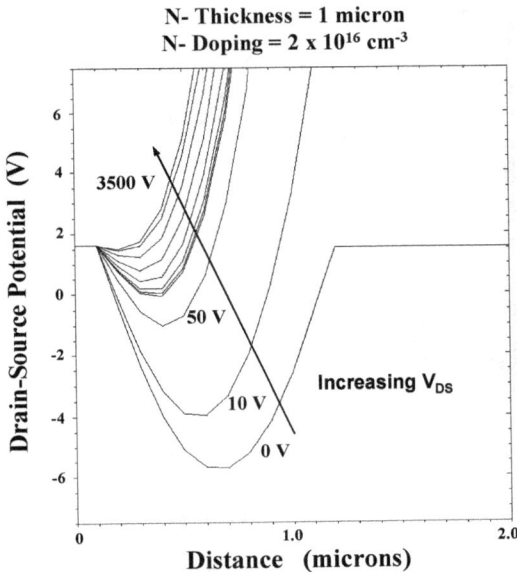

Fig. 13.31 Potential Barrier in a Charge Coupled JFET Structure.

The blocking characteristics of the charge coupled JFET structure are shown in Fig. 13.32. This structure exhibits a mixed triode-pentode behavior (see chapter 7 for more detailed discussion of the modes). The triode-like behavior occurs at the larger gate bias voltages with an exponential increase in drain current with increasing drain voltage. Although the characteristics are shown only upto 3000 volts, this JFET structure was found to be capable of supporting over 4000 volts with a gate bias of -20 volts. Using the data shown in the figure, the DC blocking gain for the charge coupled JFET structure was calculated to be 175 at a gate bias of -16 volts and a leakage current density of 1 mA/cm^2.

This is adequate for supporting high drain bias voltages with gate bias voltages below the gate-source breakdown voltage.

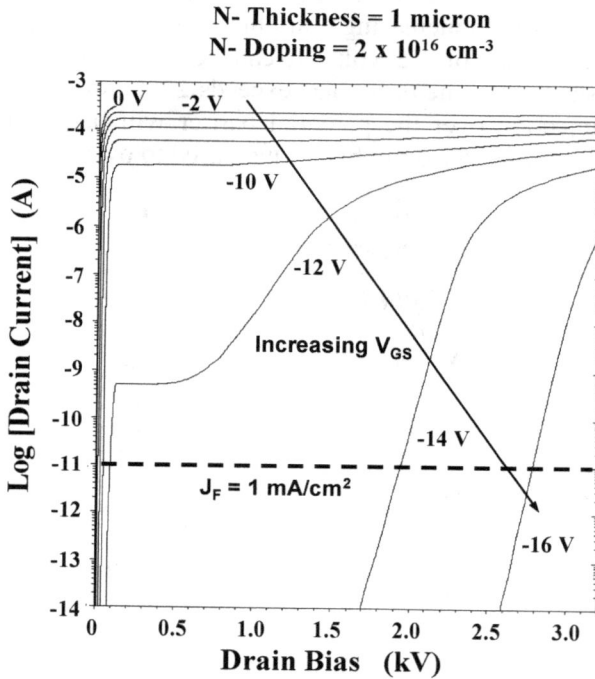

N- Thickness = 1 micron
N- Doping = 2 x 10^{16} cm^{-3}

Fig. 13.32 Blocking Characteristics of a Charge Coupled JFET Structure.

The potential distribution in the charge coupled JFET structure in the blocking mode is shown in Fig. 13.33 for the case of a drain bias of 3000 volts and a gate bias of -20 volts. The gate bias has depleted the N- region under the N$^+$ source region and produced a potential barrier to suppress drain current flow. The potential difference across this region can be seen to be small (less than the 25 volt separation between the potential contours). The drain voltage is supported below the N- region with a uniform variation of the potential along the trench. The potential distribution is very similar to that shown earlier for the enhanced charge coupled Schottky rectifier structure (see Fig. 13.14). This confirms that the charge coupled JFET structure operates in the same manner as the enhanced charge coupled Schottky rectifier structure from the stand point of the design of the N and P-type charge coupling regions.

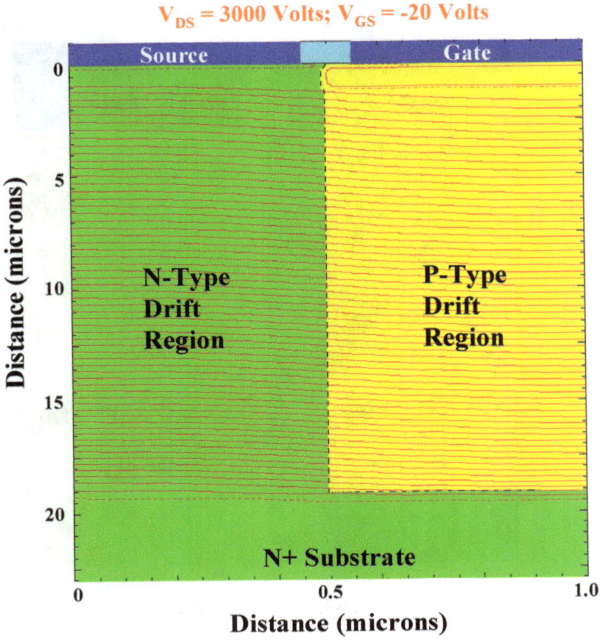

Fig. 13.33 Potential Distribution in the Charge Coupled 4H-SiC JFET Structure.

13.5.2 Charge Coupled JFET Structure: On-State Characteristics

The on-state characteristic of the charge coupled JFET structure was obtained at zero gate bias. The current distribution within the upper portion of the structure is shown in Fig. 13.34 at a drain bias of 0.1 volts. It can be observed that the current becomes constricted in the N- channel region. This is because the N- region is partially depleted by the zero-bias depletion width associated with the built-in potential of the P^+/N-junction. Once the current enters the underlying N-type drift region, it becomes uniformly distributed within it. It is worth pointing out that the zero-bias depletion width in the N-type drift region is not negligible when compared with its width. Consequently, this must also be accounted for in the model for the specific on-resistance. The absence of current flow in the P-type region confirms that current flow in the charge coupled JFET structure occurs via unipolar conduction with no injection across the P-N junction.

$$V_{DS} = 0.1 \text{ Volts}; \ V_{GS} = 0 \text{ Volts}$$

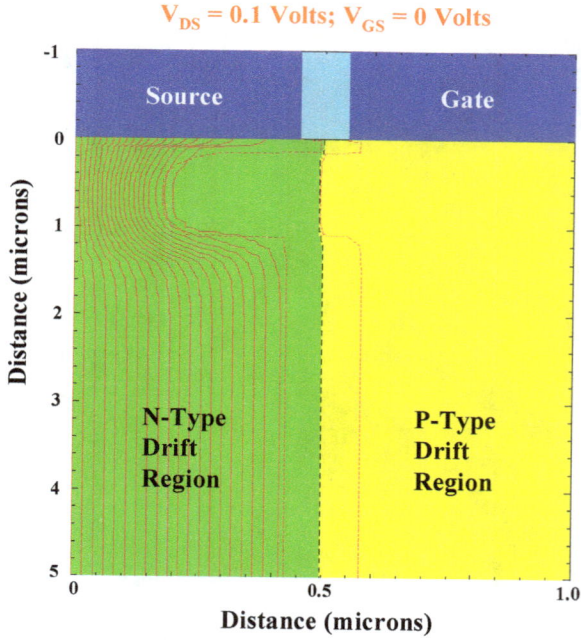

Fig. 13.34 On-State Current Distribution in the Charge Coupled JFET Structure.

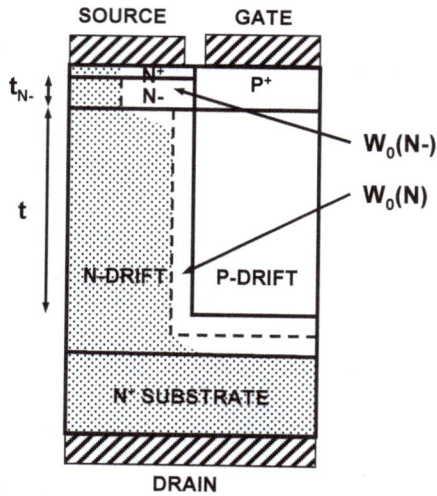

Fig. 13.35 Current Flow Pattern in the Charge Coupled JFET Structure.

The on-resistance of the charge coupled JFET structure can be modeled as the resistance of the N-type drift region with the additional resistance arising in the N- channel region including the impact of current constriction. The current flow pattern is indicated by the shaded area in Fig. 13.35. The total on-resistance for the charge coupled JFET structure is given by the resistance of all the components in the current path:

$$R_{on,sp} = R_{CH} + R_D + R_{subs} \qquad [13.8]$$

where R_{CH} is the channel resistance, R_D is the resistance of the drift region after taking into account current spreading from the channel, and R_{subs} is the resistance of the N$^+$ substrate.

The specific channel resistance is given by:

$$R_{CH} = \rho_{N-}.p.\left(\frac{t_{N-}}{W_N - W_0(N-)} \right) \qquad [13.9]$$

where ρ_{N-} is the resistivity of the N- channel region, t_{N-} is the thickness of the N- region, and $W_0(N-)$ is the zero-bias depletion width in the N-channel region. The resistivity and zero-bias depletion width used in this equation must be computed using the reduced doping concentration of the N- channel region.

The drift region spreading resistance can be obtained by using:

$$R_D = \rho_N.p.\ln\left(\frac{W_N - W_0(N)}{W_N - W_0(N-)} \right) + \rho_N.p.\left(\frac{t - W_0(N-) + 2W_0(N)}{W_N - W_0(N)} \right) \qquad [13.10]$$

where t is the thickness of the drift region below the N- channel region as shown in Fig. 13.35 and W_N is the width of the mesa region. Here, $W_0(N)$ is the zero-bias depletion width in the N-type drift region, which is given by:

$$W_0(N) = \sqrt{\frac{\varepsilon_S.V_{bi}}{q.N_D}} \qquad [13.11]$$

where N_D is the doping concentration of the N-type drift region. Note that this equation takes into account the fact that only half the built-in voltage is supported on the N-type side of the drift region because of equal concentrations on both sides of the junction.

The contribution to the resistance from the N^+ substrate is given by:

$$R_{subs} = \rho_{subs} \cdot t_{subs}$$ [13.12]

where ρ_{subs} and t_{subs} are the resistivity and thickness of the substrate, respectively. A typical value for this contribution is 4×10^{-4} Ω-cm^2.

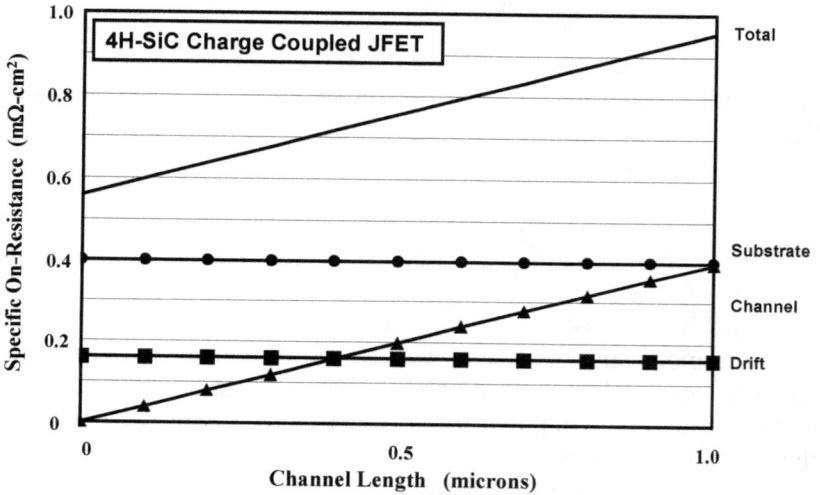

Fig. 13.36 Analytically Calculated Specific On-Resistance of the Charge Coupled JFET Structure.

The specific on-resistance calculated by using the above model is shown in Fig. 13.36 for the case of a cell pitch of 1 micron. A doping concentration of 3×10^{17} cm^{-3} was used for both the N-type and P-type drift regions. The N- channel region had a doping concentration of 2×10^{16} cm^{-3}. The analytically calculated zero-bias depletion widths in the N-channel region and N-drift region are 0.42 and 0.077 microns, respectively. Only the channel resistance contribution increases with the channel length because the cell pitch remains unchanged. The model predicts a specific on-resistance of 0.915 mΩ-cm^2 for a channel length of 0.9 microns. The specific on-resistance obtained with the simulations was 0.835 mΩ-cm^2 for the structure with N- channel region thickness of 0.9 microns. The current flow lines in Fig. 13.34 indicate that the current spills over into the edge of the depletion region reducing the on-resistance contribution from the channel. This accounts for the slightly larger value for the specific on-resistance calculated by using the model.

From Fig. 13.36, it can be seen that the N^+ substrate contribution is dominant for the charge coupled JFET structure. This is due to the extremely low specific on-resistance of the charge coupled drift region. Reduction of the contribution from the substrate can be achieved by increasing its doping concentration and aggressively reducing its thickness. The available manufacturing technology for reducing the thickness of silicon devices to less than 50 microns must be extended to silicon carbide devices.

13.5.3 Charge Coupled JFET Structure: Output Characteristics

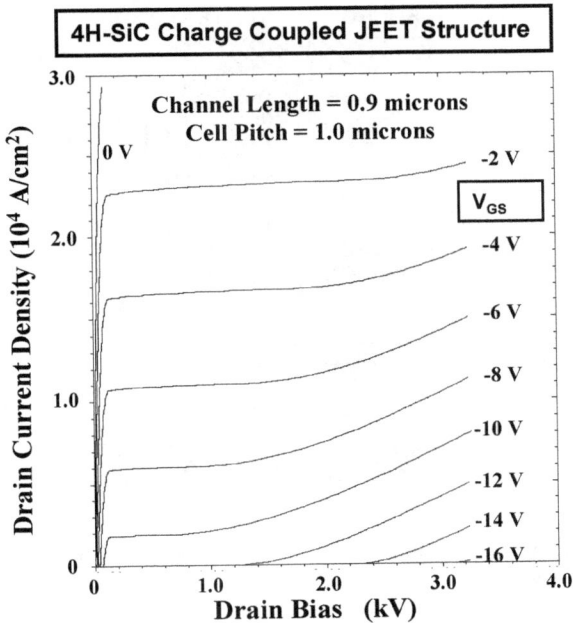

Fig. 13.37 Output Characteristics of the Charge Coupled 4H-SiC JFET Structure.

The output characteristics of the charge coupled JFET structure were obtained using two dimensional numerical simulations by sweeping the drain bias while using various negative gate bias voltages. For the structure with a N- channel length of 0.9 microns and doping concentration of 2 x 10^{16} cm^{-3}, the output characteristics show a mixed triode-pentode mode of operation. At low negative gate bias voltages, the drain current exhibits pentode-like current saturation to high drain bias voltages. At high negative gate bias voltages, a triode-like behavior is evident for the drain current-voltage characteristics. This combination of triode-like and pentode-like behavior is optimum for achieving a low specific on-resistance in the on-state and a high blocking gain for the off-state. This charge coupled JFET structure has ideal characteristics for utilization in the Baliga-Pair configuration.

13.6 Shielded Charge Coupled MOSFET Structure

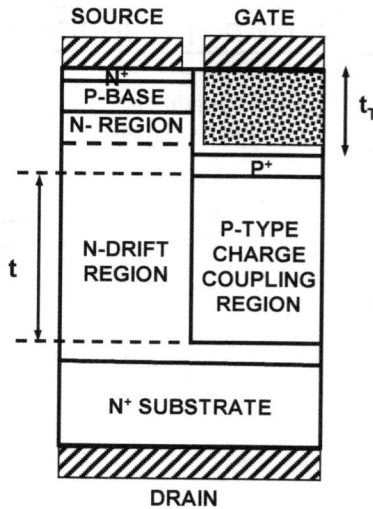

Fig. 13.38 The Shielded Charge Coupled MOSFET Structure.

In the previous chapters, where trench gate power MOSFETs were discussed, it was pointed out that a high electric field is developed in the gate oxide at the trench bottom resulting in a drastic degradation in device performance. Any power MOSFET structures proposed with the charge coupled drift regions must also take this problem into account. A structure for the charge coupled MOSFET has therefore been proposed[5] that includes a P$^+$ shielding region to protect the gate oxide. This structure is illustrated in Fig. 13.38. In addition to the N and P-type charge coupling regions, which are similar to those already shown for the Schottky rectifier, the shielded MOSFET structure contains a P^+ shielding region, located at the bottom of the trench based MOS-gate structure, which is connected to the source contact. In addition, a more lightly doped N- region is incorporated in the upper part of the MOSFET structure to reduce the electric field developed at the P-base region in order to suppress reach-through.

The shielded trench-gate charge coupled MOSFET structure can be fabricated by the growth of the N-type drift region followed by the growth of the more lightly doped N- region. The P-base and N^+ source regions can then be formed by either the growth of epitaxial layers or by ion implantation of appropriate dopants. A deep trench is then etched with a depth designed to support the desired blocking voltage. The trench is refilled with P-type silicon carbide. A second etch is performed self-aligned to the deep trench to make room for the MOS gate structure. The highly doped P-type shielding region is then formed by ion implantation of aluminum. In order to prevent doping the trench sidewalls, an oxide spacer can be formed prior to this implant step. The gate oxide is then formed on the shallow trench sidewalls followed by refill with polysilicon to form the gate electrode.

The shielded charge coupled MOSFET structure is a normally-off device which can support a large drain bias within the charge coupled drift regions. With zero gate bias, the drain voltage initially reverse biases the P-base/N- drift region junction as well as the junction between the N and P-type drift regions. As the drain bias increases, the entire N and P-type drift regions become depleted setting up a two dimensional charge coupling similar to that already discussed in detail in the section on enhanced charge coupled Schottky rectifiers. The presence of the P^+ shielding region prevents the development of a high electric field in the gate oxide.

With the application of a gate positive bias in excess of the threshold voltage, an inversion layer channel is formed along the trench sidewalls in the P-base region providing a path for the transport of electrons from the source to the drain. In addition, an accumulation layer is formed on the surface of the N- region along the trench sidewalls to reduce the resistance for current flow through this portion of the structure. A low threshold voltage can be achieved in the shielded charge coupled MOSFET structure by reducing the doping concentration of the P-base region without reach-through problems because the potential under the P-base region is screened from the drain bias by the presence of a JFET formed between the P^+ shielding region and the N-type drift region. For the same reason, the width of the P-base region can be shortened to reduce the channel length without encountering reach-through problems. This is beneficial for reducing the contribution of the channel to the on-resistance. The structure can also be modified to utilize an accumulation-mode of operation by the addition of an N-type layer along the trench sidewalls.

In order to illustrate the operating principles of the shielded charge coupled MOSFET structure, two-dimensional numerical simulations were performed with a cell pitch of 1 micron. The doping concentration of the N-type and P-type drift regions was 3×10^{17} cm^{-3} based upon the optimization discussed for the enhanced charge coupled Schottky rectifier structure. The N^+ source region had a depth of 0.1 microns, while the P-base region has a doping concentration of 8×10^{16} cm^{-3} and depth of 0.5 microns, leading to a channel length of 0.4 microns. The N- region and the shallow trench (with MOS-gate electrode) extended to a depth of 1 micron. The deep trench region had a depth of 18 microns - suitable for obtaining a breakdown voltage of 4000 volts for the charge coupled drift region. As discussed in the previous section for the enhanced charge coupled Schottky rectifier, the specific on-resistance for the N-type drift region is only 0.13 mΩ-cm^2. This extremely low value can be taken advantage of only if the resistance contributed by the channel can be made comparable while achieving the desired blocking voltage capability.

13.6.1 Shielded Charge Coupled MOSFET Structure: Blocking Characteristics

In the shielded charge coupled MOSFET structure, the blocking voltage will be determined by the length of the charge coupled drift regions if reach-through is suppressed in the P-base region. The two dimensional numerical simulations confirm that this holds true despite the relatively low P-base doping concentration and small thickness. In order to illustrate this, the potential contours within the shielded charge coupled MOSFET structure are shown in Fig. 13.39 for three drain bias values. At a drain bias of 100 volts, it can be seen on the left hand side of the figure that a depletion region has extended downwards from the P-base region and sideways from the vertical junction formed between the N and P-type drift regions. When the drain bias is increased to 1000 volts, it can be seen in the center of the figure that the two dimensional charge coupling has completely depleted the N and P-type drift regions. Note that very little potential drop is visible just below the P-base region indicating the screening of the potential in this location. This screening continues to prevail even when the drain bias is increased to 3000 volts as shown on the right hand side of the figure. The uniformity of the potential contours within the drift region below the gate structure is indicative of optimum charge coupling.

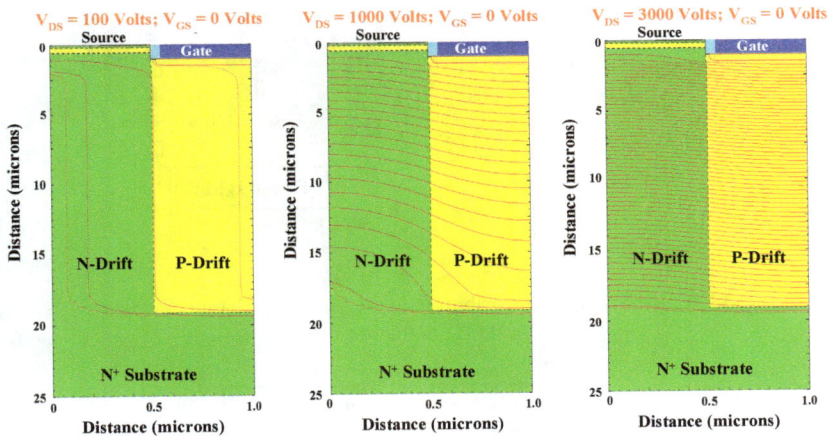

Fig. 13.39 Potential Distribution in the Shielded Charge Coupled 4H-SiC MOSFET Structure.

In the above figure, some potential crowding is noticeable at the corner of the P^+ region located under the trench-gate electrode. The enhanced electric field at this location can be clearly observed in a three dimensional view of the electric field provided in Fig. 13.40 at a drain bias of 3000 volts. At the same time, it can be seen that the electric field in the gate oxide is below 1×10^6 V/cm due to the shielding provided by the P^+ region. In addition, the electric field in the vicinity of the P-base region is also small suppressing the reach-through phenomenon. The electric field profiles taken at $x = 0$ and $x = 0.5$ microns at various drain bias voltages provide further insight into the shielding by the P^+ region. In the electric field profiles shown in Fig. 13.41 at $x = 0$ microns, it can be seen that the electric field at the P-base region is suppressed to below 7×10^5 V/cm at a drain bias of 3000 volts. This prevents reach-through in the P-base region despite its narrow width of 0.4 microns.

Drain Bias = 3000 Volts

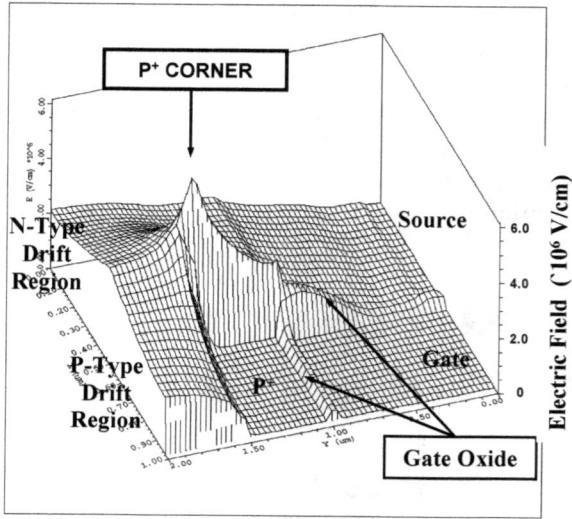

Fig. 13.40 Electric Field Distribution in the Shielded Charge Coupled
MOSFET Structure.

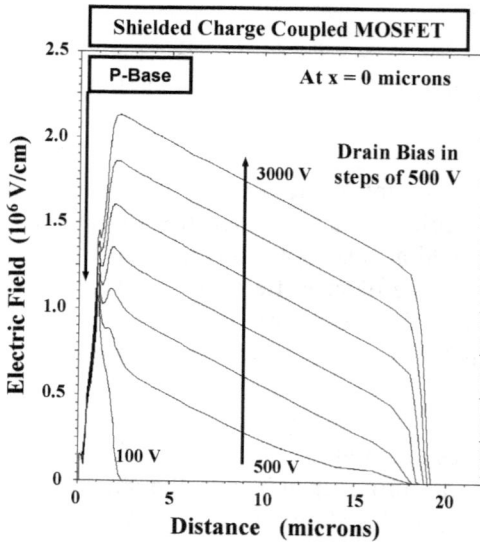

Fig. 13.41 Electric Field in the Shielded Charge Coupled MOSFET
Structure.

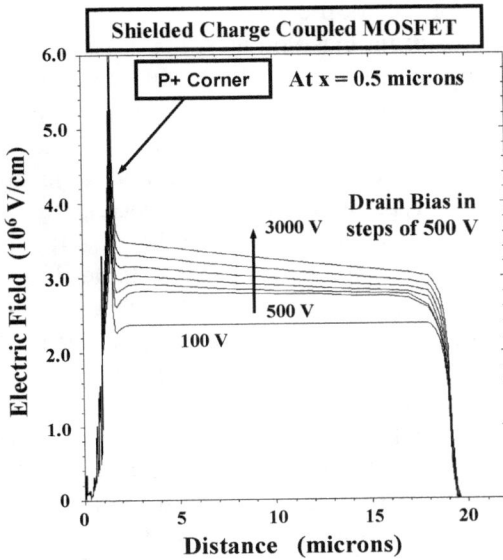

Fig. 13.42 Electric Field in the Shielded Charge Coupled MOSFET Structure.

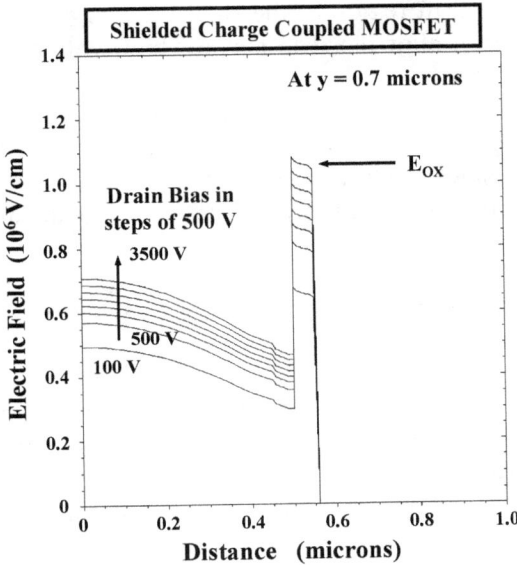

Fig. 13.43 Electric Field in the Shielded Charge Coupled MOSFET Structure.

The electric field profile taken at x = 0.5 microns is shown in Fig. 13.42 for various drain bias voltages. It can be seen that a very high electric field develops at the edge of the P$^+$ shielding region. Fortunately, the width of this spike in the electric field is very narrow limiting the impact ionization. This spike in the electric field can be reduced by rounding of the corner of the P$^+$ region and adjusting the doping concentration of the N-type region near the corner.

In the shielded charge coupled MOSFET structure, the electric field in the gate oxide is suppressed even at large drain bias voltages. This can be observed clearly in Fig. 13.43 where the electric field is shown at a depth of 0.7 microns from the top surface. This location is just below the P-base region where the gate oxide could be vulnerable to exposure to the high electric fields developed in the silicon carbide drift region. It can be seen that the electric field in the gate oxide is below 1 x 10^6 V/cm even when the drain bias is at 3000 volts. This ensures that there are no reliability or rupture problems when operating the shielded charge coupled MOSFET structure. It can also be seen in this figure (on the left hand side) that the electric field under the P-base region is very low enabling suppression of the reach-through phenomenon.

The above simulation results demonstrate that the electric field distribution in the drift region of the shielded charge coupled MOSFET structure is uniform allowing it to support high voltages. The blocking voltage for the simulated structure was found to be about 4000 volts confirming that reach-through has been suppressed in the P-base region. Since the electric field in the gate oxide has been shown above to be very low due to the shielding provided by the P$^+$ region, it is possible to operate the structure to the full breakdown voltage capability of the drift region.

13.6.2 Shielded Charge Coupled MOSFET Structure: On-State Characteristics

The on-state characteristic of the shielded charge coupled MOSFET structure was obtained by application of a gate bias above the threshold voltage. The current distribution within the upper portion of the structure is shown in Fig. 13.44 at a drain bias of 1 volt and a gate bias of 10 volts. It can be seen that the current flows through the inversion channel created by the gate bias at the surface of the P-base region and then spreads into the N- region via the accumulation layer formed at the trench sidewalls. It is distributed uniformly through the JFET region formed adjacent to the P^+ shielding region and then flows through the N-type drift region. The current is constricted by the depletion layer formed in the JFET region as well as by the P-N junction between the P and N-type drift regions. This must be accounted for in the model for the specific on-resistance.

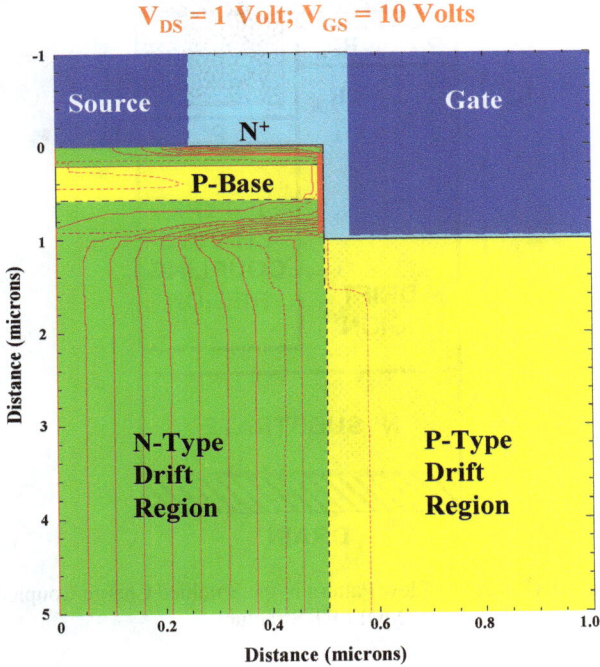

Fig. 13.44 On-State Current Distribution in the Shielded Charge Coupled MOSFET Structure.

The on-resistance of the shielded charge coupled MOSFET structure can be modeled as the resistance of the N-type drift region with the additional resistance arising in the current path due to the channel and JFET regions. The current flow pattern is indicated by the shaded area in Fig. 13.45. The total on-resistance for the shielded charge coupled MOSFET structure is given:

$$R_{on,sp} = R_{CH} + R_{N-} + R_{JFET} + R_D + R_{subs} \qquad [13.13]$$

where R_{CH} is the channel resistance, R_{N-} is the resistance of the current spreading through the N- drift region, R_{JFET} is the resistance of the JFET region, R_D is the resistance of the drift region after taking into account the depletion region from the presence of the vertical junction, and R_{subs} is the resistance of the N^+ substrate.

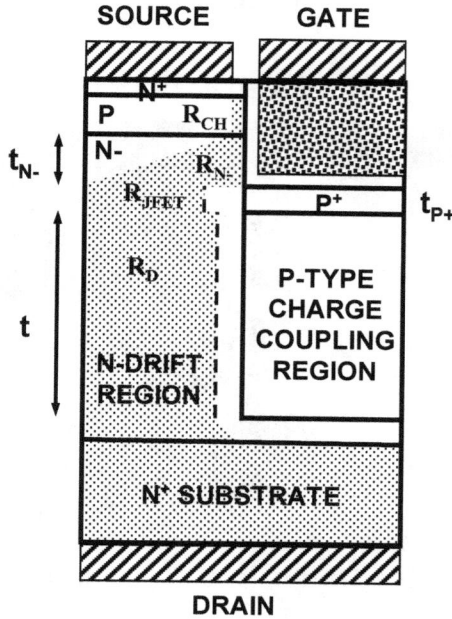

Fig. 13.45 Current Flow Pattern in the Shielded Charge Coupled
MOSFET Structure.

The specific channel resistance is given by:

$$R_{CH} = \frac{L_{CH} \cdot p}{\mu_{inv} \cdot C_{OX} \cdot (V_G - V_T)}$$ [13.14]

where L_{CH} is the channel length determined by the width of the P-base region, μ_{inv} is the mobility for electrons in the inversion layer channel, C_{ox} is the specific capacitance of the gate oxide, V_G is the applied gate bias, and V_T is the threshold voltage. The specific capacitance can be obtained using:

$$C_{ox} = \frac{\varepsilon_{ox}}{t_{ox}}$$ [13.15]

where ε_{ox} is the dielectric constant for the gate oxide and t_{ox} is its thickness.

The specific on-resistance of the N- region is given by:

$$R_{N-} = \rho_{N-} \cdot p \cdot \ln\left(\frac{W_N}{t_{CH}}\right)$$ [13.16]

where ρ_{N-} is the resistivity of the N- region and t_{CH} is the thickness of the channel from which the current spreads to a width (W_N) of the mesa region. A value of 3×10^{-7} cm for the parameter t_{CH} was found to provide good agreement with the simulations.

The specific resistance of the JFET region can be calculated using:

$$R_{JFET} = \rho_N \cdot p \cdot \left(\frac{t_{P+}}{W_N - W_0(P^+)}\right)$$ [13.17]

where ρ_N is the resistivity of the N-type drift region, t_{P+} is the junction depth of the P$^+$ shielding region, and $W_0(P^+)$ is the zero-bias depletion width at the P$^+$/N drift junction.

The drift region resistance can be obtained by using:

$$R_{JFET} = \rho_N \cdot p \cdot \left(\frac{t}{W_N - W_0(N)}\right)$$ [13.18]

where t is the thickness of the drift region below the P$^+$ shielding region and W_N is the width of the N-type drift region. Here, $W_0(N)$ is the zero-bias depletion width in the N-type drift region, which is given by:

$$W_0(N) = \sqrt{\frac{\varepsilon_S \cdot V_{bi}}{q \cdot N_D}}$$ [13.19]

where N_D is the doping concentration of the N-type drift region. Note that this equation takes into account the fact that only half the built-in voltage is supported on the N-type side of the drift region because of equal concentrations on both sides of the junction.

The contribution to the resistance from the N$^+$ substrate is given by:

$$R_{subs} = \rho_{subs} \cdot t_{subs}$$ [13.20]

where ρ_{subs} and t_{subs} are the resistivity and thickness of the substrate, respectively. A typical value for this contribution is 4×10^{-4} Ω-cm^2.

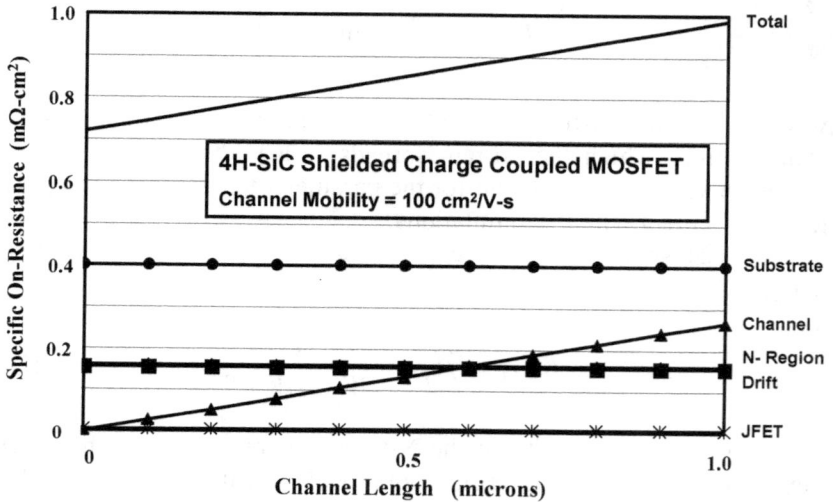

Fig. 13.46 Analytically Calculated Specific On-Resistance of the Shielded Charge Coupled MOSFET Structure.

The specific on-resistance calculated by using the above model is shown in Fig. 13.46 for the case of a cell pitch of 1 micron when an inversion layer mobility of 100 cm^2/V-s is used. A doping concentration of 3×10^{17} cm^{-3} was used for both the N-type and P-type drift regions. The N- region had a doping concentration of 2×10^{16} cm^{-3}. The P$^+$ shielding region was assumed to have a thickness of 0.5 microns. The analytically calculated zero-bias depletion widths were 0.11 and 0.077 microns for the JFET and N-drift regions, respectively. Only the channel resistance contribution increases with the channel length because the cell pitch remains unchanged. The model predicts a specific on-resistance of 0.83 mΩ-cm^2 for a channel length of 0.4 microns. From Fig. 13.46, it can be seen that the N$^+$ substrate contribution is dominant for the charge coupled JFET structure. This is due to the extremely low specific on-resistance of the charge coupled drift region. Reduction of the contribution from the substrate can be achieved by increasing its doping concentration and aggressively reducing its thickness. The available manufacturing technology for reducing the thickness of silicon devices to less than 50 microns must be extended to silicon carbide devices.

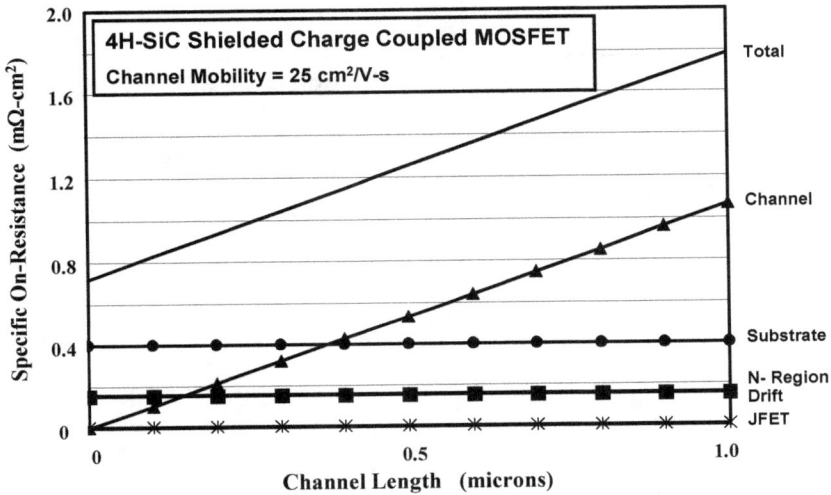

Fig. 13.47 Analytically Calculated Specific On-Resistance of the Shielded Charge Coupled MOSFET Structure.

The above results are based upon using a channel mobility of 100 cm^2/Vs. The impact of reducing the inversion layer mobility for the shielded charge coupled MOSFET structure is depicted in Fig. 13.47 for the case of mobility values of 25 cm^2/Vs. With an inversion layer mobility of 25 cm^2/Vs, the channel contribution becomes comparable to the contribution from the N^+ substrate. This results in the specific on-resistance increasing to a value of 1.15 $m\Omega$-cm^2 for a channel length of 0.4 microns.

The simulations of the specific on-resistance for the shielded charge coupled 4H-SiC MOSFET structure were performed with different values for the channel inversion layer mobility. The specific on-resistance was obtained by using a drain bias of 1 volt and sweeping the gate bias to 20 volts to obtain the transfer characteristic. A drain bias of 1 volt was used for these simulations because the on-state voltage drop for these structures is typically in this range. The transfer characteristics obtained by using this method are shown in Fig. 13.48 for the device with cell pitch of 1 micron. From the transfer characteristics, it can be seen that the threshold voltage that determines the on-resistance is approximately 4.5 volts (allowing operation of this device with a gate bias of 10 volts). The impact of changing the channel inversion layer mobility is also shown in this figure. As expected, there is a reduction in the drain current as the channel mobility is reduced.

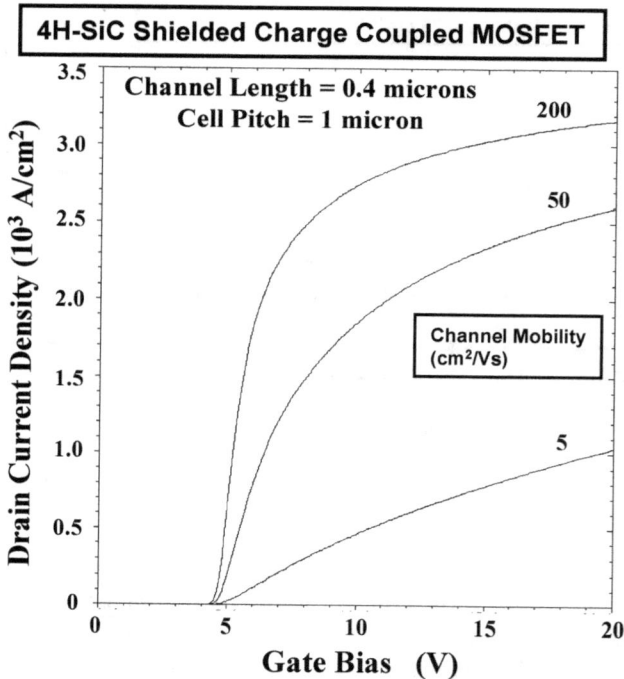

Fig. 13.48 Transfer Characteristics of the 4H-SiC Shielded Charge Coupled MOSFET Structure.

The specific on-resistances obtained from the on-state simulations are compared in Fig. 13.49 with the analytically calculated values obtained by using the model developed in the previous section. For the simulation cases, the channel mobility values were extracted from the *i-v* characteristics obtained for the lateral 4H-SiC MOSFET structure that was previously described in chapter 9 by using the same interface degradation factor as used during the simulation of the vertical 4H-SiC shielded charge coupled MOSFET structure. It can be seen that the predictions of the analytical model are in excellent agreement with the values obtained from the simulations. These results provide confidence in the ability to use the analytical model during the design of the shielded charge coupled 4H-SiC power MOSFET structure. From the figure, it can be concluded that a channel mobility of more than 50 cm^2/Vs is satisfactory for achieving an extremely low specific on-resistance for a 4 kV shielded charge coupled 4H-SiC power MOSFET structure with cell pitch of 1 micron. Further, the specific on-resistance shown in Fig. 13.49 can be reduced in half with significant improvement in the substrate contribution by reducing the wafer thickness and resistivity.

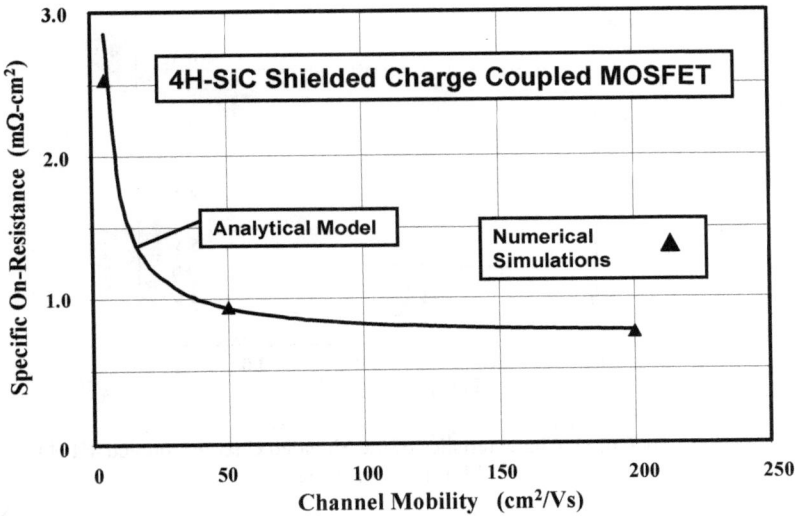

Fig. 13.49 Specific On-Resistance for the 4H-SiC Shielded Charge Coupled MOSFETs.

13.6.3 Shielded Charge Coupled MOSFET Structure: Output Characteristics

The output characteristics of the shielded charge coupled MOSFET structure were obtained using two dimensional numerical simulations by sweeping the drain bias while using various gate bias voltages. For the structure with a channel length of 0.4 microns, the output characteristics exhibit excellent current saturation as shown in Fig. 13.50. The device structure has a very wide safe-operating-area free of any short channel or base reach-through related limitations. Consequently, the loci of the drain current-voltage during switching will be constrained only by thermal considerations.

Channel Length = 0.4 microns; Cell Pitch = 1.0 microns

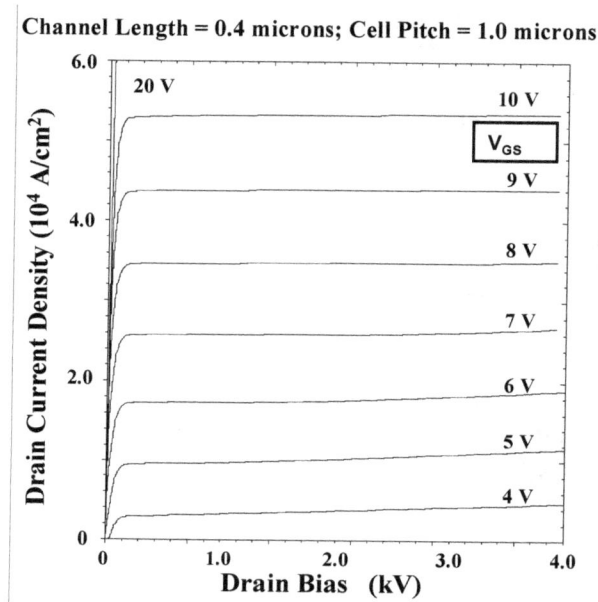

Fig. 13.50 Output Characteristics of the Shielded Charge Coupled 4H-SiC MOSFET Structure.

13.6.4 Shielded Charge Coupled MOSFET Structure: Switching Characteristics

The switching behavior of silicon power MOSFETs is often evaluated by the application of a constant current to the gate electrode to switch the device from the blocking state to the on-state[14]. The gate voltage of the 4H-SiC shielded charge coupled MOSFET structure increases in response to this stimulus as shown in Fig. 13.51. When the gate voltage exceeds the threshold voltage at time t_1, the drain current begins to rise. The simulations were designed to maintain the drain current density at a constant value once it reached a magnitude of 200 A/cm^2. At this time t_2, the drain voltage falls rapidly until time t_3 resulting in a high dV/dt at the output terminals of the FETs. The plateau in the gate voltage until time t_4

is associated with the charging of the Miller capacitance of the MOSFET.

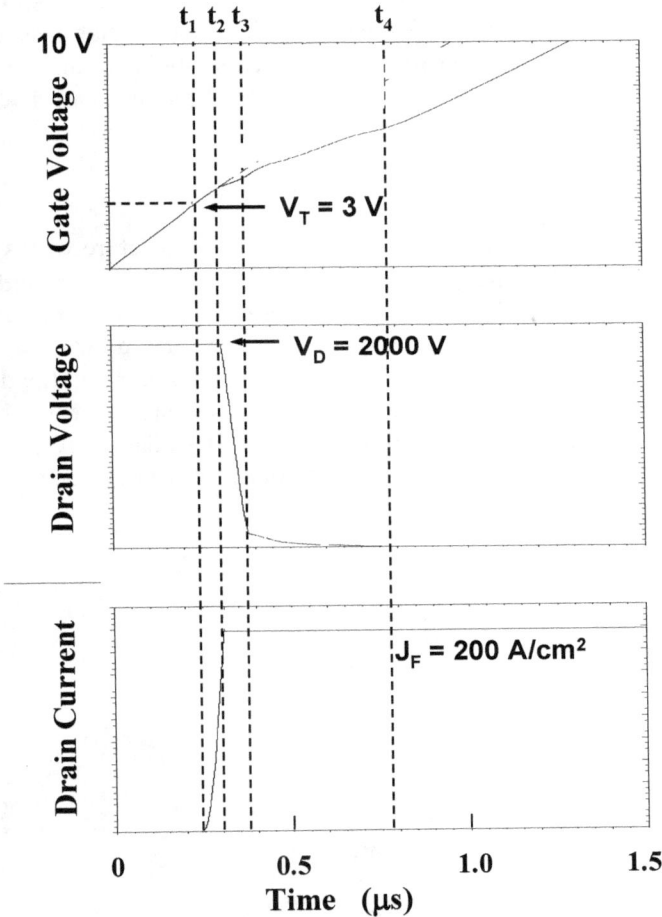

Fig. 13.51 Switching Performance of the 4H-SiC Charge Coupled
MOSFET Structure.

During the switching event for turning on the 4H-SiC shielded charge coupled MOSFET structure, high power dissipation occurs during the transition time ($t_2 - t_1$) for the drain current and the transition time ($t_3 - t_2$) for the drain voltage. The switching energy associated with these events is given by:

$$E_{off} = 0.5 * V_D * I_D * (t_3 - t_1) \qquad\qquad [13.21]$$

For the simulated case with V_D = 2000 volts and J_D = 200 A/cm^2, the calculated switching energy is 30 mJ/cm^2 - the same as that observed for the Baliga-Pair configuration in chapter 8. This would produce a power dissipation of 30 W/cm^2 at a typical operating frequency of 1 kHz. In comparison, the on-state power dissipation is given by:

$$P_{on} = R_{DSonsp} * J_D^2 \qquad\qquad [13.22]$$

For the 4H-SiC shielded charge coupled MOSFET structure with specific on-resistance of 0.8 mΩ-cm^2, the conduction power loss is found to be 30 W/cm^2. If the turn-off switching losses are assumed to be comparable to the turn-on power loss computed above, the total power loss is 90 W/cm^2, which is acceptable from the point of view of cooling the devices with available heat sink technology. This demonstrates that the 4H-SiC shielded charge coupled MOSFET structure is capable of providing excellent switching performance for high voltage power systems.

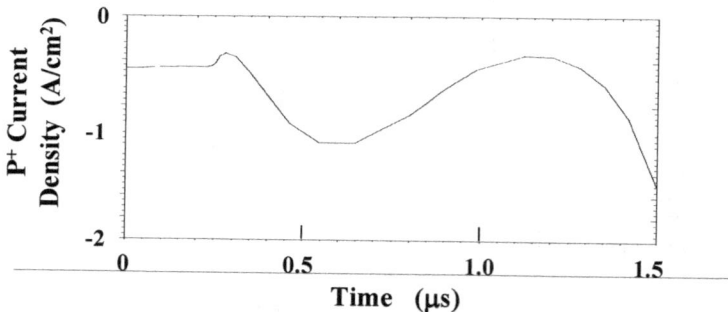

Fig. 13.52 Current Flow through the P$^+$ Shielding Region in the Charge Coupled 4H-SiC MOSFET Structure.

One of the issues that must be considered when designing the shielded charge coupled MOSFET structure is the impact of the transient current through the P$^+$ shielding region. Since this region is located below the trench gate structure, its contact must be provided at discrete locations along the cell orthogonal to the cross-section. Transient current flow through the relatively high sheet resistance of the P-type regions in silicon carbide could produce a change in the potential of the P$^+$ shielding region which could prolong the switching times. Fortunately, the current through the P$^+$ shielding region was found to be quite small, as shown in Fig. 13.52, through the entire switching event. This provides latitude when designing the contacts to the P$^+$ shielding region.

13.7 Summary

In this chapter, the charge coupling concept has been applied to silicon carbide devices by forming the P-type drift region in a trench located adjacent to the N-type drift region. It has been demonstrated that this device architecture allows supporting high voltages with very high doping concentration in the N-type drift region. As an example, devices that could support 4000 volts were achieved with N-type drift region doping concentration of 3 x 10^{17} cm^{-3} by using a trench depth of 18 microns. This design of the charge coupled drift region was found to be applicable for making high performance Schottky rectifiers, JFETs and MOSFETs. A power MOSFET structure with a shielding region incorporated within it to prevent high electric field in the gate oxide has been demonstrated to exhibit an extremely low specific on-resistance. However, two-dimensional numerical simulations indicate that the breakdown voltage for the charge coupled structures is quite sensitive to the charge balance as has been reported for silicon devices as well. This may impose a significant hurdle for commercialization of these charge coupled structures. The imposition of the same manufacturing tolerances used for commercially available silicon COOLMOS power MOSFET structures[4] will have to be incorporated into the silicon carbide arena.

References

[1]   B. J. Baliga, "The Evolution of MOS-Bipolar Power Semiconductor Technology", Proceeding of the IEEE, Vol. 74, pp. 409-418, 1988.

[2]   T. Fujihira and Y. Miyasaka, "Simulated Superior Performances of Semiconductor Super-junction Devices", IEEE International Symposium on Power Semiconductor Devices and ICs, Abstr. 12.4, pp. 423-426, 1998. (See cited patents X. Chen, U.S. Patent 5,216,275, 1993; J. Tihanyi, U.S. Patent 5,438,215, 1995)

[3]   B. J. Baliga, "Trends in Power Discrete Devices", IEEE International Symposium on Power Semiconductor Devices and ICs, Abstr. P-2, pp. 5-10, 1998. (See cited patents: M. Mehrotra and B.J. Baliga, U.S. Patent 5,365,102, 1994; B. J. Baliga, U.S. Patent 5,612,567, 1997; B.J. Baliga, U.S. Patent 5,637,898, 1997).

[4]   L. Lorenz, et al, "COOLMOS – A New Milestone in High Voltage Power MOS", IEEE International Symposium on Power Semiconductor Devices and ICs, Abstr. 1.1, pp. 3-10, 1999.

[5]   B. J. Baliga, "Silicon Carbide Power Devices having Trench-Based Silicon Carbide Charge Coupling Regions Therein", U. S. Patent 6,313,482, Issued November 6, 2001.

[6]   B. J. Baliga, "A Power Junction Gate Field Effect Transistor Structure with High Blocking Gain", IEEE Transactions on Electron Devices, Vol. 27, pp. 368-373, 1980.

[7]   B. J. Baliga, "Silicon Liquid-Phase Epitaxy", Chapter 3 in 'Epitaxial Silicon Technology', Academic Press, 1986.

[8]   K. Hara, "Vital Issues for SiC Power Devices", Silicon Carbide and Related Materials – 1997, Material Science Forum, Vol. 264-268, pp. 901-906, 1998.

[9]   Y. Li, J. A. Cooper, and M. A. Capano, "High Voltage (3kV) UMOSFETs in 4H-SiC", IEEE Transactions on Electron Devices, Vol. 49, pp. 972-975, 2002.

[10] S. Mukherjee, et al, "The Effects of SIPOS Passivation on DC and Switching Performance of High Voltage MOS Transistors", IEEE International Electron Device Meeting, pp. 646-649, 1986.

[11] V. Lauer, et al, "Spectroscopic Investigation of Vanadium Acceptor Level in 4H and 6H-SiC", Silicon Carbide and Related Materials – 1999, Material Science Forum, Vol. 338-342, pp. 635-638, 1999.

[12] R. van Dalen, C. Rochefort, and G. A. M. Hurkx, "Breaking the Silicon Limit using Semi-Insulating RESURF Layers", IEEE

Charge Coupled Structures

International Symposium on Power Semiconductor Devices and ICs, Abstract 10.1, pp. 391-394, 2001.

P. M. Shenoy, A. Bhalla, and G. M. Dolny, "Analysis of the Effect of Charge Imbalance on the Static and Dynamic Characteristics of the Super-Junction MOSFET", IEEE International Symposium on Power Semiconductor Devices and ICs, Abstract 5.1, pp. 99-102, 1999.

B. J. Baliga, "Power Semiconductor Devices, Chapter 7, pp. 387-391, PWS Publishing Company, 1996.

Chapter 14

Integral Diodes

One high volume commercial application for power devices is for the control of electric motors. For motor control, an important trend that has produced huge savings in energy loss is the introduction of adjustable speed or variable frequency drives[1]. In this approach, the speed of motors is controlled by varying the frequency of the voltage and current supplied to the motor windings. This requires the ability to generate a sinusoidal power source whose frequency can be adjusted by a control algorithm. The variable frequency sinusoidal power source is generated from a DC power source by using pulse-width-modulation (PWM) techniques with the signal applied to power devices in a totem-pole configuration, as illustrated in Fig. 14.1.

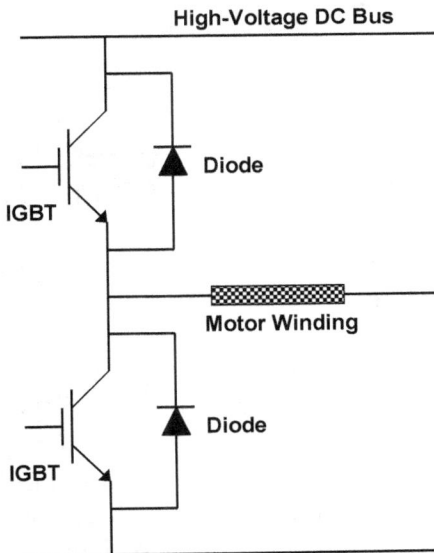

Fig. 14.1 The Basic Totem-Pole Configuration for Motor Control.

The basic totem-pole configuration illustrated in Fig. 14.1 uses IGBTs as the switches and P-i-N rectifiers as the fly-back diodes. The variable frequency sinusoidal waveform that is supplied to the motor windings can be synthesized by turning on the appropriate IGBTs in an H-bridge circuit. The fly-back diodes are required to allow current to flow through each branch of the totem-pole circuit in the reverse direction to the conduction state for IGBTs. In the future, the IGBTs could be replaced by silicon carbide MOSFETs while the silicon P-i-N rectifiers used as fly-back diodes could be replaced by silicon carbide Schottky rectifiers. However, it would be desirable to forgo the need for the fly-back diode to save cost if the body diode within the silicon carbide power MOSFET can be utilized to carry the reverse current.

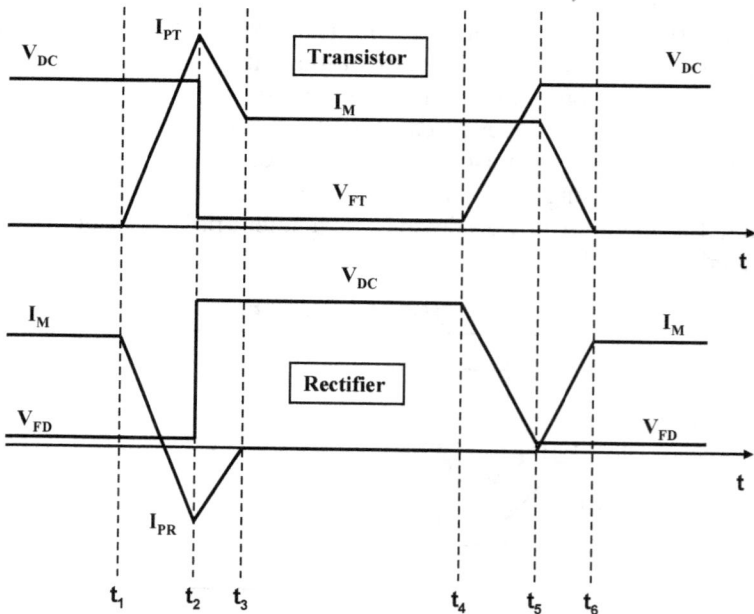

Fig. 14.2 Linearized Current and Voltage Waveforms for the Transistor and Diode in the Totem-Pole Configuration.

Typical waveforms for the current and voltage in the totem-pole configuration are shown in Fig. 14.2 for both the transistor and the fly-back rectifier. During each cycle of the PWM signal, the motor current can be assumed to remain approximately constant if the frequency of the

PWM operation is much larger than the frequency of the sinusoid that is being synthesized. The motor current is then flowing through either the transistor or the rectifier (in another branch) with transitions made between them as indicated in the figure. In the case of high voltage silicon P-i-N rectifiers, a large reverse-recovery current (I_{PR}) occurs due to the stored charge within the structure. This produces significant power loss at each switching event not only in the rectifier but also in the transistor because the peak current (I_{PT}) in the transistor is the sum of the motor current and the reverse recovery current. The switching losses become dominant in the case of silicon devices as the operating frequency exceeds 1 kHz[2]. However, if the silicon P-i-N rectifier is replaced with a silicon carbide rectifier with small reverse recovery current, the on-state conduction power loss becomes more significant. It is then more important to reduce the on-state voltage drop of the rectifier, as well as that of the power transistor.

If the body-diode within the silicon carbide power MOSFET structures can be utilized as the fly-back rectifier, it would reduce the number of components and decrease the cost of the system. An obvious detriment to this approach is the high on-state voltage drop of the P-N junction in silicon carbide. In this chapter, it is demonstrated that current conduction in the third quadrant can be obtained through the power MOSFET structure in a unipolar mode with lower on-state voltage drop than that for the silicon carbide P-i-N rectifier even without the application of a gate bias voltage. Even lower on-state voltage drop can of course be achieved by turning on the channel in the MOSFET with a positive gate bias. However, this requires synchronization of the gate signal to the reverse current conduction period.

14.1 Trench-Gate MOSFET Structure

Two dimensional numerical simulations of the trench-gate MOSFET structure were performed with negative bias applied to the drain electrode to obtain the characteristics for the structure in the third quadrant of operation. The device structure is the same as the one described in chapter 11 (see Fig. 11.7 for the device cross-section). The simulated structure had a drift region with doping concentration of 1×10^{16} cm^{-3} and thickness of 20 microns. The P$^+$ and P-base regions had a depth of 1 micron. The mesa and trench widths were 0.5 microns, leading to a cell pitch of 1 micron.

The *i-v* characteristic for the structure in the third quadrant of operation is shown in Fig. 14.3. At first glance, it is similar in appearance to that of the 4H-SiC P-i-N rectifiers previously discussed in chapter 4. However, at a current density of 100 A/cm^2, the on-state voltage drop is found to be 2.9 volts, which is slightly smaller than that observed for the 4H-SiC P-i-N rectifier with the same blocking voltage capability. The reason for this was found to be associated with a unique mode of operation within the structure that has not been observed and reported for silicon devices.

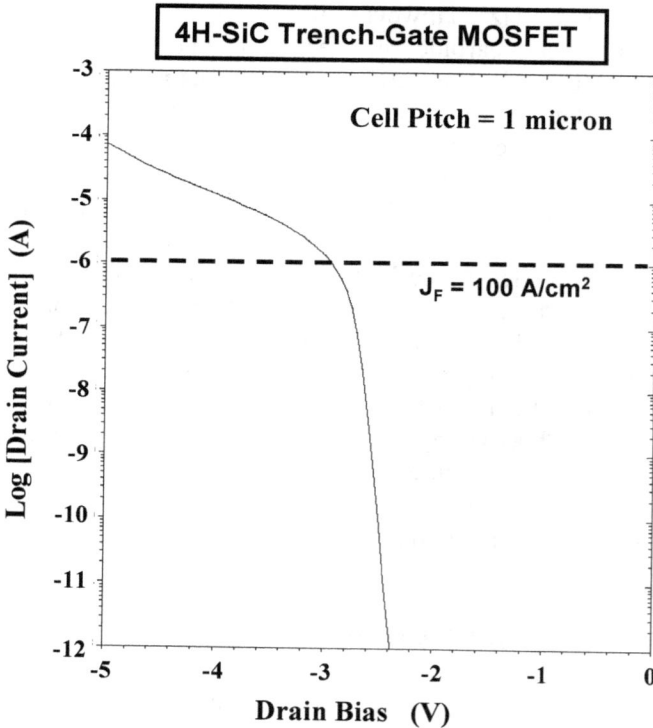

Fig. 14.3 Characteristics of the Body-Diode in the 4H-SiC Trench-Gate MOSFET Structure.

The current flow within the trench-gate MOSFET structure obtained with the two dimensional numerical simulations is shown in Fig. 14.4 at drain bias of -3 and -5 volts, with the gate held at zero volts. At a drain bias of -3 volts, it can be observed that the current does not flow across the P-N junction formed by the P-base/N-drift region. Instead, the current flows along the sidewall of the trench gate region into the N+ source region in a mode of operation not previously reported for silicon devices. This pattern for the current flow is similar to that observed during on-state current flow in the first quadrant when a positive gate bias above the threshold voltage is applied to the MOSFET.

When the drain bias is increased to -5 volts, the current flow occurs across the P-N junction within the structure as observed in silicon devices. In this case, the current flows into the contact to the P-base region located on the left hand side of the cross-section and also into the N+ source region after transition through the P-base region. This current pattern has been reported previously in silicon MOSFETs as well[3].

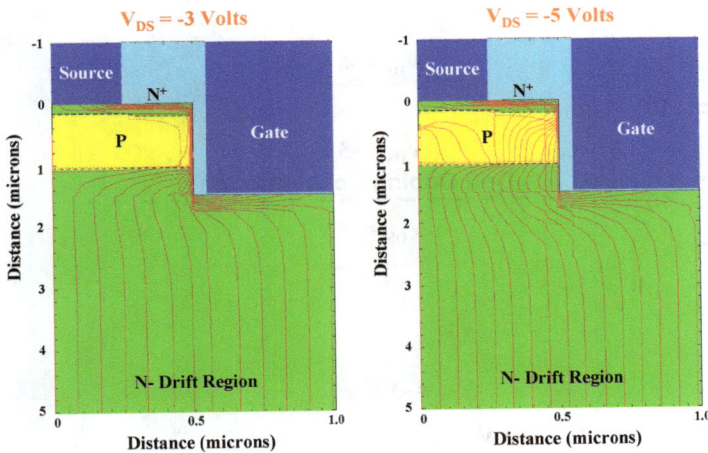

Fig. 14.4 Current Distribution in the Trench-Gate 4H-SiC MOSFET Structure.

The activation of the P-N junctions within the MOSFET structure is known to produce stored minority carrier charge that must be removed during the reverse recovery transient. The injected hole concentration within the N-type drift region is an indication of this stored charge. The hole concentration in the drift region can be analyzed by examining the profile taken at x = 0.1 microns, as shown in Fig. 14.5. It can be seen that high-level minority carrier injection - when the injected hole concentration exceeds the background doping concentration - occurs only when the drain bias exceeds -3 volts. The low concentration of injected holes at a drain bias of -3 volts indicates that the reverse recovery for the body-diode of the 4H-SiC charge coupled MOSFET structure will occur with small recovery charge leading to low switching losses. It is therefore feasible to use the body-diode in the 4H-SiC charge coupled MOSFET structure as the fly-back rectifier in H-bridge circuits if the losses associated with a relatively high on-state voltage drop are tolerable.

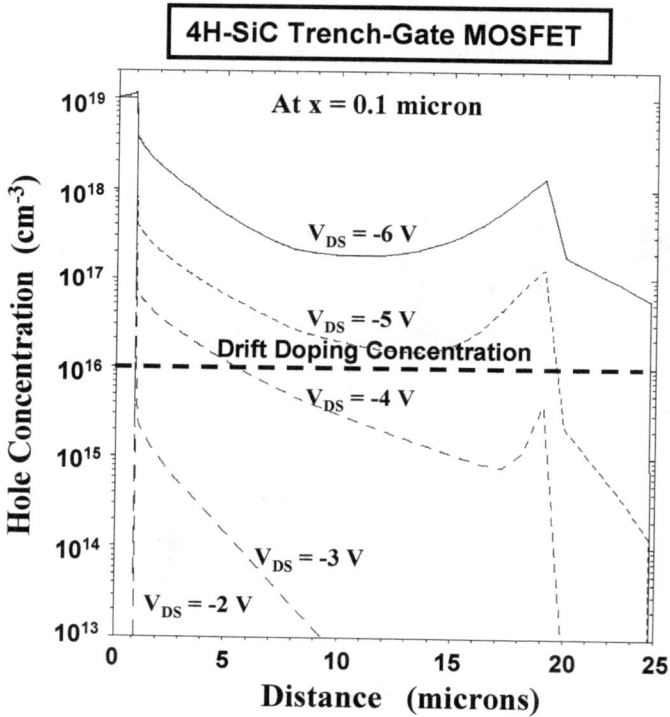

4H-SiC Trench-Gate MOSFET

At x = 0.1 micron

$V_{DS} = -6$ V

$V_{DS} = -5$ V

Drift Doping Concentration

$V_{DS} = -4$ V

$V_{DS} = -3$ V

$V_{DS} = -2$ V

Hole Concentration (cm^{-3})

Distance (microns)

Fig. 14.5 Hole Distribution in the Trench-Gate 4H-SiC MOSFET Structure.

In order to understand the new mode of current conduction within the silicon carbide trench-gate MOSFET in the third quadrant of operation, the simulations were repeated with different doping concentrations for the P-base region. The highly doped P$^+$ region was retained in all the structures with a doping concentration of 1×10^{19} cm^{-3}. The resulting i-v characteristics in the third quadrant are shown in Fig. 14.6. It can be seen that the i-v characteristic shifts to the right hand side when the P-base doping concentration is reduced from 1 x 10^{17} to 1 x 10^{16} cm^{-3} and to the left hand side when the P-base doping concentration is increased to 1 x 10^{18} cm^{-3}. It is worth pointing out that the on-state voltage drop for the structure in the third quadrant can be reduced to 2.3 volts with the smaller P-base doping concentration. Although this may result in reach-through breakdown problems for the conventional trench-gate structure, this is a feasible approach for the shielded trench-gate structure (discussed in chapter 12) as shown in the next section of this chapter.

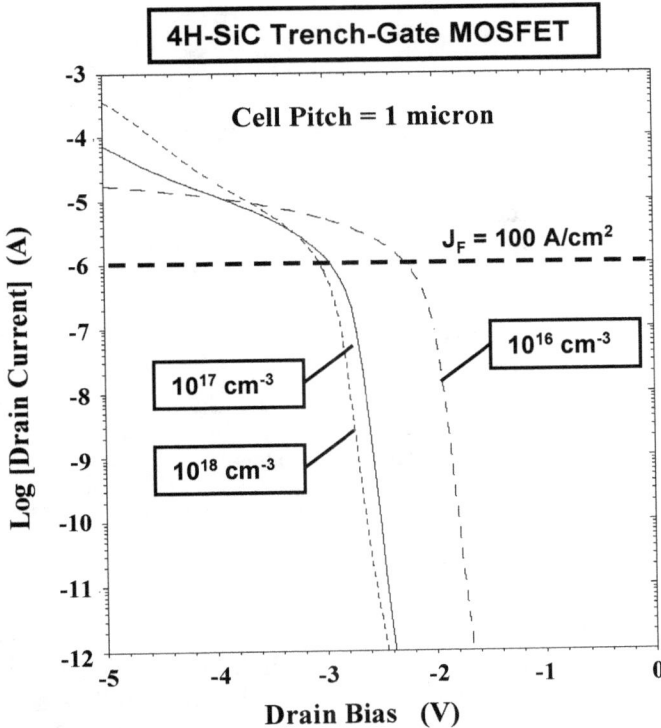

Fig. 14.6 Characteristics of the Body-Diode in the 4H-SiC Trench-Gate MOSFET Structures.

A further understanding of the new mode of current conduction in the third quadrant for the trench-gate 4H-SiC MOSFET structure can be obtained by examination of the potential barrier for transport of electrons through the P-base region. The potential distribution for various negative drain bias voltages is shown in Fig. 14.7 at x = 0 microns (through the P$^+$ region), for the structure with P-base doping concentration of 1 x 10^{17} cm^{-3}. A potential barrier of 3 volts for the transport of electrons exists at zero drain bias. The potential barrier becomes smaller as the negative drain bias is increased but can be observed at all negative drain bias voltages at the P$^+$ region.

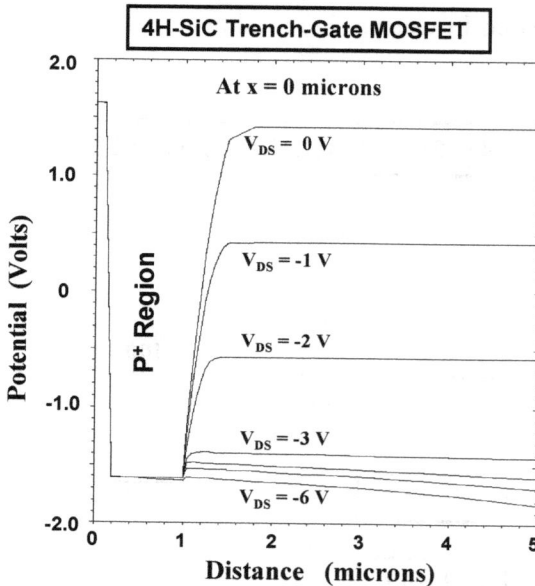

Fig. 14.7 Potential Distribution in the 4H-SiC Trench-Gate MOSFET Structure.

The potential barrier at the P-base region (x = 0.5 microns) is shown in Fig. 14.8 at various negative drain bias voltages for the structure with P-base doping concentration of 1 x 10^{17} cm^{-3}. A potential barrier of 2.6 volts for the transport of electrons exists at zero drain bias. This barrier is smaller due to the movement of the Fermi level closer to the mid-gap for the more lightly doped P-base region when compared with the P$^+$ region. The potential barrier is completely removed when the drain bias is increased from -2 V to -3 V, which is consistent with the on-set of current flow at a drain bias of about -2.5 volts in Fig. 14.3. This provides an opportunity to reduce the on-state voltage drop during current flow in the third quadrant of operation for the MOSFET.

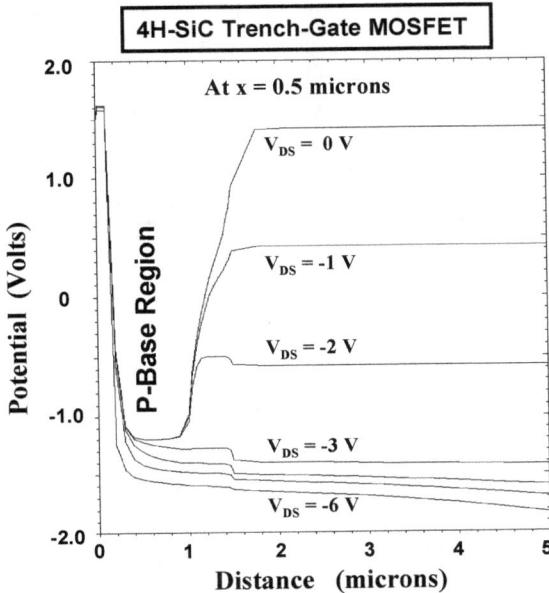

Fig. 14.8 Potential Distribution in the 4H-SiC Trench-Gate MOSFET Structure.

14.2 Shielded Trench-Gate MOSFET Structure

In the previous section, it was demonstrated that current flow in the third quadrant occurs via a new mode of operation through the P-base region adjacent to the trench-gate region. It was also shown that the on-state voltage drop during this current flow can be reduced by reducing the doping concentration of the P-base region. In the conventional trench-gate MOSFET structure, a reduction of the P-base doping concentration can lead to poor voltage blocking capability due to the reach-through phenomenon. However, the reach-through effect can be suppressed in the shielded trench gate MOSFET structure allowing taking advantage of the new mode.

Two dimensional numerical simulations of the shielded trench-gate MOSFET structure were performed with negative bias applied to the drain electrode to obtain the characteristics for this structure in the third quadrant of operation. The device structure is the same as the one described in chapter 12 (see Fig. 12.1 for the device cross-section). The simulated structure had a drift region with doping concentration of 1 x

10^{16} cm^{-3} and thickness of 20 microns. The P-base region had a depth of 0.5 microns with a N$^+$ source region depth of 0.1 microns, leading to a channel length of 0.4 microns. The mesa and trench widths were 1.25 microns and 0.55 microns, leading to a cell pitch of 1.8 microns.

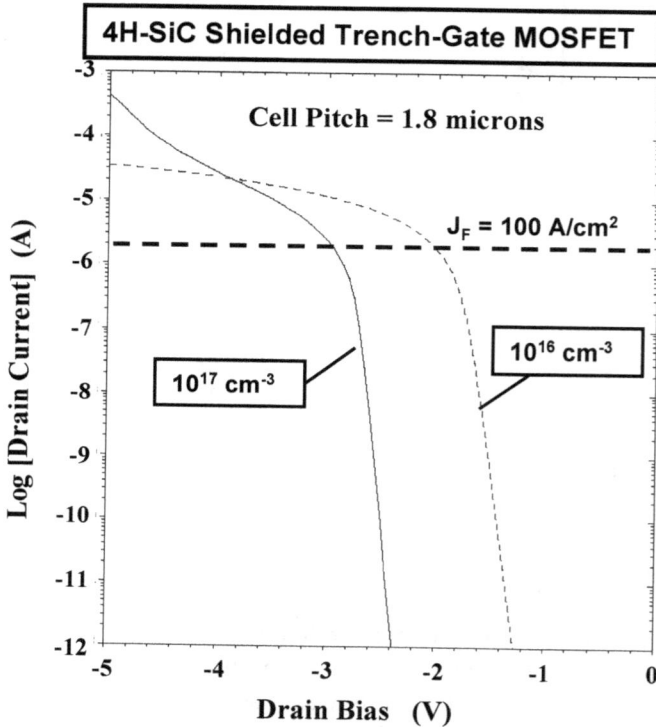

Fig. 14.9 Characteristics of the Body-Diode in the 4H-SiC Shielded Trench-Gate MOSFET Structures.

The *i-v* characteristics for the shielded trench-gate MOSFET structures in the third quadrant of operation are shown in Fig. 14.9 for two values of the P-base doping concentration. At a current density of 100 A/cm^2, the on-state voltage drop is found to be only 1.9 volts for the structure with P-base doping concentration of 1 x 10^{16} cm^{-3}. This is much smaller than that observed for the 4H-SiC P-i-N rectifier with the same blocking voltage capability. These results indicate that the new mode of current conduction in the third quadrant is relevant for the shielded trench-gate MOSFET structure.

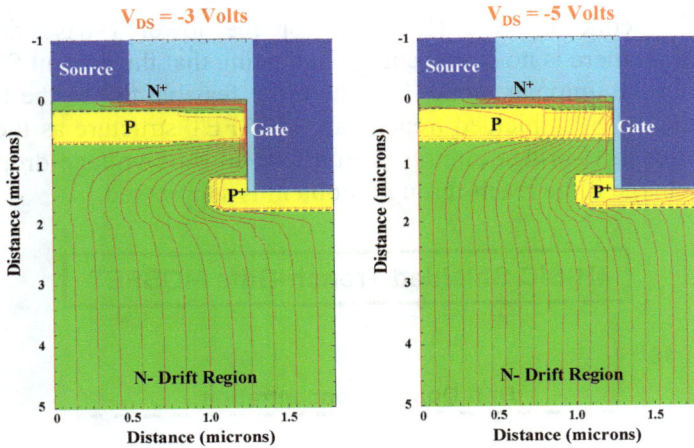

Fig. 14.10 Current Distribution in the Shielded Trench-Gate 4H-SiC MOSFET Structure.

In order to confirm the existence of the new mode of operation in the shielded trench-gate MOSFET structure, the current flow within the shielded trench-gate MOSFET structure was obtained with the two dimensional numerical simulations, as shown in Fig. 14.10, with the gate held at zero volts. At a drain bias of -3 volts, it can be observed that the current does not flow across the P-N junction formed by the P^+ shielding region or the P-base region. Instead, the current flows along the sidewall of the trench gate region into the N^+ source region as observed for the trench-gate structure. When the drain bias is increased to -5 volts, the current flow occurs across both the P-N junctions within the structure. In this case, the current flows into the contact to the P^+ shielding region located on the right hand side of the cross-section and also into the N^+ source region after transition through the P-base region.

The activation of the P-N junctions within the MOSFET structure is known to produce stored minority carrier charge that must be removed during the reverse recovery transient. The injected hole concentration within the N-type drift region is an indication of this stored charge. The hole concentration injected into the drift region in the shielded trench-gate MOSFET structure can be analyzed by examining the profile taken at x = 0.1 microns, as shown in Fig. 14.11. Here, the hole concentration is shown for the cases of P-base doping concentration of 1×10^{17} and 1×10^{16} cm^{-3} for the same on-state current density of 100 A/cm^2. Minority carrier injection is observed for the case of the larger P-

base doping concentration. However, for the lower P-base doping
concentration, there is no stored charge indicating that the current flow is
occurring in the unipolar mode. It is therefore feasible to use the body-
diode in the 4H-SiC shielded trench-gate MOSFET structure as the fly-
back rectifier in H-bridge circuits with a low on-state voltage drop and
no stored charge if the reach-through problem is suppressed.

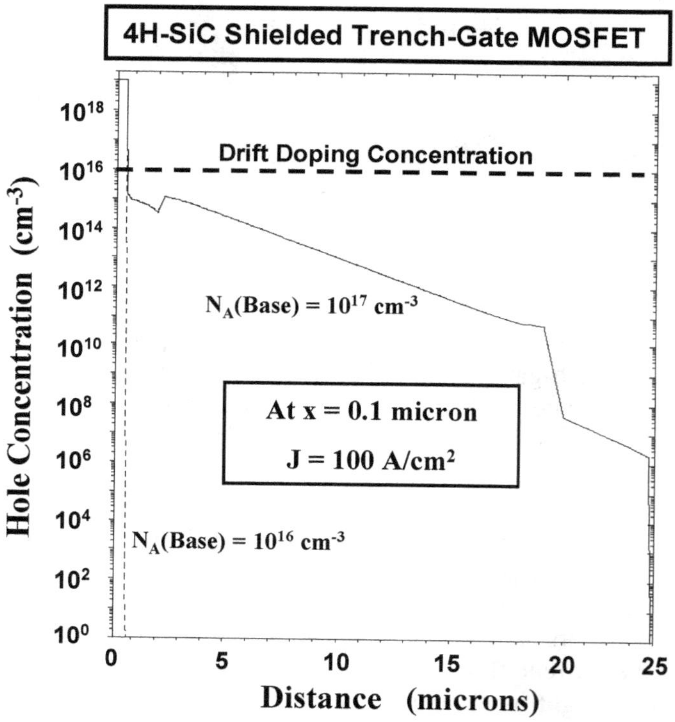

Fig. 14.11 Hole Distribution in the Shielded Trench-Gate 4H-SiC MOSFET
Structures.

In order to confirm good blocking voltage capability and safe
operating area for the shielded trench-gate MOSFET structure, two
dimensional numerical simulations were performed for the structure with
P-base doping concentration of 1×10^{16} cm^{-3} by sweeping the drain bias
while using various positive gate bias voltages. The resulting blocking
characteristics are shown in Fig. 14.12. In comparison with the
characteristics for the shielded trench-gate 4H-SiC MOSFET with P-base

doping concentration of 1×10^{17} cm^{-3} previously shown in chapter 12 (see Fig. 12.6), a slightly enhanced leakage current is observed at high drain bias voltages for the case of zero gate bias. However, the absolute magnitude of the leakage current density is less than 1 mA/cm^2 even when the drain bias approaches 3000 volts. This implies that the reach-through effect has been adequately suppressed allowing the use of a low P-base doping concentration. From this figure, it can also be concluded that the threshold voltage has been reduced from about 4 volts to 1 volt by the reduced P-base doping concentration.

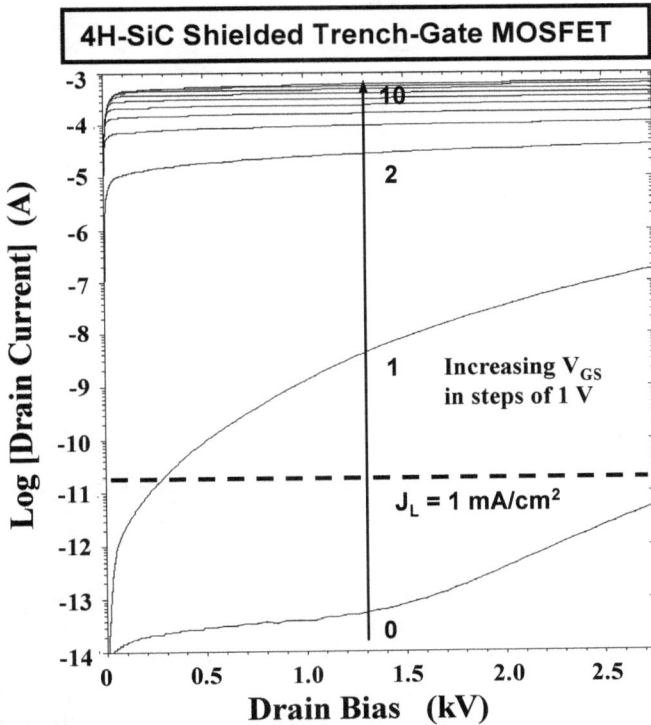

Fig. 14.12 Blocking Characteristics of the 4H-SiC Shielded Trench-Gate MOSFET Structure.

The suppression of the reach-through effect also enables this shielded trench-gate MOSFET structure to exhibit excellent drain current saturation up to large drain bias voltages when the gate bias is increased above the threshold voltage. The output characteristics of the shielded trench-gate MOSFET structure with P-base doping concentration of 1 x

10^{16} cm^{-3} are shown in Fig. 14.13. It can be seen that the device exhibits excellent gate controlled drain current saturation upto 3000 volts with high output resistance. The safe-operating-area for the structure is very broad enabling a wide range for the loci of the current-voltage during switching as long as the power dissipation is not excessive.

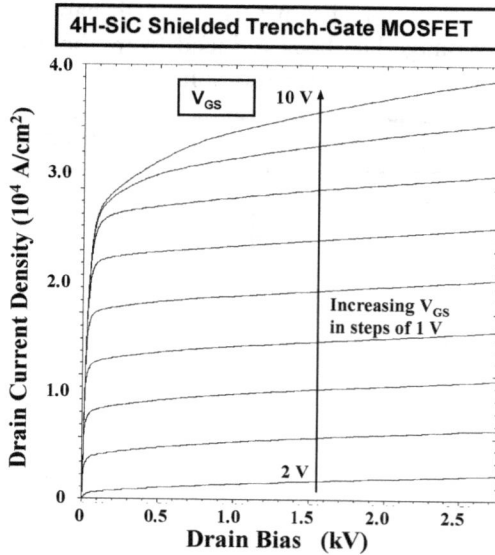

Fig. 14.13 Output Characteristics of the 4H-SiC Shielded Trench-Gate MOSFET Structure.

A smaller P-base doping concentration is known to produce a lower threshold voltage. The impact of this on the on-state resistance of the shielded trench-gate MOSFET structure was determined by performing two-dimensional numerical simulations of the structure with a drain bias of 1 volt and sweeping the gate bias up to 20 volts. The channel mobility was varied during these simulations by using the channel degradation coefficient. The results, shown in Fig. 14.14, clearly show the reduction of the threshold voltage due to the smaller P-base doping concentration. This is beneficial for reducing the channel resistance contribution and hence the specific on-resistance of the device structure. The specific on-resistance for the 3 kV shielded trench-gate MOSFET structure with P-base doping concentration of 1 x 10^{16} cm^{-3} was found to be only 1.5 mΩ-cm^2 at a gate bias of 10 volts.

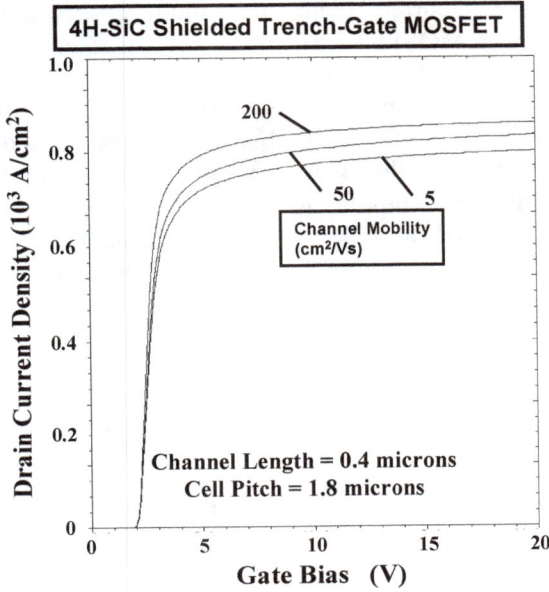

Fig. 14.14 Transfer Characteristics of the 4H-SiC Shielded Trench-Gate
MOSFET Structure.

14.3 Planar Shielded Accumulation Mode MOSFET Structure

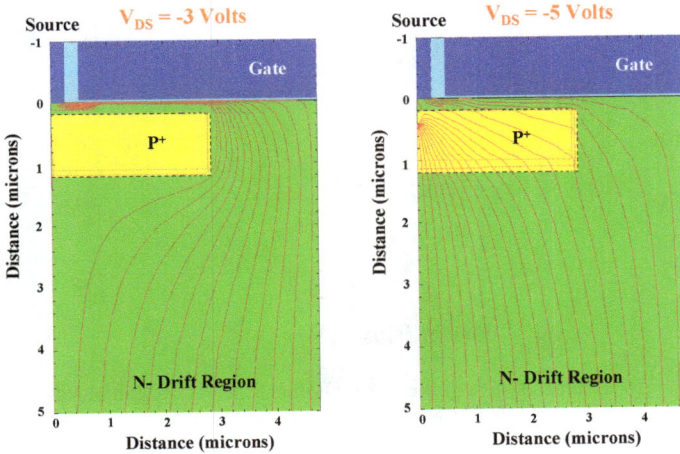

Fig. 14.15 Current Distribution in the Planar Shielded Accumulation Mode
4H-SiC MOSFET Structure.

The unusual current conduction mode observed in the trench gate 4H-SiC power MOSFET structure was also found to occur in the planar shielded accumulation mode MOSFET structure. The current flow-lines are shown in Fig. 14.15 for this structure for the cases of a drain bias of -3 and -5 volts. With a bias of -3 volts applied to the drain, the current flows through the channel region bypassing the P-N junction. This confirms that the unipolar mode of operation also occurs in the planar accumulation mode 4H-SiC MOSFET structure. When the drain bias is increased to -5 volts, the current flows through the P$^+$ region as well as into the N$^+$ source region. Thus, the usual bipolar mode is observed at larger drain bias levels.

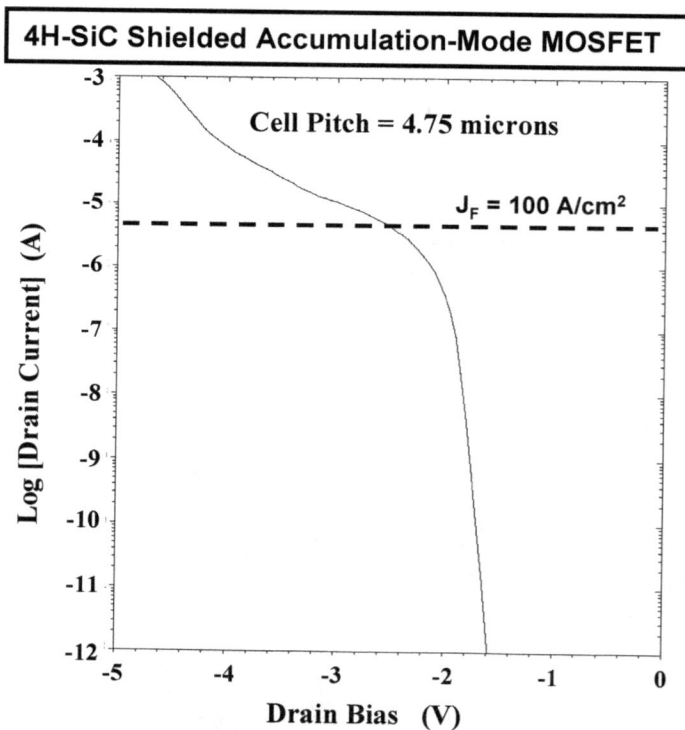

Fig. 14.16 Characteristics of the Body-Diode in the 4H-SiC Shielded Accumulation-Mode MOSFET Structure.

The *i-v* characteristic for the shielded 4H-SiC accumulation-mode MOSFET structures in the third quadrant of operation is shown in

Fig. 14.16. At a current density of 100 A/cm^2, the on-state voltage drop is found to be 2.5 volts for the structure. This is slightly (about 0.5 volts) smaller than that observed for the 4H-SiC P-i-N rectifier with the same blocking voltage capability.

14.4 Charge Coupled MOSFET Structure

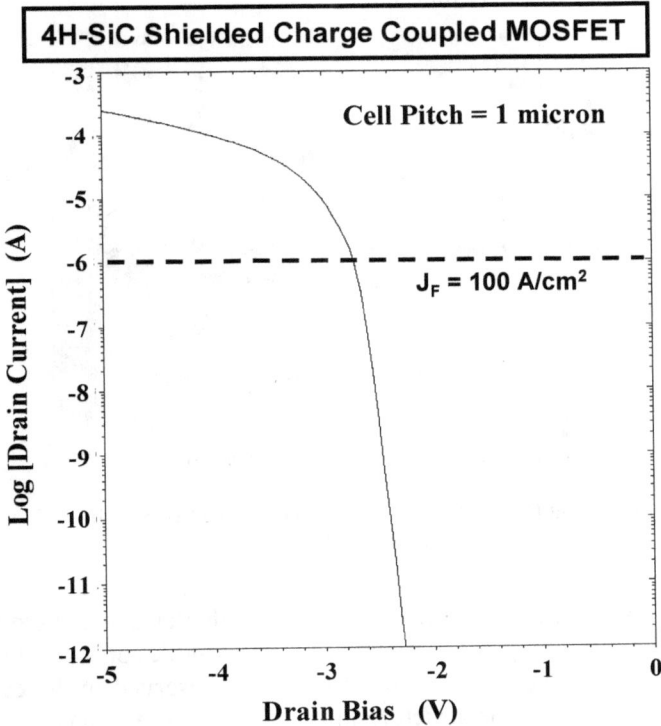

Fig. 14.17 Characteristics of the Body-Diode in the Shielded Charge Coupled 4H-SiC MOSFET Structure.

In the previous chapter, it was demonstrated that two-dimensional charge coupling in the drift region can be employed to drastically reduce the specific on-resistance. However, since the charge coupling is obtained by the formation of a vertical P-N junction that extends through the entire drift region, it is not obvious that the current transport in the third quadrant can bypass the junction. In order to find out if the new mode of operation is relevant to the charge coupled 4H-SiC MOSFET structure, two dimensional numerical simulations of the shielded charge coupled MOSFET structure were performed with negative bias applied to the drain electrode to obtain the characteristics for the structure in the third quadrant of operation. The *i-v* characteristic is shown in Fig. 14.17. At a current density of 100 A/cm^2, the on-state voltage drop is found to be about 2.7 volts, which is less than that observed in the 4H-SiC P-N junction.

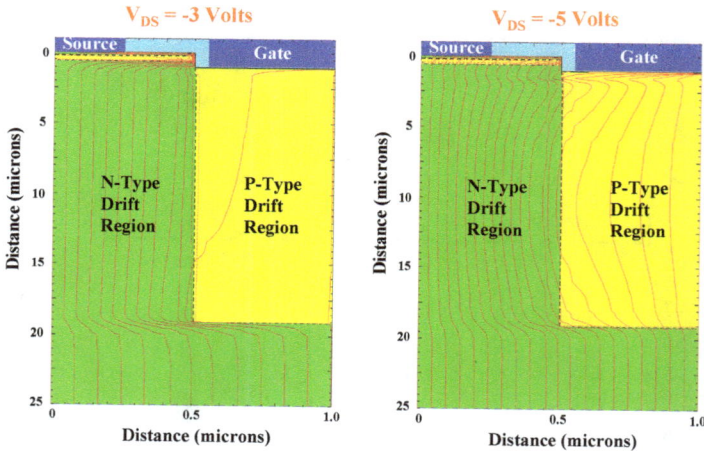

Fig. 14.18 Current Distribution in the Shielded Charge Coupled 4H-SiC MOSFET Structure.

The current flow pattern within the shielded charge coupled 4H-SiC MOSFET structure reveals that the new mode of operation is applicable for this structure as well. This can be observed in the current flow-lines shown within the structure in Fig. 14.18 at drain bias of -3 and -5 volts. At a drain bias of -3 volts, it can be observed that the current does not flow across either the P-N junction formed by the N and P-type drift regions or the P-N junction formed by the P-base/N-drift region. Instead, the current flows along the sidewall of the trench gate region and exclusively through the N-type drift region. This pattern for the current flow is similar to that observed during on-state current flow in the first quadrant when a positive gate bias above the threshold voltage is applied to the MOSFET. However, when the drain bias is increased to -5 volts, the current flow occurs across the P-N junctions within the structure.

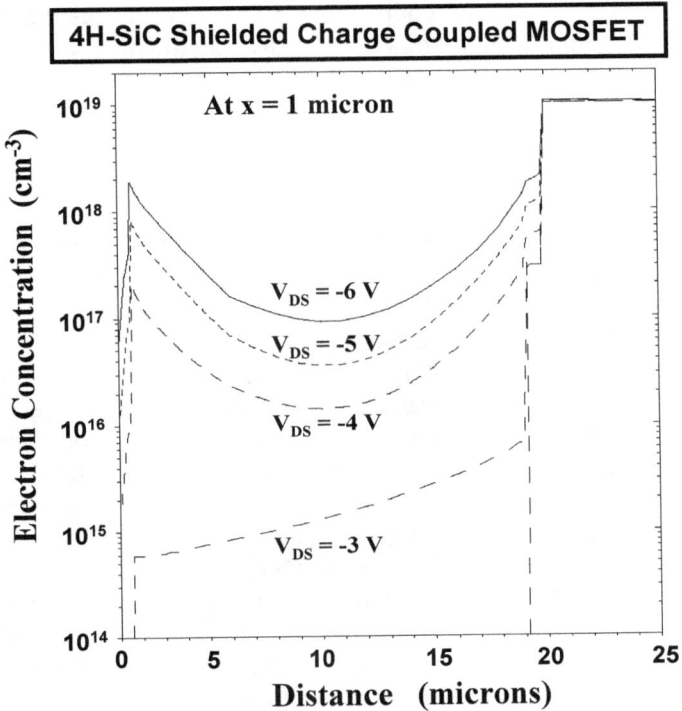

Fig. 14.19 Electron Distribution in the Shielded Charge Coupled 4H-SiC MOSFET Structure.

The activation of the P-N junctions within the MOSFET structure is known to produce stored minority carrier charge that must be removed during the reverse recovery transient. The electron distribution within the P-type drift region is an indication of this stored charge. The electron concentration in the P-type drift region can be analyzed by examining the profile taken at x = 1 micron, as shown in Fig. 14.19. It can be seen that significant minority carrier (electrons in the P-type drift region) injection occurs when the drain bias exceeds -4 volts. Due to high doping concentration (3×10^{17} cm^{-3}) in the P-type drift region, high level injection does not take place until the drain bias exceeds -5 volts. The low concentration of injected electrons at a drain bias of -3 volts indicates that the reverse recovery for the body-diode of the 4H-SiC charge coupled MOSFET structure will occur with small recovery charge leading to low switching losses. It is therefore feasible to use the body-diode in the 4H-SiC charge coupled MOSFET structure as the fly-back rectifier in H-bridge circuits if the losses associated with a relatively high on-state voltage drop are tolerable.

14.5 Summary

In this chapter, the current flow within various 4H-SiC power MOSFET structures has been investigated in the third quadrant of operation. Unlike silicon power MOSFETs, it has been demonstrated that a new unique mode of operation occurs with unipolar current conduction at low drain bias voltages. The on-state voltage drop in the third quadrant of operation can be reduced by taking advantage of this new mode of operation by reducing the doping concentration of the P-base region. Proper shielding of the P-base region must be provided to suppress reach-through breakdown problems that can arise due to the low doping concentration in the P-base region. With appropriate design, it has been found that the integral diode within the 4H-SiC power MOSFETs can be utilized as the fly-back rectifier in totem-pole circuits commonly used for motor control applications. This allows elimination of the external flyback diode commonly used at present in H-bridge circuits with IGBTs for motor control. In the case of systems operating at lower DC-bus voltages, it has been possible to use the integral diode within silicon power MOSFETs after performing lifetime control with electron irradiation to reduce the stored charge[4]. The development of 4H-SiC power MOSFETs can extend this monolithic approach to higher voltage systems.

References

[1]   B. K. Bose, "Power Electronics and Variable Frequency Drives", IEEE Press Book, New York, 1997.

[2]   B. J. Baliga, "Power Semiconductor Devices for Variable Frequency Drives", Proceeding of the IEEE, Vol. 82, pp. 1112-1122, 1994.

[3]   R. Sunkavalli and B. J. Baliga, "Integral Diodes in Lateral DI Power Devices", IEEE International Symposium on Power Semiconductor Devices and ICs, Abstract 8.45, pp. 385-390, 1995.

[4]   B. J. Baliga and J. P. Walden, "Power MOSFET Integral Diode Reverse Recovery Tailoring using Electron Irradiation", IEEE Transactions on Electron Devices, Vol. 29, pp. 1685-1686, 1982.

Chapter 15

Lateral High Voltage FETs

The ability to integrate high voltage power device structures with low voltage CMOS and bipolar devices has enabled the evolution[1] of silicon *smart power technology*. Due to the high on-resistance of silicon lateral high voltage structures, silicon MOS-bipolar structures must be utilized when the operating voltages exceed 200 volts[2]. Since the bipolar current flow in these structures can interfere with the CMOS logic circuits when implemented using junction isolation, it has been necessary to develop dielectrically isolated lateral high voltage configurations for the power transistors[3]. It has been demonstrated that high performance structures can be made with this approach with breakdown voltages upto 1200 volts. However, the cost of the dielectrically isolated silicon wafers has been substantially higher than junction isolated wafers, which has limited their applications.

The silicon lateral high voltage structures utilized for smart power applications are usually based upon the *REduced SURface electric Field (RESURF) concept*. In this approach[4], two dimensional charge-coupling between the N-drift layer and a P-type substrate is utilized to modify the lateral electric field so that the breakdown voltage is enhanced. The charge coupling allows increasing the doping concentration of the N-drift region enabling reduction of the on-resistance for the lateral high voltage devices. It is necessary to optimize the charge within the drift region to maximize the breakdown voltage.

In this chapter, a basic equation is first developed[5] for the specific on-resistance for the lateral high voltage structures that use the charge coupling concept to homogenize the lateral electric field. This equation predicts that the specific on-resistance of the drift region for the high voltage lateral device is 4 times smaller than that for the vertical device. This provides motivation for the development of lateral high voltage structures. More relevant to this book, the equation predicts that the specific on-resistance is inversely proportional to the cube of the critical electric field for breakdown in the semiconductor. This provides strong justification for exploring lateral silicon carbide high voltage structures. After a brief discussion of the first high voltage lateral devices that were based upon the charge coupling concept, the analysis of lateral power MOSFET structures is undertaken in this chapter. It is demonstrated that field plates must be used to ameliorate the generation of high electric fields at the edges of the drift region as previously found in silicon devices. More importantly, it is demonstrated that the P-base region must be shielded to prevent reach-through breakdown if the high blocking voltage capability of the drift region is to be fully utilized. With proper design precautions, high performance lateral silicon carbide structures are feasible with very high breakdown voltages and low on-resistance. The results of experimental work on lateral high voltage silicon carbide MOSFET structures are summarized at the end of the chapter to provide a perspective on the state of the art.

15.1 Fundamental Analysis

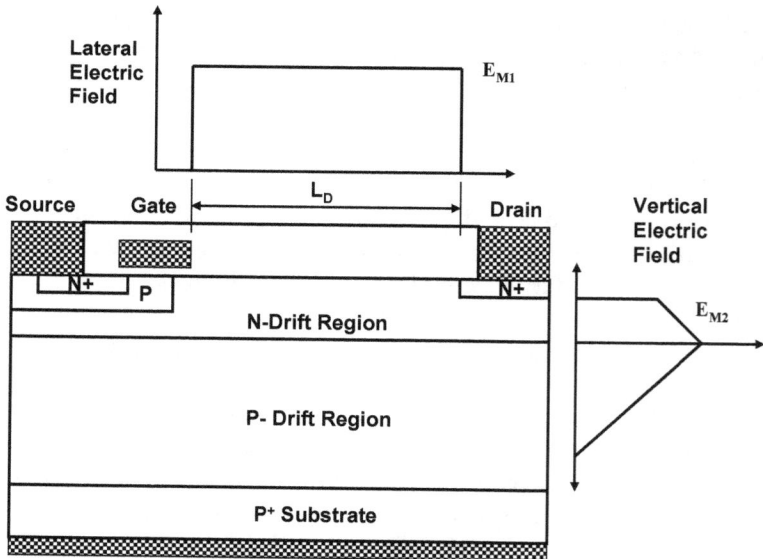

Fig. 15.1 Electric Field Profiles within a Lateral Charge-Coupled MOSFET Structure.

For performing a fundamental analysis of the specific on-resistance for the lateral high voltage FET structure with charge coupling, a MOSFET structure will be considered. As shown in Fig. 15.1, the voltage is supported in the structure across the N-type drift region and the P-type substrate. With optimum charge coupling, the electric field can be assumed to uniform in the lateral direction for the ideal case as illustrated in the figure. The applied drain voltage is also supported across the junction between the N-drift region and the P-type substrate because it is customary to connect the substrate to the source terminal. This electric field profile is basically triangular in nature as shown in the figure. The breakdown voltage for the lateral MOSFET structure can be limited by either of the maxima (E_{M1} or E_{M2}) in the electric field becoming equal to the critical electric field for breakdown for the semiconductor. In order to maximize the benefits of the charge coupling phenomenon, the doping concentration of the P-type substrate region is made sufficiently low so that E_{M2} is less than E_{M1}. With this design, the breakdown of the lateral structure occurs when E_{M1} becomes equal to the critical electric field for breakdown for the semiconductor. The breakdown voltage of the structure is then given by:

$$BV = L_D . E_C \qquad \text{[15.1]}$$

where L_D is the length of the drift region and E_C is the critical electric field for breakdown for the semiconductor.

The resistance for the drift region is given by:

$$R_D = \frac{L_D}{q.\mu_N.N_D.Z.t_N} \qquad \text{[15.2]}$$

where μ_N is the electron mobility corresponding to the doping concentration N_D in the drift region, t_N is the thickness of the drift region, and Z is the width of the structure orthogonal to the cross-section. In the idealized case where the space contributed by the other portions of the structure is neglected, the surface area for the structure is ($Z.L_D$). The specific on-resistance for the drift region can be then obtained by taking the product of this surface area and the resistance:

$$R_{D,sp} = \frac{L_D^2}{q.\mu_N.N_D.t_N} \qquad \text{[15.3]}$$

In an optimized charge coupled structure, the optimum charge ($Q_{Optimum}$) in the drift region is equal to the product of the dielectric constant and the critical electric field for breakdown for the semiconductor:

$$Q_{optimum} = q.N_D.t_N = \varepsilon_S.E_C \qquad [15.4]$$

where q is the charge of an electron (1.6×10^{-19} Coulombs), N_D is the doping concentration of the N-Type drift region, t_N is the thickness of the N-type drift region, ε_S is the dielectric constant of the semiconductor, and E_C is the critical electric field for breakdown in the semiconductor. This equation indicates that the optimum RESURF dose for silicon carbide devices is an order of magnitude larger than for silicon devices due to the larger critical electric field for breakdown. Using this relationship in the previous equation and eliminating L_D by using Eq. [15.1]:

$$R_{D,sp} = \frac{BV^2}{\mu_N.\varepsilon_S.E_C^3} \qquad [15.5]$$

This is a fundamental relationship for the specific on-resistance of the drift region within lateral high voltage devices. By co-incidence, the denominator in this equation is the same *Baliga's Figure of Merit* derived in chapter 1. It indicates that the specific on-resistance will increase as the square of the breakdown voltage. More relevant to the discussion of silicon carbide in this book, the equation predicts a strong reduction of the specific on-resistance due to the much larger critical electric field for breakdown when compared with silicon. In comparison to the specific on-resistance for vertical high voltage devices derived in chapter 1 (see Eq. [1.4]), the lateral device has a four times smaller value. This provides strong motivation for the development of high voltage lateral structures from silicon carbide. However, it must be kept in mind that the space taken up by the metallization in lateral structures can be significant enlarging the area much beyond that taken up by just the drift region.

15.2 Lateral High Voltage Buried Junction FET Structure

The first lateral high voltage silicon carbide FET structure[6] that utilized the charge-coupling concept to enhance the breakdown voltage was demonstrated at PSRC and reported in 1995. The device structure, illustrated in Fig. 15.2, was fabricated by ion implantation of nitrogen to form the N-type drift region in a P-type epitaxial layer grown on a P$^+$ substrate. The P$^+$ substrate was then used as the gate to modulate the current flow between the source and drain regions formed in the N-type drift region by shallow high dose ion implantation of nitrogen. The device was fabricated with just three mask layers.

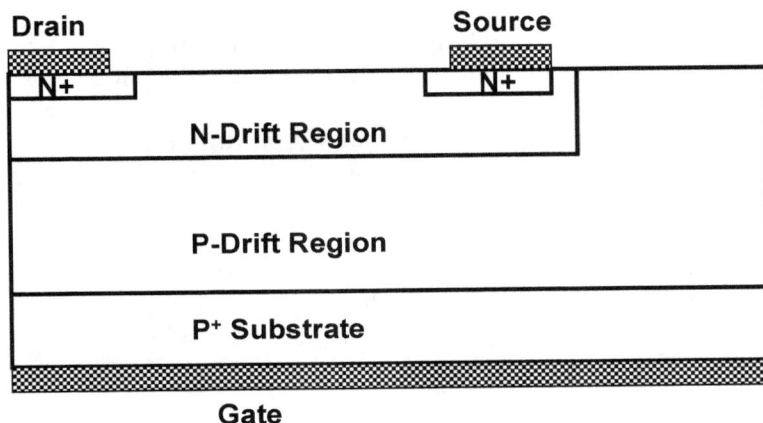

Fig. 15.2 The Buried Gate Lateral Junction Field Effect Transistor.

Due to the low doping concentration in the P-type drift region, the gate bias voltage in the BG-FET structure is supported across both the N-drift layer and the P-drift layer. Consequently, a high gate bias is required to pinch off the N-drift layer because a large part of the applied gate bias voltage appears across the more lightly doped P-drift region. It was found that a drain bias of 450 volts could be supported by the structure when a negative bias of 440 volts was applied to the gate. Under these bias conditions, the junction between the N-type and P-type drift regions is supporting 840 volts. This is possible due to the charge coupling phenomenon obtained by using a dose of 1.3×10^{13} cm^{-2} in the N-drift region.

The specific on-resistance for this structure was found to be 75 mΩ-cm^2 at zero gate bias. The breakdown voltage and specific on-resistance were found to scale down with reduction of the length of the drift region. These results provided the first confirmation that the charge coupling or RESURF effect was applicable to silicon carbide lateral structures with an order of magnitude increase in the dose of the N-drift region when compared with silicon structures. However, the large gate bias voltages required for the BG-FET structure makes it impractical for applications.

15.3 Lateral High Voltage MESFET Structure

The first surface gate lateral high voltage silicon carbide FET structure[7] based upon charge coupling was developed at PSRC and reported in 1996. The device structure, illustrated in Fig. 15.3, was fabricated by ion implantation of nitrogen to form the N-type drift region in a P-type epitaxial layer grown on a P^+ substrate. The N^+ source and drain regions were then formed in the N-type drift region by shallow high dose ion implantation of nitrogen. The Schottky gate contact was then fabricated by deposition of titanium. By using titanium as the metallization, it was possible to simultaneously form both the gate Schottky contact and the source/drain ohmic contacts. This enabled fabrication of the device structure with just three mask layers.

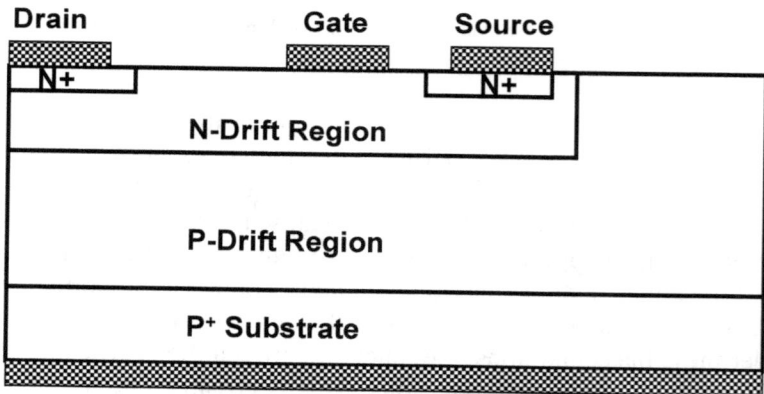

Fig. 15.3 The Lateral High Voltage MESFET Structure.

In the lateral MESFET structure, drain current flow can occur at zero gate bias due to the existence of an un-depleted region under the gate electrode. With a dose of 1.5×10^{13} cm^{-2} for the N-drift region, a specific on-resistance of 21 mΩ-cm^2 was anticipated for the structure. The measured specific on-resistance was 4 times larger due to high contributions from the ohmic contacts. The device, with N-drift region length of 15 microns, was able to support 450 volts when a negative gate bias of 20 volts was applied. This demonstrated the operation of the charge coupling effect in silicon carbide with a RESURF dose ten times larger than in silicon.

A lateral high voltage 4H-SiC JFET structure has also been recently demonstrated[8]. This structure has a cross-section similar to the one shown in Fig. 15.3 with the Schottky gate replaced by a P-N junction. The authors obtained a breakdown voltage of 800 volts at a negative gate bias of -4 volts by using a drift region with doping concentration of 2×10^{17} cm^{-3} and thickness of 0.4 microns, corresponding to a charge of 8×10^{12} cm^{-2}. At zero gate bias, the device had a very high specific on-resistance of 6 Ω-cm^2 which could be reduced to 0.24 Ω-cm^2 by application of a positive gate bias of 2.5 volts. These results indicate that the channel design was not adequately optimized for normally-on operation.

15.4 Lateral High Voltage MOSFET Structure

The basic lateral high voltage MOSFET structure was illustrated in Fig. 15.1. As shown in that figure, the voltage is supported across the N-drift region between the gate electrode and the N$^+$ drain region. The charge in the N-drift region must be optimized to obtain the maximum breakdown voltage. When the charge in the drift region is too small, it becomes depleted at a low drain bias voltage. In this case, a high electric field is developed at the drain side within the N-drift layer resulting in a reduction of the breakdown voltage. When the charge in the drift region is too high, the N-drift region cannot be depleted by the drain bias and the two-dimensional charge coupling effect is lost. A high electric field develops at the gate side within the N-drift region resulting in a low breakdown voltage. At the optimum charge in the drift region, the electric field is balanced between the drain side and the gate side within the N-drift region producing the highest breakdown voltage.

When a positive gate bias is applied, a current path is created between the drain and source by the formation of an inversion layer channel. Current transport between the source and the drain can now occur via unipolar conduction. The on-resistance for the structure is determined by the sum of the channel resistance and the drift region resistance if the contact resistance is sufficiently low. However, the specific on-resistance can be adversely impacted by the area consumed by the source and drain contacts, as well as enlargement of the cell pitch to accommodate the high current source and drain metal fingers.

Two dimensional numerical simulations of the 4H-SiC lateral high voltage MOSFET structure were performed with zero bias applied to the gate electrode to obtain the blocking characteristics. The device structure had the following structural parameters: a P-drift region with doping concentration of 1×10^{15} cm^{-3} and thickness of 20 microns; a N-drift region with doping concentration of 5×10^{17} cm^{-3} and thickness of 0.1 microns; a drift region length of 21.5 microns between the edge of the gate electrode and the N$^+$ drain region; a gate oxide thickness of 0.05 microns with a gate length of 2 microns; a P-base region with doping concentration of 1×10^{17} cm^{-3} with channel length of 1 micron. The parameters for the P-drift region were chosen so that the breakdown voltage of the vertical diode between the N$^+$ drain and the P$^+$ substrate exceeded 3000 volts, the anticipated breakdown voltage of the N-drift region.

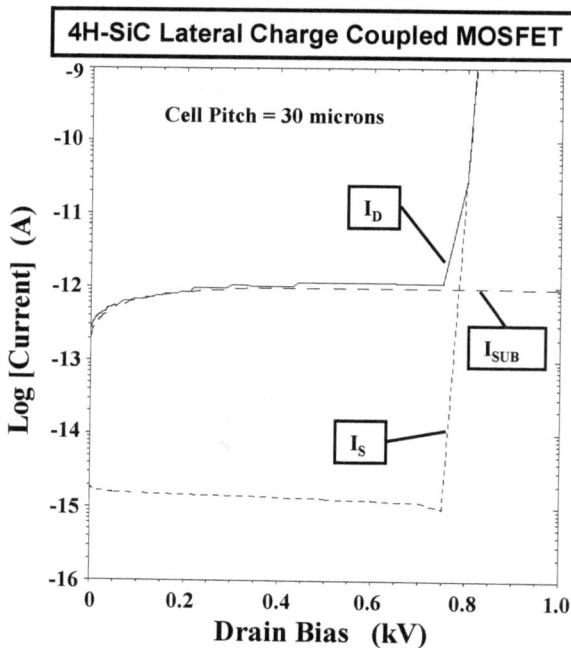

Fig. 15.4 Blocking Characteristics of the Lateral High Voltage MOSFET Structure.

The blocking characteristic obtained with the numerical simulations for the lateral 4H-SiC charge coupled MOSFET structure is shown in Fig. 15.4. Here, the substrate current is included with the drain and source currents to elucidate the breakdown phenomenon. It can be seen that the drain current flow matches the substrate current at lower drain bias voltages. However, the source current abruptly increases at a drain bias of 750 volts and exceeds the substrate current at a drain bias of 800 volts. The maximum blocking voltage for the structure is limited by this abrupt increase in the source current.

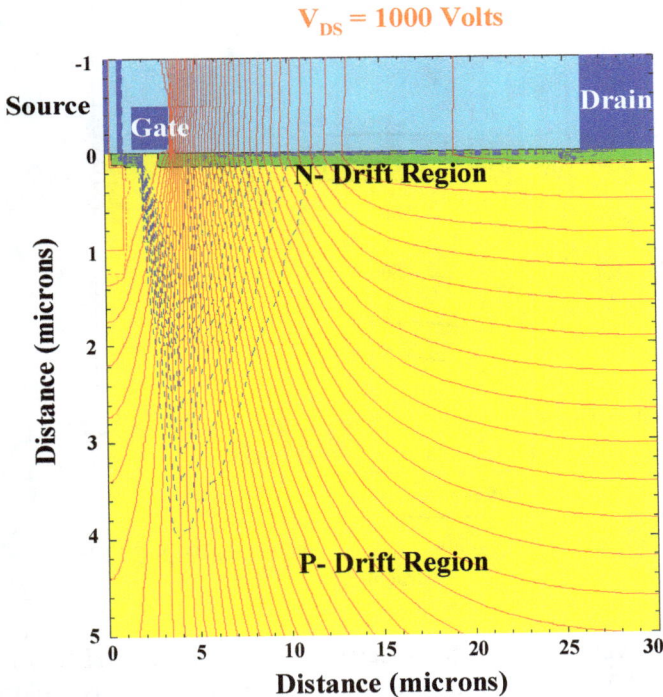

$V_{DS} = 1000$ Volts

Fig. 15.5 Potential and Current Distribution in the Lateral 4H-SiC Charge-Coupled MOSFET Structure.

The potential distribution within the 4H-SiC lateral charge coupled MOSFET structure is shown in Fig. 15.5 at a drain bias of 1000 volts together with the current flow pattern (indicated by the dashed lines). The potential lines are crowded at the gate edge indicating a high electric field at this location. The high electric field in the vicinity of the

P-base/N-drift junction promotes depletion of the P-base region leading
to reach-through breakdown problems.

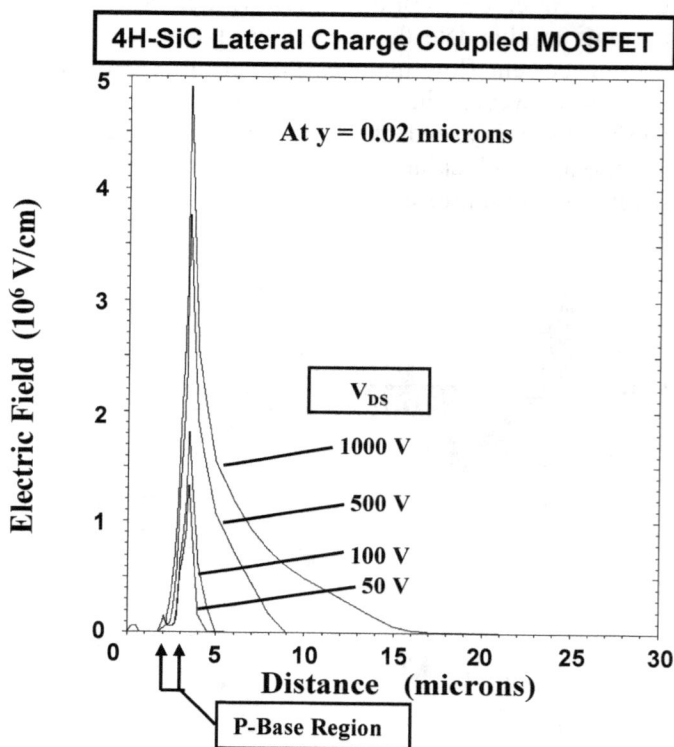

Fig. 15.6 Electric Field along the Surface in the Lateral 4H-SiC MOSFET
Structure.

The development of the electric field within the 4H-SiC lateral
charge coupled MOSFET structure is shown in Fig. 15.6 just below the
semiconductor surface along the x-direction. The location of the P-base
region is also indicated in this figure. When the drain bias exceeds 500
volts, the electric field spans the P-base region due to its high value at the
gate edge (located at 3.5 microns). This confirms that the sharp rise in
the source current, responsible for limiting the breakdown voltage, is a
consequence of the reach-through phenomenon.

In the case of silicon lateral MOSFETs that utilize the RESURF
effect, it has been found that the electric field at the gate and drain sides
of the N-drift region can be reduced by incorporation of field plates[9]. The

utility of electric field plates for ameliorating reach-through in the 4H-SiC charge coupled MOSFET structure is discussed in the next section.

15.5 Lateral High Voltage MOSFET Structure with Field Plates

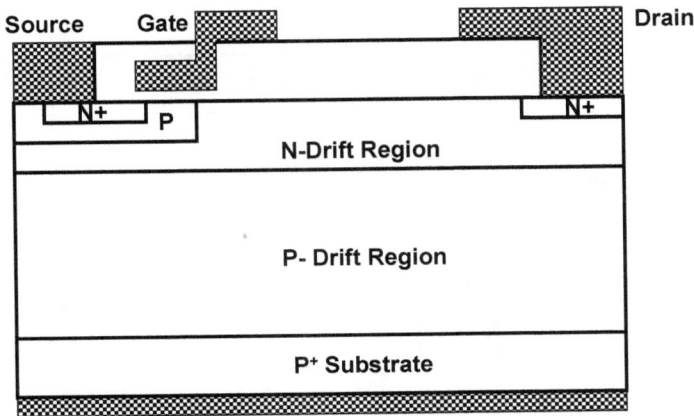

Fig. 15.7 The Lateral Charge-Coupled MOSFET Structure with Field Plates.

The lateral charge coupled MOSFET structure with field plates at the gate and drain ends of the N-drift region is illustrated in Fig. 15.7. The drain field plate can be constructed by simply extending the drain contact metal towards the gate side over the field oxide. On the gate side of the N-drift region, the field plate can be formed by stepping the gate electrode over the field oxide. Alternately, the field plate on the gate side can be formed by extending the source metal over the gate electrode to overlap the N-drift region.

In order to examine the influence of the field plates on the blocking characteristics, two-dimensional numerical simulations were performed using the same structural parameters reported in the previous section with the addition of the field plates. On the gate side, the field plate was extended over the field oxide to a distance of 1.5 microns. On the drain side, the field plate was extended on the field oxide by a distance of 2 microns from the edge of the N^+ drain region. The blocking characteristic obtained from the simulations is shown in Fig. 15.8. Here, the substrate current is included with the drain and source currents to elucidate the breakdown phenomenon. It can be seen that the drain current flow matches the substrate current at lower drain bias voltages. However, the source current abruptly increases at a drain bias of 1000 volts and exceeds the substrate current at a drain bias of 1100 volts. The maximum blocking voltage for the structure is limited by this abrupt increase in the source current. When compared with the lateral MOSFET structure discussed in the previous section, there is an increase in the blocking voltage capability by 300 volts.

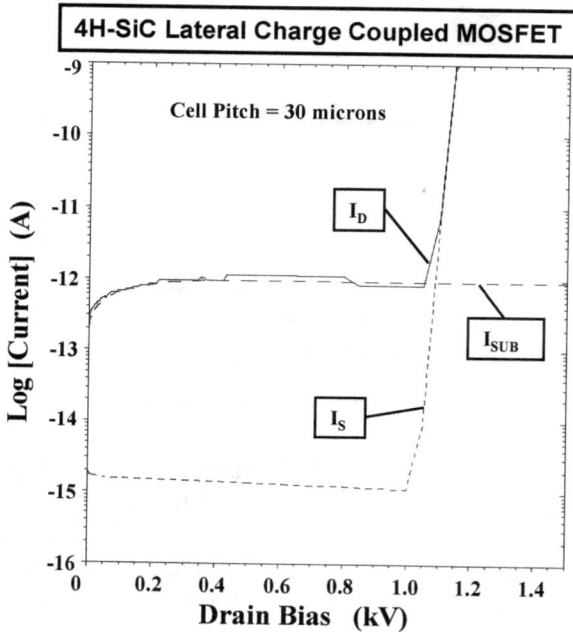

Fig. 15.8 Blocking Characteristics of the Lateral High Voltage MOSFET Structure with Field Plates.

The potential distribution within the 4H-SiC lateral charge coupled MOSFET structure with field plates is shown in Fig. 15.9 at a drain bias of 1000 volts together with the current flow pattern (indicated by the dashed lines). The presence of the field plate on the gate side reduces the crowding of the potential lines at the gate edge. The electric field in the vicinity of the P-base/N-drift junction is consequently reduced leading to the suppression of the reach-through breakdown problem. The current flow pattern indicates that the drain current flows into the substrate rather than into the source electrode, which is consistent with the results shown in the previous figure.

$$V_{DS} = 1000 \text{ Volts}$$

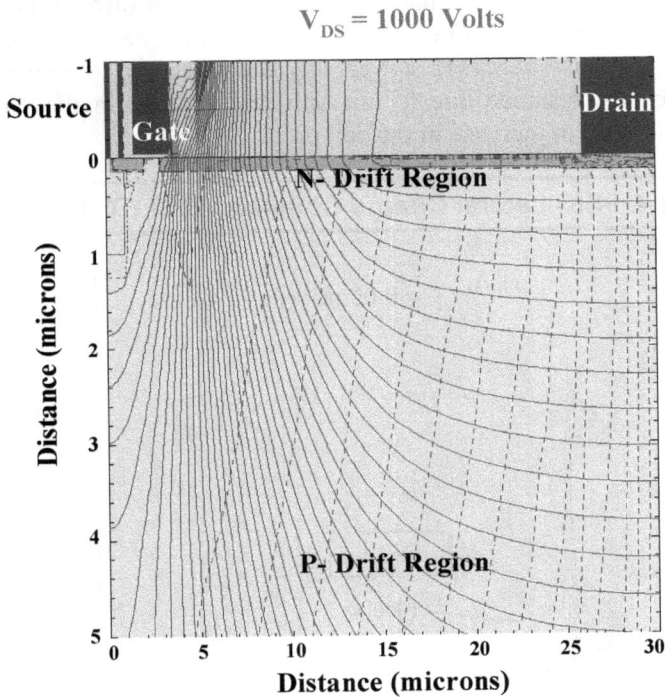

Fig. 15.9 Potential and Current Distribution in the Lateral 4H-SiC MOSFET Structure with Field Plates.

The development of the electric field within the 4H-SiC lateral charge coupled MOSFET structure with field plates is shown in Fig. 15.10 just below the semiconductor surface along the x-direction. The location of the P-base region is also indicated in this figure. In comparison with the structure without the field plates, it can be seen that the electric field at the P-base region has been suppressed by the presence of the field plate. However, when the drain bias exceeds 1000 volts, the electric field spans the P-base region because its value is still high at the gate edge (located at 3.5 microns). Thus, the blocking voltage capability of the lateral 4H-SiC charge coupled MOSFET structure is limited by the reach-through phenomenon even after inclusion of the field plate on the gate side. This demonstrates that, unlike silicon structures, the reach-through problem cannot be solved by using a field plate on the gate side because the magnitude of the electric field in silicon carbide is much larger than in silicon. The reach-through problem in the lateral 4H-SiC MOSFET structure could be mitigated by increasing the P-base width. However, this is detrimental to achieving a low specific on-resistance due to not only an increase in the channel resistance but also an increase in the cell pitch.

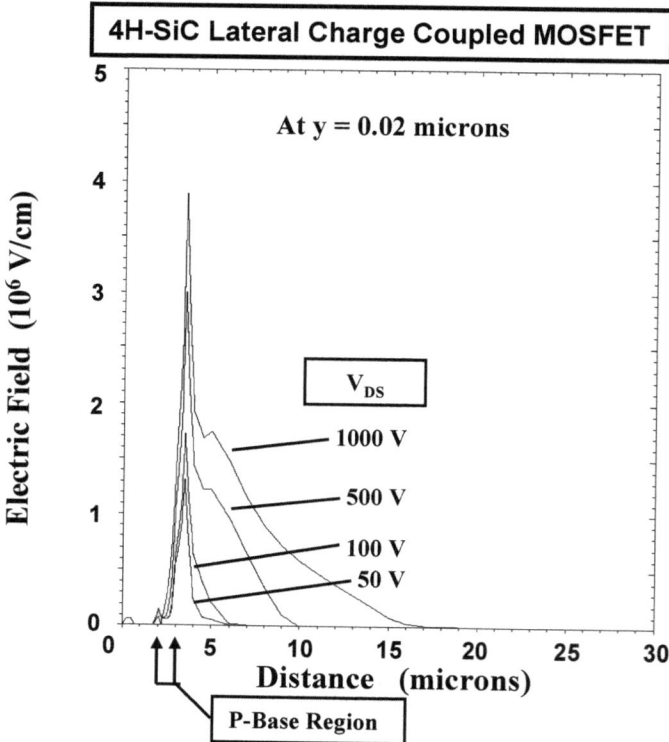

Fig. 15.10 Electric Field along the Surface in the Lateral 4H-SiC MOSFET Structure with Field Plates.

Based upon the results of the analysis of the lateral 4H-SiC MOSFET structure provided in this section, it can be concluded that it is necessary to solve the reach-through problem by incorporation of shielding of the P-base region from the high electric field developed in the N-drift region. A lateral MOSFET structure containing a sub-surface

P^+ shielding region located below the P-base region to address these issues is discussed in the next section.

15.6 Lateral High Voltage Shielded MOSFET Structure

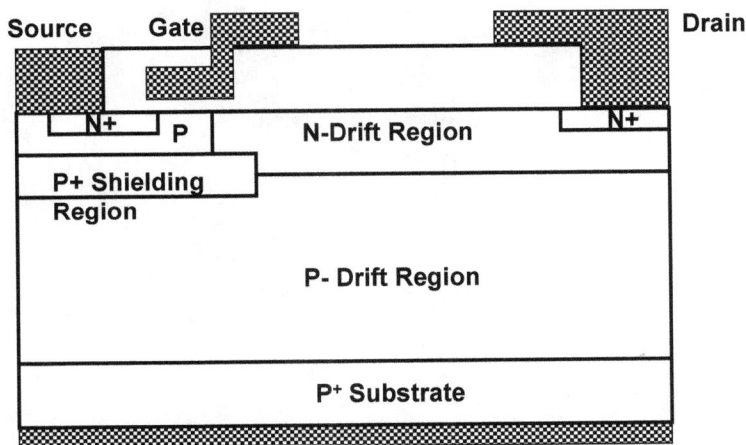

Fig. 15.11 The Lateral Shielded Charge-Coupled MOSFET Structure.

The lateral shielded charge coupled MOSFET structure with field plates at the gate and drain ends of the N-drift region is illustrated in Fig. 15.11. The structure is similar to one discussed in the previous section with the addition of the sub-surface P^+ shielding region which extends beyond the edge of the P-base region towards the N-drift region. The sub-surface P^+ shielding region can be heavily doped because the threshold voltage of the MOSFET is determined by the doping concentration of the P-base region.

In order to examine the influence of the sub-surface P^+ shielding region on the blocking characteristics, two-dimensional numerical simulations were performed using the same structural parameters reported in the previous section with incorporation of the sub-surface P^+ shielding region. The sub-surface P^+ shielding region extended from the left hand edge of the structure up to 4 microns placing it 1 micron beyond the extension of the P-base region. The blocking capability for this structure was found to exceed 3000 volts with no indication of reach through current flow into the source electrode.

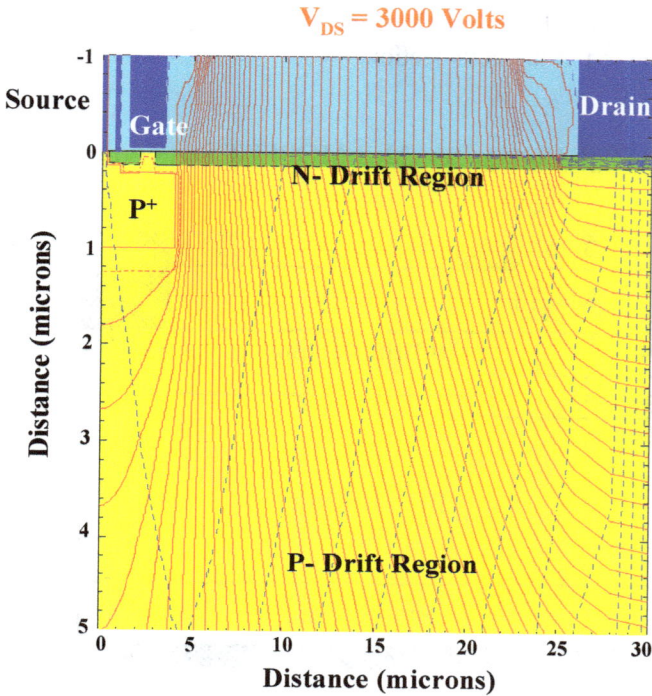

Fig. 15.12 Potential and Current Distribution in the Lateral 4H-SiC Shielded
MOSFET Structure with Field Plates.

The potential distribution within the 4H-SiC lateral shielded
charge coupled MOSFET structure with field plates is shown in Fig.
15.12 at a drain bias of 3000 volts together with the current flow pattern
(indicated by the dashed lines). The presence of the sub-surface P^+
shielding region on the gate side prevents the encroachment of the
potential into the P-base region. This completely suppresses the reach-
through problem allowing the structure to support a much larger drain
voltage. The current flow pattern indicates that the drain current flows
into the substrate rather than into the source electrode, which provides
confirmation that the reach-through phenomenon does not occur even at
a drain bias of 3000 volts.

The development of the electric field within the 4H-SiC lateral
shielded charge coupled MOSFET structure with field plates is shown in
Fig. 15.13 just below the semiconductor surface along the x-direction.

The location of the P-base region and the edge of the P⁺ shielding region are also indicated in this figure. In comparison with the structure without the field plates, it can be seen that the electric field at the P-base region has been drastically suppressed by the presence of the P⁺ shielding region. The high electric field is confined to the N-drift region beyond the edge of the P⁺ shielding region. When the drain bias exceeds 1000 volts, the N-drift region becomes completely depleted and the electric field spans the entire length between the gate and drain edges. At a drain bias of 3000 volts, the electric field is symmetrical indicating an optimum charge within the N-drift region. The magnitude of the electric field at a drain bias of 3000 volts is insufficient to generate impact ionization induced breakdown.

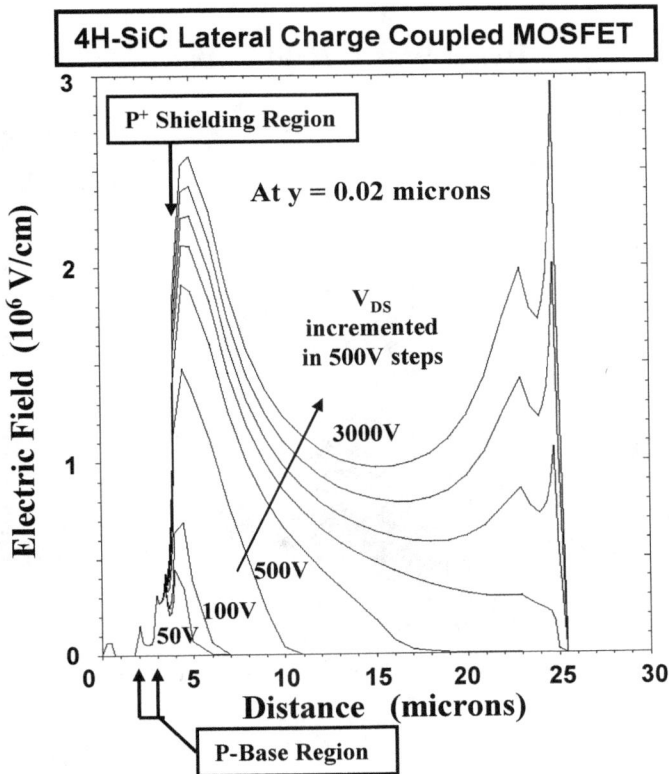

Fig. 15.13 Electric Field along the Surface in the Lateral 4H-SiC Shielded MOSFET Structure with Field Plates.

A three-dimensional view of the electric field distribution is shown in Fig. 15.14 at a drain bias of 3000 volts for the lateral 4H-SiC shielded charge coupled MOSFET structure with field plates. It can be seen that a high electric field occurs at the source and drain side of the drift region not only within the semiconductor but also in the field oxide. The electric field in the field oxide is below its rupture strength but sufficiently high to warrant reliability concerns. This problem can be ameliorated by replacing the field oxide with silicon nitride[10] to reduce the electric field. Silicon nitride has a larger dielectric constant (permittivity of 7.5) than silicon dioxide (permittivity of 3.85) resulting in reducing the electric field within the dielectric by half. Since silicon nitride is commonly used for manufacturing silicon devices, mature chemical vapor deposition technology is available for application to silicon carbide devices as well.

Drain Bias = 3000 Volts

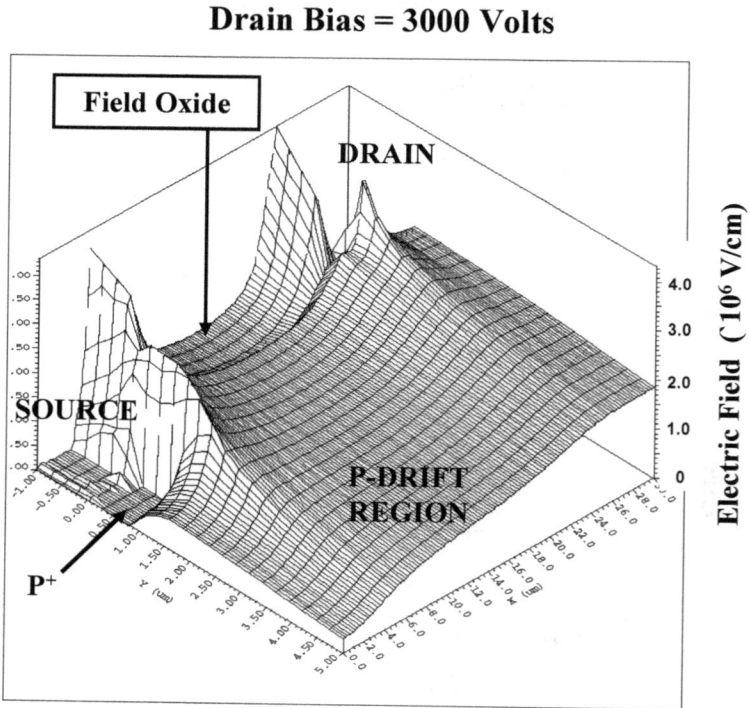

Fig. 15.14 Electric Field Distribution within the Lateral 4H-SiC Shielded MOSFET Structure with Field Plates.

15.7 Lateral High Voltage MOSFET Structure Optimization

As previously mentioned, the charge in the N-drift region of the lateral MOSFET structure must be optimized in order to maximize the blocking voltage capability. When the charge in the drift region is too small, it becomes depleted at a low drain bias voltage. In this case, a high electric field is developed at the drain side within the N-drift layer resulting in a reduction of the breakdown voltage. When the charge in the drift region is too high, the N-drift region cannot be depleted by the drain bias and the two-dimensional charge coupling effect is lost. A high electric field develops at the gate side within the N-drift region resulting in a low breakdown voltage. At the optimum charge in the drift region, the electric field is balanced between the drain side and the gate side within the N-drift region producing the highest breakdown voltage. It is important to have a precise knowledge of the optimum charge because the doping concentration of the N-drift region must be maximized to reduce the specific on-resistance.

Two dimensional numerical simulations of the lateral 4H-SiC charge coupled MOSFET structure were performed with various doping concentrations for the N-drift region to ascertain the breakdown voltage and the on-resistance. The field plates on the gate and drain side were retained to reduce the electric field. The P^+ shielding region was included in order to suppress the reach-through problem. The electric field distribution within the N-drift region was monitored as a good indication of the on-set of breakdown.

The electric field profiles at various drain bias voltages are shown in Fig. 15.15 for the case of a doping concentration of 1×10^{17} cm^{-3} (corresponding to a charge of 1×10^{12} cm^{-2}) in the N-drift region. It can be seen that N-drift region becomes depleted at a relatively low drain bias of about 100 volts leading to a high electric field at the drain side. In contrast, the electric field profiles at various drain bias voltages are shown in Fig. 15.16 for the case of a doping concentration of 2×10^{18} cm^{-3} (corresponding to a charge of 2×10^{13} cm^{-2}) in the N-drift region. It can be seen that N-drift region no longer becomes depleted even at a relatively high drain bias of 1000 volts leading to a high electric field at the gate side. At an optimum doping concentration of 5×10^{17} cm^{-3} (corresponding to a charge of 5×10^{12} cm^{-2}) in the N-drift region, the electric field profile becomes symmetrical as already shown in Fig. 15.13. Thus, the breakdown voltage goes through a maximum when the doping concentration exceeds the optimum value.

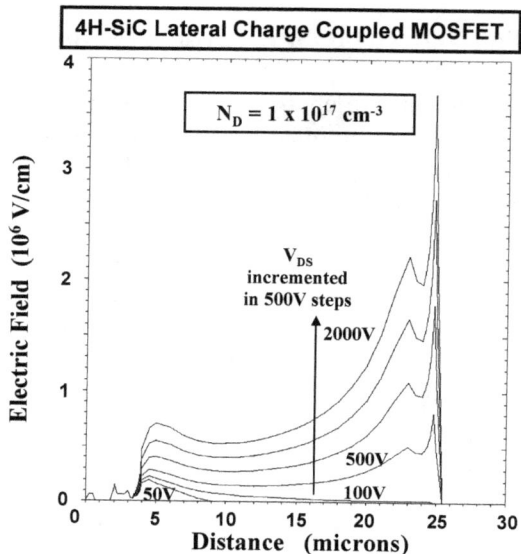

Fig. 15.15 Electric Field along the Surface in the Lateral 4H-SiC
Shielded MOSFET Structure with Field Plates.

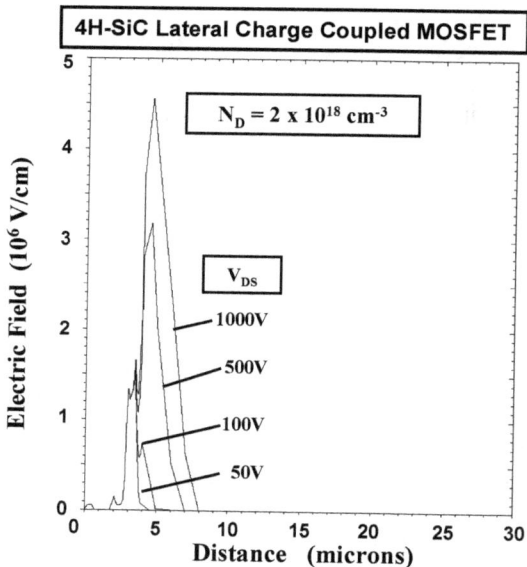

Fig. 15.16 Electric Field along the Surface in the Lateral 4H-SiC
Shielded MOSFET Structure with Field Plates.

The breakdown voltage of the lateral 4H-SiC shielded charge coupled MOSFET structure is plotted in Fig. 15.17 as a function of the charge in the N-drift region. The breakdown voltage increases with increasing charge in the N-drift region until it reaches a maximum value at a dose of about 5×10^{12} cm^{-2}. At above a dose of 8×10^{12} cm^{-2}, the breakdown voltage decreases rapidly with increasing dose in a manner similar to that observed for silicon devices. The optimum dose of 5×10^{12} cm^{-2} is significantly (about 5 times) larger than that observed for silicon lateral devices. This results in a reduction of the on-resistance for the 4H-SiC lateral MOSFET structure.

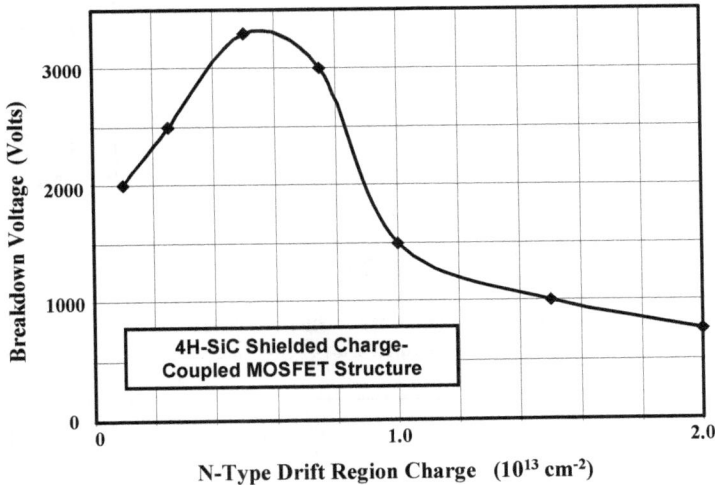

Fig. 15.17 Breakdown Voltages for the Lateral 4H-SiC Shielded MOSFET Structures with Field Plates.

The on-resistance of the lateral 4H-SiC shielded charge-coupled MOSFET structure was obtained by using two dimensional numerical simulations of each of the cases with different doping concentration in the N-drift region. In these simulations, the drain bias was held at 1 volt while sweeping the gate bias up to 20 volts. The resulting composite transfer characteristics for all the structures are shown in Fig. 15.18. Using the drain current extracted for each case at a gate bias of 20 volts, the on-resistance was calculated and plotted in Fig. 15.19. It can be seen that the on-resistance decreases with increasing doping concentration in the N-drift region.

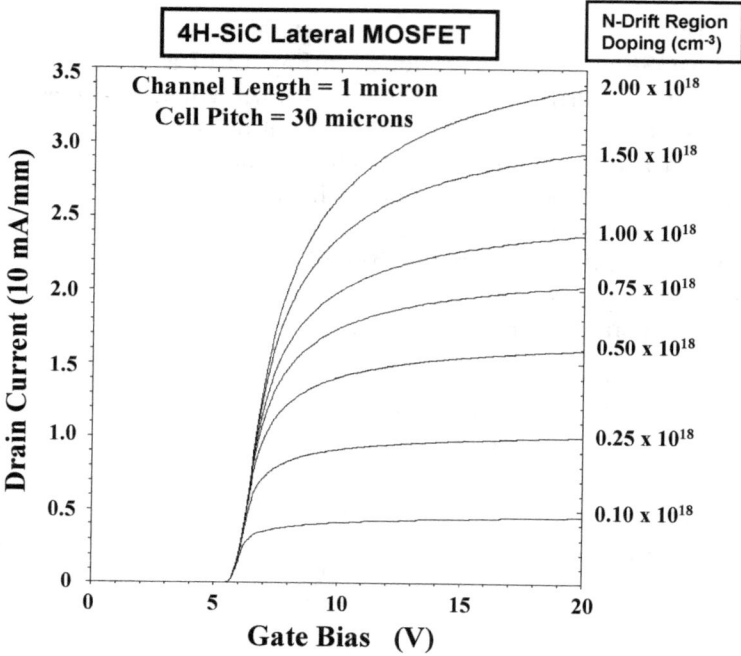

Fig. 15.18 Transfer Characteristics for the Lateral 4H-SiC Shielded MOSFET Structures with Field Plates.

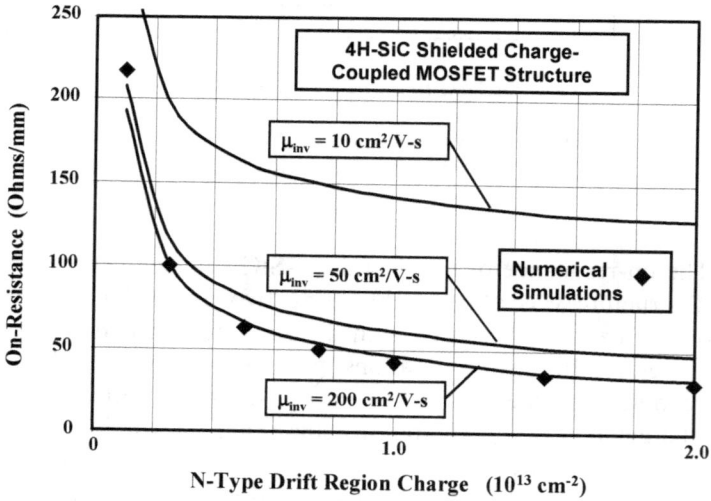

Fig. 15.19 On-Resistance for the Lateral 4H-SiC Shielded MOSFET Structures with Field Plates.

A simple model for the on-resistance for the lateral shielded charge coupled MOSFET structure can be posited by adding the channel and drift region contributions:

$$R_T = R_{CH} + R_D \qquad [15.6]$$

The channel resistance is given by:

$$R_{CH} = \frac{L_{CH}}{\mu_{inv}.C_{OX}.(V_G - V_T).Z} \qquad [15.7]$$

where L_{CH} is the channel length determined by the width of the P-base region, μ_{inv} is the mobility for electrons in the inversion layer channel, C_{ox} is the specific capacitance of the gate oxide, V_G is the applied gate bias, V_T is the threshold voltage, and Z is the device width orthogonal to the cross-section. The specific capacitance can be obtained using:

$$C_{ox} = \frac{\varepsilon_{ox}}{t_{ox}} \qquad [15.8]$$

where ε_{ox} is the dielectric constant for the gate oxide and t_{ox} is its thickness. It is worth pointing out that this is not the specific channel resistance but the channel resistance for a width Z of the lateral device structure.

The on-resistance contribution from the N-drift region is given by:

$$R_D = \frac{L_D}{q.\mu_n.N_D.t_N.Z} \qquad [15.9]$$

where L_D is the length of the N-drift region, μ_n is the mobility for electrons in the drift region appropriate to its doping concentration (N_D), and t_N is the thickness of the N-drift region. Here, the depletion of the N-drift region at its junction with the P-drift region has been neglected due to the much larger doping concentration of the N-drift region. This equation is not the specific drift region resistance but the drift region resistance for a width Z of the lateral device structure.

The analytically calculated values for the on-resistance for the lateral 4H-SiC shielded charge coupled MOSFET structure are plotted in Fig. 15.19 for three cases of inversion layer mobility. The structural parameters used for the calculations were the same as those used during the two-dimensional numerical simulations, namely: a gate oxide thickness of 0.05 microns, a channel length of 1 micron, a drift region length of 21 microns and thickness of 0.1 microns. The electron mobility in the N-drift region was adjusted based upon its doping concentration. It can be seen that there is an excellent match between the analytically calculated resistance per mm of device width and the values extracted from the numerical simulations when an inversion layer mobility of 200 cm^2/V-s was assumed in the analytical model. This provides validation of the simple analytical model for predicting the on-resistance of the lateral 4H-SiC shielded charge coupled MOSFET structure.

The impact of reduction of channel mobility on the on-resistance of the lateral 4H-SiC shielded charge coupled MOSFET structure was obtained by changing the inversion layer mobility value in the analytical model. The results are shown in Fig. 15.19 for the case of channel mobility values of 50 and 10 cm^2/V-s. It can be concluded that a mobility of 50 cm^2/V-s is adequate for obtaining good performance in the lateral 4H-SiC shielded charge coupled MOSFET structure.

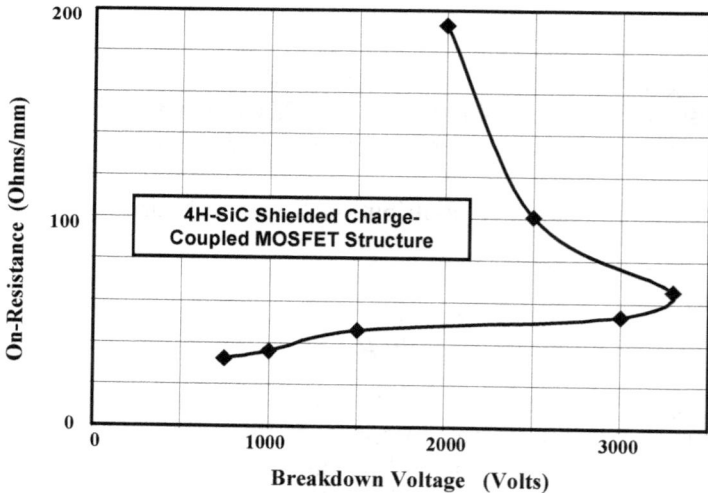

Fig. 15.20 Optimization of the Lateral 4H-SiC Shielded MOSFET Structure.

Further optimization of the lateral 4H-SiC shielded charge coupled MOSFET structure can be performed by plotting the on-resistance of the device as a function of the breakdown voltage using the doping concentration of the N-drift region as a parametric variable. This plot, shown in Fig. 15.20, indicates that the highest breakdown voltage of just above 3000 volts can be obtained with an on-resistance of 65 ohms per mm of device width. This value translates to a specific on-resistance of 19 mΩ-cm^2 for the lateral 4H-SiC shielded charge coupled MOSFET structure, which is substantially larger than about 1 mΩ-cm^2 obtained by using Eq. [15.5] for the ideal 3000 volt case with a critical electric field of 3 x 10^6 V/cm for 4H-SiC. The reason for the discrepancy is related to the non-uniform electric field in the N-drift region along the lateral direction which degrades the breakdown voltage and enlarges the lateral spacing. This problem can be partially overcome by employing multiple zones for the N-drift region with different doping concentrations[11]. An improvement in the on-resistance can also be achieved by using a thicker N-drift region because this improves the field distribution and reduces the doping concentration resulting in a larger mobility in the drift region.

15.8 Lateral High Voltage MOSFET Structure: Experimental Results

Experimental results on the first[12] high voltage lateral 4H-SiC MOSFET structures were reported in 1998 with blocking voltage of 2.6 kV. However, these devices were fabricated on semi-insulating substrates and did not utilize the two-dimensional charge coupling effect. The specific on-resistance for the devices was extremely large (550 Ω-cm^2) because of the very poor inversion layer mobility (7.5 x 10^{-3} cm^2/V-s) and long channel length used in the design, as well as the lack of the charge coupling effect.

The first experimental results[13] on high voltage lateral 6H-SiC MOSFETs were reported in 1999 with blocking voltage of 475 volts. These devices utilized the charge coupling (RESURF) effect in the drift region to enhance the breakdown voltage. The blocking voltage of 475 volts was obtained using a drift region with length of 10 microns and charge of 2.5 to 5 x 10^{12} cm^{-2}. A high specific on-resistance of 0.25 to 0.77 Ω-cm^2 was measured at a gate bias of 10 volts. This was attributed by the authors to poor activation of the implanted dopant in the N-drift region. These results were subsequently extended[14] to a breakdown voltage of 600 volts with a specific on-resistance of 57 mΩ-cm^2 by increasing the implant annealing temperature from 1200 °C to 1400 °C. The authors also fabricated devices using 4H-SiC but found very poor current conduction due to extremely low channel mobility.

The problems with poor channel mobility in 4H-SiC lateral charge-coupled MOSFETs has continued to plague[15,16] the development of high performance devices until recently. Although blocking voltages in the vicinity of 1000 volts were obtained, the specific on-resistance for these devices was very poor (about 1 Ω-cm^2). Significant improvement in the channel inversion layer mobility (to a value of 25 cm^2/V-s) was obtained by the authors with a NO annealing process[17]. Devices fabricated with two RESURF zones were found to exhibit a breakdown voltage of 930 volts with a specific on-resistance of 170 mΩ-cm^2 at a gate bias of 24 volts.

Similar results have been reported by other groups. It was found that the inversion layer mobility in 4H-SiC could be improved from 5 to 41 cm^2/V-s by going from the silicon-face to the carbon-face[18]. This allowed the fabrication of charge coupled lateral MOSFETs with a breakdown voltage of 460 volts and specific on-resistance of 79 mΩ-cm^2 at a gate bias of 25 volts. The successful fabrication of lateral MOSFETs from 4H-SiC was also reported[19] using a two-zone RESURF N-drift region with a breakdown voltage of 1230 volts and specific on-resistance of 138 mΩ-cm^2 at a gate bias of 20 volts.

The relatively high specific on-resistance of the experimentally fabricated lateral 4H-SiC MOSFETs based upon the charge coupling effect can be attributed to the absence of adequate shielding for preventing reach-through breakdown. The fabricated devices have utilized large channel lengths that enhance the channel contribution and enlarge the cell pitch which contributes to a high specific on-resistance. In order to avoid reach-through limited breakdown, the fabricated devices have long drift region lengths and lower than optimal charge in the N-drift region. This implies that significantly better lateral high voltage lateral MOSFETs can be developed from 4H-SiC with the use of the P-base shielding approach described in this chapter.

15.9 Summary

A fundamental analysis of lateral charge coupled FET structures has been shown to yield an ideal specific on-resistance four times smaller than that of vertical structures with the same breakdown voltage. This analysis assumes that the electric field is uniform between the source and drain along the surface. The charge coupling (or RESURF) concept can provide this uniform distribution in principle but two-dimensional numerical simulations demonstrate that the actual electric field has pronounced peaks located at the edges of the N-drift region in the vicinity of the gate and drain regions. This non-uniform electric field degrades the breakdown voltage and enlarges the length of the drift region to achieve a given breakdown voltage. In addition, the high electric field developed in the drift region for silicon carbide promotes premature breakdown due to the reach-through phenomenon. It has been demonstrated in this chapter that the reach-through problem can be overcome by screening the P-base region with a sub-surface P^+ region located beneath it. With the screening, it is possible to operate the structure with an optimum charge in the N-drift region with an on-resistance of about 65 Ohms/mm of device width for a device capable of supporting 3000 volts. However, the experimental results reported on lateral MOSFETs have been far inferior to the intrinsic capability of the structure because of lack of screening of the P-base region in the fabricated structures.

References

[1] B. J. Baliga, "Smart Power Technology: An Elephantine Opportunity", IEEE International Electron Devices Meeting, Abstract 1.1.1, pp. 3-6, 1990.

[2] B. J. Baliga, "An Overview of Smart Power Technology", IEEE Transactions on Electron Devices, Vol. 38, pp. 1568-1575, 1991.

[3] Y-S. Huang and B. J. Baliga, "Extension of the RESURF Principle to Dielectrically Isolated Power Devices", International Symposium on Power Semiconductor Devices and ICs, Abstract 2.2, pp. 27-30, 1991.

[4] J. A. Appels and H. M. J. Vaes, "High Voltage Thin Layer Devices (RESURF Devices)", IEEE International Electron Devices Meeting, pp. 238-241, 1979.

[5] B. J. Baliga, "Prospects for Development of SiC Power Devices", Silicon Carbide and Related Materials - 1995, Institute of Physics Conference Series, Vol. 142, pp. 1-6, 1996.

[6] D. Alok and B. J. Baliga, "High Voltage (450V) 6H-SiC Buried Gate FET (BG-FET)", Silicon Carbide and Related Materials - 1995, Institute of Physics Conference Series, Vol. 142, pp. 749-752, 1996.

[7] D. Alok and B. J. Baliga, "A High Voltage (450V) 6H-SiC Lateral MESFET Structure", Electronics Letters, Vol. 32, pp. 1929-1931, 1996.

[8] K. Fujikawa, et al, "800V 4H-SiC RESURF-Type Lateral JFETs", IEEE Electron Device Letters, Vol. 25, pp. 790-791, 2004.

[9] T. Yamaguchi and S. Morimoto, "Process and Device Design of a 1000-V MOS IC", IEEE Transactions on Electron Devices, Vol. 29, pp. 1171-1178, 1982.

[10] P. Mehrotra and B. J. Baliga, "4H-SiC Lateral Single Zone RESURF Diodes", PSRC Technical Report TR-98-23, 1998.

[11] P. Mehrotra and B. J. Baliga, "4H-SiC Lateral RESURF Devices", PSRC Technical Report TR-99-010, 1999.

[12] J. Spitz, et al, "2.6 kV 4H-SiC Lateral DMOSFET's", IEEE Electron Device Letters, Vol. 19, pp. 100-102, 1998.

[13] N. S. Saks, et al, "A 475V High Voltage 6H-SiC Lateral MOSFET", IEEE Electron Device Letters, Vol. 20, pp. 431-433, 1999.

[14] A. K. Agarwal, et al, "Investigation of Lateral RESURF 6H-SiC MOSFETs", Silicon Carbide and Related Materials - 1999, Materials Science Forum, Vols. 338-342, pp. 1307-1310, 2000.

[15] K. Chatty, et al, "High Voltage Lateral RESURF MOSFETs on 4H-SiC", IEEE Electron Device Letters, Vol. 21, pp. 356-358, 2000.

[16]   S. Banerjee, et al, "Improved High Voltage Lateral RESURF MOSFETs in 4H-SiC", IEEE Electron Device Letters, Vol. 22, pp. 209-211, 2001.

[17]   W. Wang, et al, "930V, 170 mΩ-cm2 Lateral Two-Zone RESURF MOSFETs in 4H-SiC with NO Annealing", IEEE Electron Device Letters, Vol. 25, pp. 185-187, 2004.

[18]   M. Okamoto, et al, "Lateral RESURF MOSFET Fabricated on 4H-SiC (0001) C-Face", IEEE Electron Device Letters, Vol. 25, pp. 405-407, 2004.

[19]   T. Kimoto, et al, "Design and Fabrication of RESURF MOSFETs on 4H-SiC and 6H-SiC", IEEE Transactions on Electron Devices, Vol. 52, pp. 112-117, 2005.

Chapter 16

Synopsis

The applications for power devices were described in chapter 1 with the aid of several illustrations. The suitability for silicon carbide structures for these applications is indicated in Fig. 16.1 by the box drawn using dashed lines. The basis for the selected area is the satisfactory performance of silicon MOSFETs and IGBTs when designed for operating at below 500 volts, and the availability of sophisticated monolithically integrated silicon solutions for applications with current levels below 10 amperes.

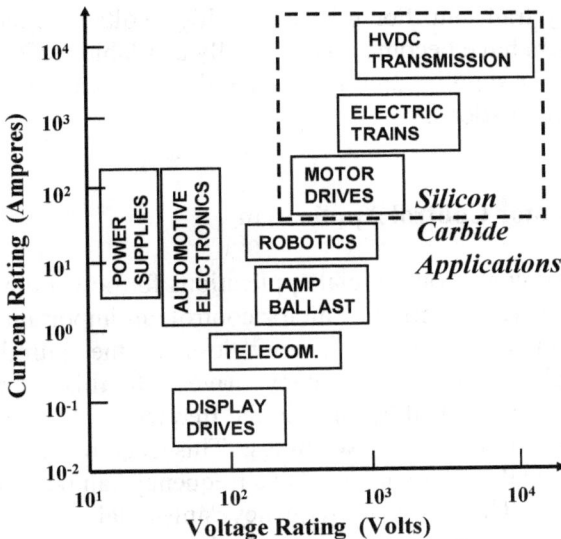

Fig. 16.1 Applications Opportunity for Silicon Carbide Power Devices.

Due to the parasitic resistance within silicon carbide structures (especially that associated with the substrate), their superior specific on-resistance when compared with silicon devices becomes an outstanding advantage only when the blocking voltage requirement exceeds 1000 volts. In the commercial arena, these applications are for high power motor control (e.g. in steel mills), for traction (e.g. electric street cars and locomotives), and the distribution of power (HVDC transmission).

In the military arena, there is considerable interest in silicon carbide technology because of the possibility to operate the power devices at elevated temperatures when compared with silicon. The success of this application is predicated upon utilization of the low intrinsic carrier concentration for silicon carbide associated with its large energy band gap. However, the potential for high temperature operation of silicon carbide structures must be tempered by reliability considerations associated with the surface passivation and the gate oxide.

It is prudent to focus initial utilization of silicon carbide devices on applications that can take advantage of the unique attributes of silicon carbide structures from the point of view of improving the efficiency of the power electronic system. The superior switching behavior of high voltage silicon carbide Schottky rectifiers when compared with silicon P-i-N structures can be utilized to reduce power losses associated with the reverse recovery transient. For this reason, high voltage silicon carbide Schottky rectifiers have become commercially available as O-ring diodes in computer power supplies and as fly-back diodes used with IGBTs for motor control applications.

16.1 Typical Motor Control Application

One of the high volume commercial applications for power devices is for the control of electric motors. For motor control, an important trend that has produced huge savings in energy loss is the introduction of adjustable speed or variable frequency drives[1]. In this approach, the speed of motors is controlled by varying the frequency of the voltage and current supplied to the motor windings. This requires the ability to generate a sinusoidal power source whose frequency can be adjusted by a control algorithm. The variable frequency sinusoidal power source is generated from a DC power source by using pulse-width-modulation (PWM) techniques with the signal applied to power devices in a totem-pole configuration, as illustrated in Fig. 14.1.

The basic totem-pole configuration illustrated in Fig. 14.1 uses IGBTs as the switches and P-i-N rectifiers as the fly-back diodes. The

variable frequency sinusoidal waveform that is supplied to the motor
windings can be synthesized by turning on the appropriate IGBTs in an
H-bridge circuit. The fly-back diodes are required to allow current to
flow through each branch of the totem-pole circuit in the reverse
direction to the conduction state for IGBTs. In the future, the IGBTs
could be replaced by silicon carbide MOSFETs while the silicon P-i-N
rectifiers used as fly-back diodes could be replaced by silicon carbide
Schottky rectifiers.

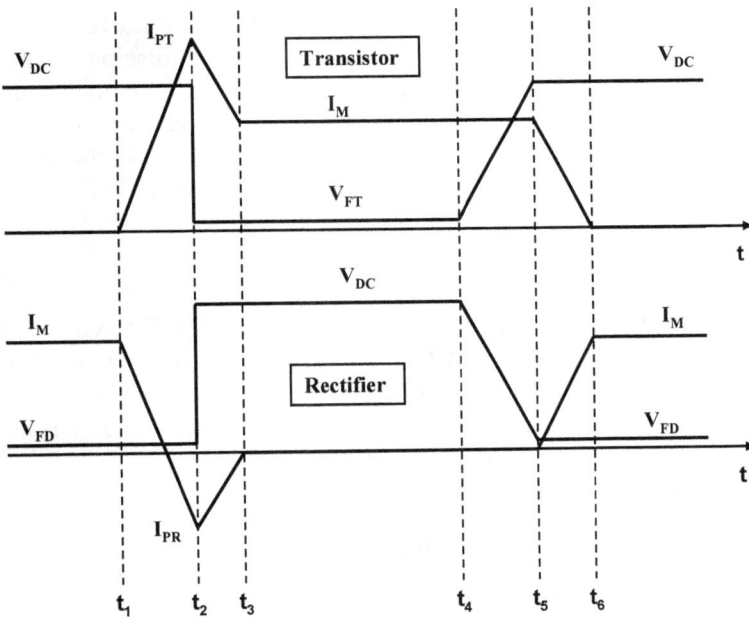

Fig. 16.2 Linearized Current and Voltage Waveforms for the Transistor and
Diode in the Totem-Pole Configuration.

Typical waveforms for the current and voltage in the totem-pole
configuration are shown in Fig. 16.2 for both the transistor and the fly-
back rectifier. During each cycle of the PWM signal, the motor current
(I_M) can be assumed to remain approximately constant if the frequency of
the PWM operation is much larger than the frequency of the sinusoid that
is being synthesized. The motor current is then flowing through either the
transistor or the rectifier (in another branch) with transitions made
between them as indicated in the figure. In the case of high voltage

silicon P-i-N rectifiers, a large reverse-recovery current (I_{PR}) occurs due to the stored charge within the structure. This produces significant power loss at each switching event not only in the rectifier but also in the transistor because the peak current (I_{PT}) in the transistor is the sum of the motor current and the reverse recovery current. However, if the silicon P-i-N rectifier is replaced with a silicon carbide rectifier with small reverse recovery current, the on-state conduction power loss becomes more significant. It is then more important to reduce the on-state voltage drop of the rectifier, as well as that of the power transistor.

The power losses in the transistor and the fly-back rectifier must be added together to determine the total power dissipated by the semiconductor devices. For each device, the power losses occur during the on-state, the turn-on phase, the turn-off phase, and during the blocking mode. For the waveforms illustrated in Fig. 16.2, the transistor is in the on-state during the time interval from t_3 to t_4. This is usually expressed in terms of the duty cycle:

$$\delta = (t_4 - t_3)/T \qquad [16.1]$$

where T is the time period (the reciprocal of the PWM operating frequency). For the analysis in this chapter, it will be assumed that the transistor has a duty cycle of 50 percent. If the transition times during the switching events are small when compared with the time period, the rectifier is also in its on-state for 50 percent of the time.

The power loss in the transistor during its on-state is given by:

$$P_{LT}(on) = \delta I_M V_{FT} \qquad [16.2]$$

where V_{FT} is the on-state voltage drop of the transistor.

The total switching power loss in the transistor is given by the sum of the turn-on power loss and the turn-off power loss incurred during the two switching events for each period. The turn-on power loss for the transistor is given by:

$$P_{LT}(turnon) = [0.5(t_2 - t_1)I_{PT}V_{DC} + 0.5(t_3 - t_2)(I_{PT} + I_M)V_{FT}]f \qquad [16.3]$$

where V_{DC} is the DC bus voltage, I_{PT} is peak current experienced by the transistor during turn-on, and f is the PWM operating frequency. The peak current experienced by the transistor is given by:

$$I_{PT} = I_M + I_{PR} \qquad [16.4]$$

where I_{PR} is peak reverse recovery current of the rectifier.

The turn-off power loss for the transistor is given by:

$$P_{LT}(turnoff) = [0.5(t_6 - t_4)I_M V_{DC}]f \qquad [16.5]$$

The power loss during the blocking state will be assumed to be negligible due to the low leakage current for the transistor. The total power losses incurred in the transistor can be obtained as a function of the operating frequency by adding these power loss components.

In the same manner, the power loss in the rectifier can be obtained by summing the power losses incurred during the on-state, the turn-on phase, the turn-off phase, and during the blocking mode. The power loss in the rectifier during its on-state is given by:

$$P_{LD}(on) = (1 - \delta)I_M V_{FD} \qquad [16.6]$$

where V_{FD} is the on-state voltage drop of the rectifier.

The total switching power loss in the rectifier is given by the sum of the turn-on power loss and the turn-off power loss incurred during the two switching events for each period. The turn-on power loss for the rectifier is given by:

$$P_{LD}(turnon) = [0.5(t_6 - t_5)I_M V_{FD}]f \qquad [16.7]$$

The turn-off power loss for the rectifier is given by:

$$P_{LD}(turnoff) = [0.5(t_3 - t_2)I_{PR} V_{DC}]f \qquad [16.8]$$

The power loss during the blocking state will be assumed to be negligible due to the low leakage current for the rectifier. The total power losses incurred in the rectifier can be obtained as a function of the operating frequency by adding these power loss components.

A quantitative evaluation of the benefits of replacing silicon power devices with their silicon carbide counterparts can be obtained by performing the analysis of a typical system delivering power to a motor at a bus voltage of 2000 volts. The motor current will be assumed to be 100 amperes during the switching event, implying a delivered power of 200 kW. For comparison purposes, the IGBT will be used as the silicon power switch and the P-i-N rectifier as the silicon fly-back diode. A silicon carbide 4H-SiC MOSFET with specific on-resistance of 2 mΩ-cm^2 will be assumed to be representative of silicon carbide power switches, while a 4H-SiC Schottky rectifier will be used as the fly-back diode.

The electrical parameters used during the analysis for these devices are given in Fig. 16.3. The on-state voltage drops were determined by assuming an on-state current density of 100 A/cm^2 for all the devices. The switching times for the silicon device case are typical for motor control applications[2]. It was assumed that the silicon carbide devices could be switched ten times faster because of the absence of the reverse recovery current in the silicon carbide Schottky rectifier.

	Silicon IGBT	Silicon P-i-N	4H-SiC MOSFET	4H-SiC SBD
V_f (V)	3.0	2.0	0.2	1.5
I_{PR} (A)	-	200	-	10
$(t_2 - t_1)$ (μs)	0.1	0.1	0.01	0.01
$(t_3 - t_2)$ (μs)	0.1	0.1	0.01	0.01
$(t_5 - t_4)$ (μs)	0.1	0.1	0.01	0.01
$(t_6 - t_5)$ (μs)	0.1	0.1	0.01	0.01

Fig. 16.3 Electrical Parameters used for the Transistor and Diode in the Totem-Pole Configuration.

In the following sections, three cases will be analyzed for comparison purposes. The first case is a bench-mark case based upon using silicon devices, namely the silicon IGBT together with the silicon P-i-N rectifier. In the second case, the silicon P-i-N rectifier is replaced with the silicon carbide Schottky rectifier but the silicon IGBT is retained as the power switch. This is a near term option for the industry because silicon carbide Schottky rectifiers have already become commercially available. The third case is based upon using only silicon carbide devices, namely the 4H-SiC MOSFET as the switch and the 4H-SiC Schottky rectifier as the fly-back diode.

16.2 Motor Control Application: Silicon IGBT and P-i-N Rectifier

Silicon IGBTs and P-i-N rectifiers are commercially available with blocking voltage capability in excess of 4.5 kV and high current modules have been developed by several companies to handle large loads. The on-state voltage drops and switching properties of these devices, given in Fig. 16.3, were based on the available data in the literature. The switching speed during the turn-on of the IGBT at time t_1 is limited by the reverse recovery behavior of the P-i-N rectifier. Although a shorter turn-on time could reduce the power dissipated during the switching event, this results in a larger di/dt for the turn-off of the P-i-N rectifier. This produces an increase in the peak reverse current in the P-i-N rectifier which can exceed its ratings. To make matters worse, when the their turn-off di/dt exceeds a critical value, silicon P-i-N rectifiers undergo "snappy" recovery with an abrupt drop in the current which produces a high voltage spike due to the parasitic inductances in the circuit. This voltage spike is superimposed upon the DC bus voltage and must be withstood by all the components. The higher blocking voltage rating demanded by this larger voltage is detrimental to the on-state voltage drop of the devices leading to larger power losses. Consequently, it is prudent to limit the turn-on speed of the IGBT to 0.1 microseconds.

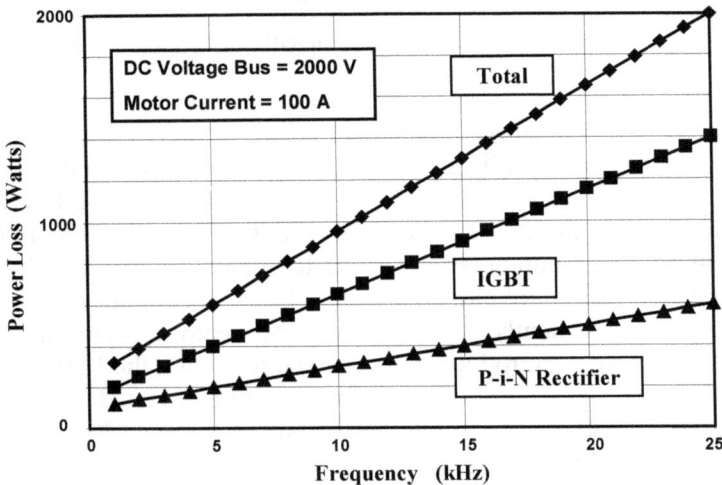

Fig. 16.4 Power Losses for the Silicon IGBT and P-i-N Diode in the Totem-Pole Configuration.

The total power loss (transistor plus diode) calculated using the above methodology is shown in Fig. 16.4 as a function of the switching frequency. It can be seen that the total power loss increases rapidly with frequency indicating that the switching losses are dominant in this case. The total power losses (on-state plus switching) within the IGBT and the P-i-N rectifier are also shown in this plot. It can be seen that the power losses in the IGBT are about twice as large as those incurred in the P-i-N rectifier. It is worth pointing out that the power losses in the IGBT are enhanced by the larger reverse recovery current of the P-i-N rectifier.

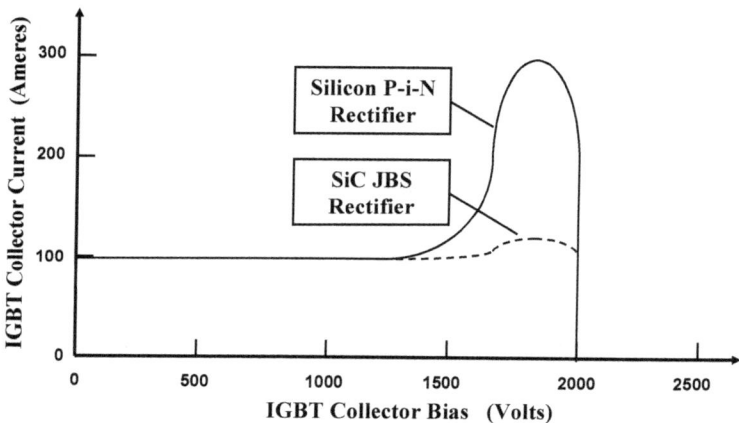

Fig. 16.5 Switching Loci for the Silicon IGBT in the Totem-Pole Configuration.

The electrical stress in the IGBT is enhanced by the large reverse recovery current in the silicon P-i-N rectifier. This is illustrated in Fig. 16.5 by graphing the collector current-voltage locus for the IGBT. As indicated by the solid line, the current in the IGBT reaches 300 amperes with a high voltage impressed across its terminals (at time t_2). This can lead to destructive failure due to latch-up of the parasitic thyristor within the IGBT[3]. Consequently, the large reverse recovery current of the silicon P-i-N rectifier introduces very high stress within the IGBT.

The cost for the power switch and the associated heat-sink is much larger than that for the rectifier in power systems. It is therefore advantageous to reduce the total power dissipation in the power switch in order to reduce its size and hence its cost. A good understanding of the power dissipation components within the IGBT is useful for finding approaches to reducing the power losses in the IGBT. These power loss components are plotted in Fig. 16.6 as a function of frequency.

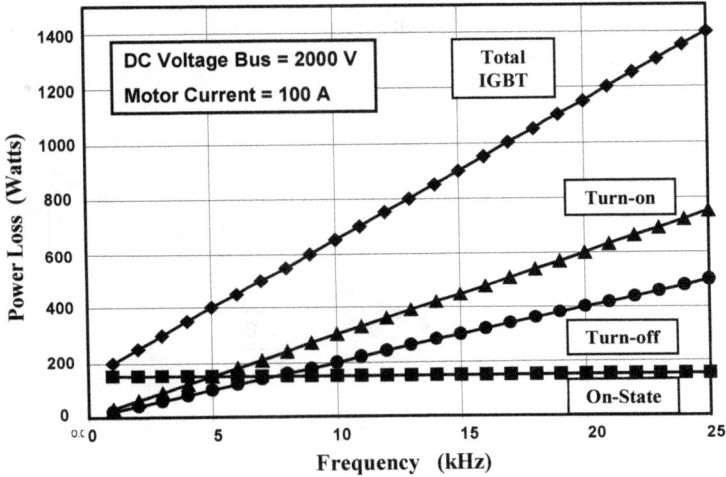

Fig. 16.6 Power Loss Components for the Silicon IGBT in the Totem-Pole Configuration with a Silicon P-i-N Diode.

It can be seen in Fig. 16.6 that, when the frequency exceeds 5 kHz, the switching losses overtake the on-state power loss. For motor control, it is desirable to raise the PWM operating frequency to above the acoustic range for human hearing in order to eliminate discomfort in residential installations. This requires operating at above 15 kHz. At these frequencies, the turn-on losses in the IGBT exceed the turn-off losses. Since the turn-on losses within the IGBT are enhanced by the large peak current (I_{PT}), significant improvements in the power losses within the IGBT can be achieved by reduction of the reverse recovery current in the P-i-N rectifier[4]. In silicon structures, this has conventionally been achieved by reduction of the minority carrier lifetime in the drift region[5]. However, a reduction of the minority carrier lifetime in the drift region of P-i-N rectifiers produces an increase in the on-state voltage drop. This necessitates a trade-off between on-state and switching losses. A significant improvement in the trade-off has more recently been achieved by using the merged-P-i-N/Schottky (MPS) rectifier structure[6]. This can enable a 25 percent improvement in power losses in the motor control application. An even greater improvement is possible by replacement of the silicon P-i-N rectifier with a silicon carbide Schottky rectifier, as discussed in the next section.

16.3 Motor Control Application: Silicon IGBT and Silicon Carbide Schottky Rectifier

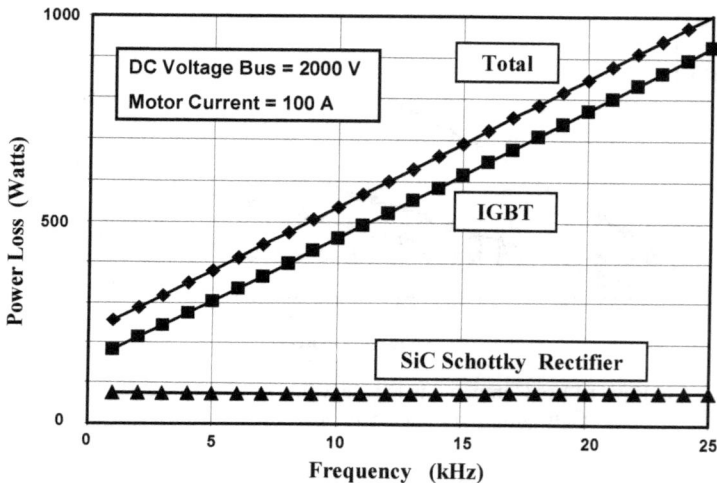

Fig. 16.7 Power Losses for the Silicon IGBT and 4H-SiC Schottky Rectifier in the Totem-Pole Configuration.

The improved performance of silicon carbide Schottky rectifiers, especially the JBS rectifier discussed in chapter 6, offers the opportunity to greatly diminish the reverse recovery current while retaining a low on-state voltage drop. Although the unipolar current flow in the silicon carbide Schottky rectifier in the on-state eliminates the stored charge, a small reverse recovery current is still observed during its turn-off due to the displacement current associated with formation of the depletion layer within the drift region[2]. The electrical parameters for the 4H-SiC JBS rectifier capable of supporting 3000 volts are given in Fig. 16.2. Using these parameters, the total power loss in the power devices for the motor control system is given in Fig. 16.7. In comparison with the all-silicon approach shown in the previous section, the total power loss has been reduced in half. The power losses in the IGBT and the power rectifier are also plotted in Fig. 16.7. It can be seen that the power loss in the IGBT is now dominant and that its power losses still increase rapidly with frequency indicating that switching losses are much larger than the on-state losses.

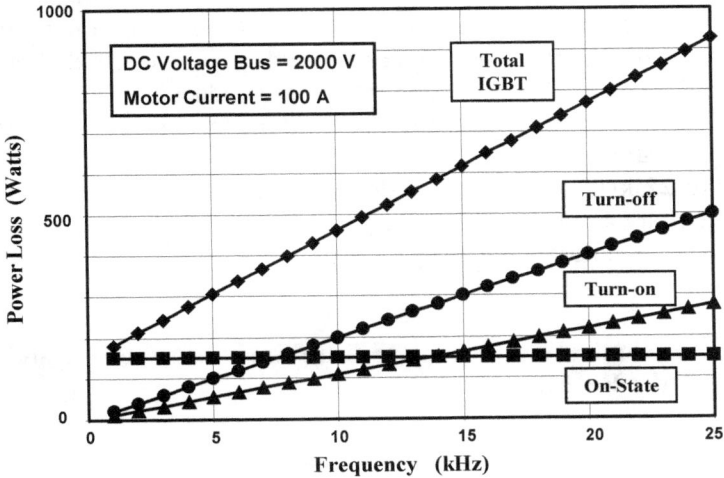

Fig. 16.8 Power Loss Components for the Silicon IGBT in the Totem-Pole Configuration with a 4H-SiC JBS Rectifier.

The total power loss (on-state plus switching) within the IGBT is reduced by about 50 percent with the replacement of the silicon P-i-N rectifier with the 4H-SiC JBS rectifier. For example, at an operating frequency of 15 kHz, the power loss is 600 watts in the IGBT when compared with 900 watts when the silicon P-i-N rectifier is used. In order to elucidate the power dissipation for the IGBT in this case, the power loss components within the IGBT are plotted in Fig. 16.8 as a function of frequency. It can be seen that the switching losses overtake the on-state power loss when the frequency exceeds a higher frequency (10 kHz) in this case. As previously pointed out, it is desirable to raise the PWM operating frequency to above the acoustic range for human hearing in order to eliminate discomfort in residential installations. At above 15 kHz, the turn-off losses in the IGBT exceed the turn-on losses for the case with the 4H-SiC JBS rectifier. In addition, the locus for the turn-off transition is greatly improved as shown in Fig. 16.5 with the dashed lines. The smaller reverse recovery current in the 4H-SiC JBS rectifier reduces the stress in the IGBT eliminating the prospects for destructive failure due to latch-up.

This case with the silicon IGBT and the silicon carbide Schottky rectifier has been experimentally demonstrated to provide significant improvement in power losses[2] by using 600 volt rated devices. A 57 percent reduction in power loss in the IGBT was observed by replacement of the silicon P-i-N rectifier with a silicon carbide Schottky rectifier. This was mainly attributable to the reduced reverse recovery current which resulted in smaller IGBT turn-on losses. The reduced power dissipation enabled increasing the PWM operating frequency from 10 kHz to 22 kHz while keeping the same junction temperature for the IGBT. The larger operating frequency was found to reduce the acoustical noise from the motor windings by 7 dB in the 30-40 Hz range.

16.4 Motor Control Application: Silicon Carbide MOSFET and Schottky Rectifier

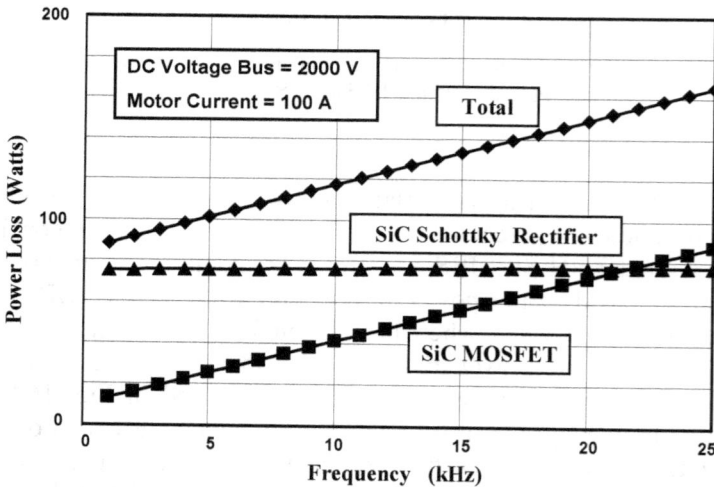

Fig. 16.9 Power Losses for the 4H-SiC MOSFET and Schottky Rectifier in the Totem-Pole Configuration.

The low specific on-resistance of 4H-SiC power MOSFETs discussed earlier in this book together with their fast switching speed provides an additional opportunity to improve the efficiency of motor control systems. The electrical parameters for the 4H-SiC power MOSFET capable of supporting 3000 volts, given in Fig. 16.2, are based upon a specific on-resistance of 2 mΩ-cm^2 and an on-state current density of 100 A/cm^2. It is assumed that the gate drive current for these silicon carbide MOSFETs can be adjusted so that they can be switched in one-tenth of the time used for the IGBTs. This faster switching time can only be utilized in conjunction with the use of silicon carbide Schottky rectifiers. Using the parameters for the silicon carbide devices, the calculated total power loss in the power devices for the motor control system is provided in Fig. 16.9. In comparison with the all-silicon approach shown in the previous section, the total power loss has been reduced by an order of magnitude. The power loss in the 4H-SiC power MOSFET and Schottky rectifier are also plotted in Fig. 16.9. It can be seen that the power loss in the rectifier is now dominant at lower frequencies. The power loss in the 4H-SiC Schottky rectifier is almost independent of frequency because of its small reverse recovery current and high on-state voltage drop. In contrast, the power losses in the 4H-SiC power MOSFET increase rapidly with frequency indicating that switching losses are much larger than the on-state losses. The power losses in the SiC power switch overtakes that in the rectifier only above 20 kHz.

16.5 Motor Control Application: Comparison of Cases

In order to compare the above cases, it is beneficial to look at the total power loss incurred at the same operating frequency. This can be done using the plot provided in Fig. 16.10. A good perspective on progressive improvements that can be achieved in motor control technology can be discerned from this graph. The replacement of silicon P-i-N rectifiers with silicon carbide Schottky diodes provides the most near term opportunity to reduce the power losses by a factor of two. This improvement is primarily due to elimination of the large reverse recovery current observed in silicon P-i-N rectifiers. The commercial availability of silicon carbide Schottky rectifiers makes this approach viable today. As the silicon carbide power MOSFET technology matures or by utilization of the Baliga-Pair configuration discussed in chapter 8, it is possible to obtain a huge reduction (by a factor of ten) in the power losses. This has important implications in the thermal design of the system enabling reduction of the size and weight of the heat sink.

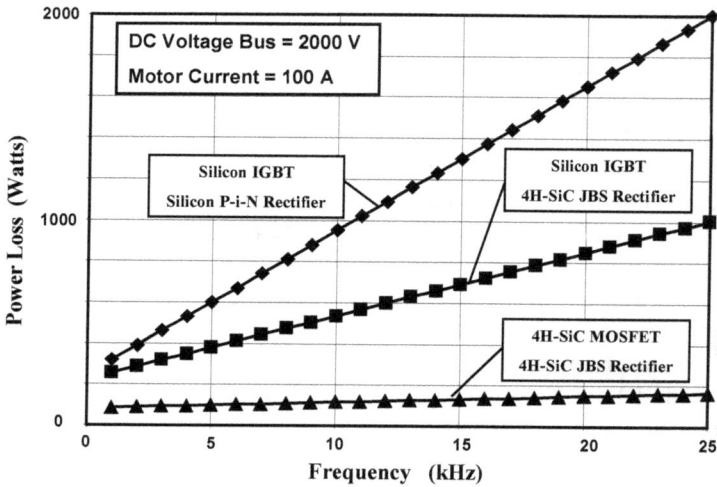

Fig. 16.10 Total Power Loss for the Devices in the Totem-Pole Configuration.

16.6 Summary

The benefits of introduction of the silicon carbide power device technology in the motor control applications arena has been explained in this chapter with the aid of a specific example. These results would be generally applicable for medium and high power motor control. Since the differential in the performance between silicon and silicon carbide devices enlarges with increase in the operating voltage, the advantages of silicon carbide will be most compelling at higher operating voltages used in higher power systems. Thus, the prognostication made using a simple analysis based upon fundamental principles[7] twenty-five years ago has been validated by the experimental demonstration of commercially viable silicon carbide power switches and rectifiers. As the cost for manufacturing these devices declines due to availability of larger diameter wafers with fewer defects, the displacement of silicon devices with their silicon carbide counterparts will accelerate resulting in higher efficiency power systems that benefit society by reduction of fossil fuel consumption and reduced environmental pollution.

References

[1]   B. K. Bose, "Power Electronics and Variable Frequency Drives", IEEE Press Book, New York, 1997.
[2]   M. O'Neill, "SiC puts new Spin on Motor Drives", Power Electronics Technology Magazine, pp. 14-22, January 2005.
[3]   B. J. Baliga, "Power Semiconductor Devices", PWS Publishing Company, 1996.
[4]   B. J. Baliga, "Power Semiconductor Devices for Variable Frequency Drives", Proceedings of the IEEE, Vol. 82, pp. 1112-1122, 1994.
[5]   B. J. Baliga and E. S. Sun, "Comparison of Gold, Platinum, and Electron Irradiation for Controlling Lifetime in Power Rectifiers", IEEE Transactions on Electron Devices, Vol. 24, pp. 685-688, 1977.
[6]   B. J. Baliga, "Analysis of a High Voltage Merged P-i-N/Schottky (MPS) Rectifier", IEEE Electron Device Letters, Vol. 8, pp. 407-409, 1987.
[7]   B. J. Baliga, "Semiconductors for High Voltage Vertical Channel Field Effect Transistors", Journal of Applied Physics, Vol. 53, pp. 1759-1764, 1982.

Index

www.ingramcontent.com/pod-product-compliance
Lightning Source LLC
Chambersburg PA
CBHW070740220326
41598CB00026B/3714